D. Ebert
J. M. Favre
R. Peikert (eds.)

Data Visualization 2001

Proceedings of the Joint Eurographics –
IEEE TCVG Symposium on Visualization
in Ascona, Switzerland,
May 28–30, 2001

Eurographics

Springer-Verlag Wien GmbH

Prof. Dr. David S. Ebert
School of Electrical and Computer Engineering
Purdue University, West Lafayette, USA

Dr. Jean M. Favre
Data Mining & Visualization, CSCS
Manno, Switzerland

Dr. Ronald Peikert
Institut für wissenschaftliches Rechnen
ETH-Zentrum, IFW
Zürich, Switzerland

© 2001 Springer-Verlag Wien
Originally published by Springer-Verlag Wien New York in 2001

Typesetting: Camera-ready by authors

Graphic design: Ecke Bonk

Printed on acid-free and chlorine-free bleached paper

SPIN: 10837726

With 212 partly coloured Figures

ISSN 0946-2767
ISBN 978-3-211-83674-3   ISBN 978-3-7091-6215-6 (eBook)
DOI 10.1007/978-3-7091-6215-6

Preface

These proceedings contain the papers presented at VisSym '01, the third Joint Visualization Symposium of the Eurographics Association and the Technical Committee on Visualization and Graphics (TCVG) of the IEEE Computer Society. The symposium is being held from May 28 to May 30, 2001, in Ascona, Switzerland.

The number of submitted papers has again increased to reach 68 this year and their high quality made it possible to extend the program to 33 presentations. VisSym has grown from a mostly European workshop to a visualization event of global significance, which is best illustrated by the 13 contributions from outside Europe.

We observed that the spectrum of topics has been quite stable over the last few years, with volume visualization, surface extraction, information visualization and visualization of vector fields and terrains as major areas of interest. A continuing trend appearing in many of these areas is the shift to multi-resolution techniques. Looking at applications, biomedical visualization is this year predominant, but also noticeable are the three automotive applications which could indicate rapidly growing interest in visualization in this area.

We thank the Centro Stefano Franscini (CSF), Advanced Visual Systems Inc. (AVS) and the Numerical Algorithms Group (NAG) for financially supporting VisSym '01. Special thanks go to the CSF for providing a professional local organization. We thank the reviewers for their invaluable work. And last but not least we thank the authors, not only for providing us with their exciting scientific contributions but also for their cooperation in producing a uniformly typeset and visually attractive book.

May 2001

David Ebert
Jean M. Favre
Ronald Peikert

Chairs, IPC, and Reviewers

Symposium Co-Chairs

David Ebert, Purdue University, West Lafayette, Indiana, U.S.A.

Jean M. Favre, Swiss Center for Scientific Computing, Manno, Switzerland

Ronald Peikert, Swiss Federal Institute of Technology, Zürich, Switzerland

International Program Committee

D. Bartz,	H. Hagen,	R. Peikert,
G.-P. Bonneau,	P. Hanrahan,	H. Pfister,
K. Brodlie,	C. Hansen,	F. Post,
S. Coquillart,	D. Keim,	B. Ribarsky,
R. Crawfis,	D. Kenwright,	M. Rumpf,
L. De Floriani,	D. Laidlaw,	G. Scheuermann,
D. Ebert,	W. Lefer,	R. Scopigno,
S. Eick,	R. van Liere,	H.-W. Shen,
T. Ertl,	R. Machiraju,	D. Silver,
J. Favre,	N. Max,	P. Slavík,
M. Grave,	R. Moorhead,	O. Staadt,
E. Gröller,	G. Nielson,	A. Varshney,
M. Gross,	R. Pajarola,	J. van Wijk

Additional Reviewers

D. Bauer,	A. Kalaiah,	N. Polapally,
C. Botha,	M. Kirby,	P. Rheingans,
P. Cignoni,	A. Knig,	S. Roettger,
P. de Bruin,	R. Koch,	R. Scoggins,
K. Engel,	M. Kraus,	N. Shareef,
S. Frisken-Gibson,	R. Lütolf,	T. Theußl,
N. Gagvani,	M. Magallon,	A. Vilanova Bartrolí,
X. Hao,	P. Magillo,	S. Würmlin,
M. Hopf,	C. Montani,	L. Yang,
J. Huang,	L. Mroz,	M. Zwicker
S. Iserhardt-Bauer,	M. Pauly,	

Table of Contents

Volume Rendering

Information Visualization Applications

Automotive Applications

Invited Speaker

Hanspeter Pfister

Research Scientist
MERL - Mitsubishi Electric Research Laboratories
201 Broadway, Cambridge, MA 02139, USA

Point-Based Graphics and Visualization

Point-based models are ideally suited to acquire, transmit, and display complex, real-life three dimensional objects as efficiently as possible. Our objective is to develop 3D graphics technology for cell phones or PDAs, eCommerce, and 3D games.

The fundamental approach of our research is to represent 3D objects as a dense set of unconnected surfels (or surface elements). In collaboration with MIT, we have built a system for acquiring high-quality 3D models from a series of captured images. Our system builds an approximate three-dimensional model based on the image-based visual hull upon which a view-dependent radiance function is mapped. We have digitized hundreds of models using our system, and we have built three generations of digitizers based on the same approach.

In collaboration with ETH Zurich, we have developed several point rendering methods to visualize our models. We developed novel screen space filtering techniques called surface splatting and EWA volume splatting. Our rigorous mathematical analysis extends anisotropic texture mapping to irregularly spaced point samples and voxels. Surface and EWA volume splatting are efficient rendering algorithms for point samples and volume data with high image quality, progressive rendering, anti-aliasing, and transparency.

Biography

Hanspeter Pfister is a Research Scientist at MERL - Mitsubishi Electric Research Laboratories - in Cambridge, MA. He is the chief architect of VolumePro, Mitsubishi Electric's real-time volume rendering system for PC-class computers. His research interests include computer graphics, scientific visualization, and computer architecture. Hanspeter Pfister received his Ph.D. in Computer Science in 1996 from the State University of New York at Stony Brook. In his doctoral research he developed Cube-4, a scalable architecture for real-time volume rendering. He received his M.S. in Electrical Engineering from the Swiss Federal Institute of Technology (ETH) Zurich in 1991. He is a member of the ACM, IEEE, the IEEE Computer Society, and the Eurographics Association.

A Case Study in Multi-Sensory Investigation of Geoscientific Data

Chris Harding[1,3], Ioannis A. Kakadiaris[1,2], John F. Casey[1,3] and R. Bowen Loftin[4]

[1] Virtual Environments Research Institute, Univ. of Houston, Houston, TX, USA

harding77019@yahoo.com

[2] Department of Computer Science, University of Houston, Houston, TX, USA

ioannisk@uh.edu

[3] Department of Geosciences, University of Houston, TX, USA

jfcasey@uh.edu

[4]VMASC, Old Dominion University, Suffolk, VA, USA

bloftin@odu.edu

Abstract: In this paper, we report our ongoing research into multi-sensory investigation of geoscientific data. Our Geoscientific Data Investigation System (GDIS) integrates three-dimensional, interactive computer graphics, touch (haptics) and real-time sonification into a multi-sensory Virtual Environment. GDIS has been used to investigate geological structures on the high-resolution bathymetry data from the Mid-Atlantic Ridge. Haptic force feedback was used to precisely digitize line features on three-dimensional morphology and to feel surface properties via varying friction settings; additional, overlapping data can be perceived via sound (sonification). We also report on the results of a psycho-acoustic study about the absolute recognition of sound signals, and on the actual feedback that we have received from a number of geoscientists during a recent major geoscience conference.

1. Introduction and Background

In recent years, the natural resource industry has recognized three-dimensional visualization and modeling of geoscientific data as playing an important part in the exploration and development of natural resources. Several research projects have demonstrated that the use of Virtual Environments has the potential to improve productivity and lower costs in areas such as petroleum exploration [3], [5]. Most current virtual environments focus entirely on improving the user's comprehension of geoscientific data by using true three-dimensional (stereo graphic) environments, large screens and cooperation between different disciplines. However, the fields of haptic force feedback devices and real-time sound synthesis have matured sufficiently in recent years to allow research of the integration of touch and sound into visual virtual environments. Both technologies have seen some application in the geoscientific domain [1], [2].

The aim of our research is to integrate the ability to feel and interact with data via touch, and to analyze data via hearing into interactive three-dimensional systems to give the user the advantage of working with multiple, overlapping data properties simultaneously. Although the visual sense is still the main channel, presenting other aspects of data simultaneously through touch and sound could lead to enormous benefits if we succeed in mapping data from its scientific domain into a useful representation of touch and sound (the reader is referred to [4] for further information). This paper focuses on a novel sonification technique and its integration with visualization and force feedback to create highly interactive virtual environments for investigating geoscientific data.

In our experience, a demonstration of GDIS to potential users is a vital research tool. After geoscientists gathered first hand experience with this new technology, their feedback lead to valuable research contributions. The system has been very well received by participating geoscientists. Their input helps to open up new areas of potential geoscientific applications for mapping, displaying and perceiving complex multi-attribute data sets simultaneously.

2. The Geoscientific Data Investigation System

We have created a demonstration prototype called GDIS (Geoscientific Data Investigation System, which allows the multi-sensory investigation of geoscientific surface data, on which several geophysical properties has been mapped (e.g., gravimetric and magnetic data). The term *investigation* is used to emphasize that three major senses (stereo vision, touch and sound) are used for what is usually called *vi*sualization and modeling. GDIS uses all three senses to simultaneously explore different, overlapping surface properties and accurately digitize lines on the surface.

Easy access to the development of human-computer interaction via touch has become possible in the last five years with the development of haptic force feedback systems such as Sensable Technology's PHANToM. We employ the Desktop PHANToM in our system (Figure 1), which can project a point force of up to 6.5 N within its three-dimensional workspace (16 cm x 13 cm x 13 cm). The PHANToM creates the illusion of touching solid objects with a virtual fingertip and the feel of physical effects such as attraction, repulsion, friction and viscosity. The stylus at the end of the arm gives the user force feedback and direct interaction with the data in three dimensions. Using this "three-dimensional force feedback mouse" allows not only the interrogation of three-dimensional data within a virtual space, but also provides input constraints for fine-grained interaction. In our research, we have concentrated on the haptic rendering of *surface data* (which is a common data type across the geosciences) onto which three-dimensional lines can be digitized. This allows the user to explore minute features of the surface's morphology and to use the tactile feedback to model line and polygonal structures. In addition, we translate surface properties into different friction values in order to make it noticeably easy or difficult to move the stylus tip over certain parts of the surface.

The use of non-speech sound to interactively explore data (scientific sonification -"the use of data to control a sound generator" [6]), has been under investigation since the early 1990s. Although recent advances in real-time sound synthesis have made this technology more widely available to researchers, there are only very few guidelines on mapping data into sounds, and there seems to be relatively little research that is related to geoscientific data [3]. It seems to be clear, however, that sound needs to not only be effective but also pleasant (or at least not disturbing). As the way different people react to sound can be very different, sound mapping in an application may need to be highly customizable to a specific user. The human hearing system is very well suited to detect even small changes in sounds (e.g., has very good temporal and pitch resolution) but has difficulties determining absolute values.

One advantage of sonification is that the user's eyes are free to process visual data while hearing a different set of data. We have integrated a novel "sound map" into the visual rendering of surfaces, giving the user the ability to listen to a local surface property while simultaneously visually observing other properties.

3. Absolute Recognition of sound—a psycho-acoustic user study

We conducted a psychological study into the use of sound to convey data in an absolute way. We use the term *absolute* signal (as opposed to a *relative* signal) to refer to a signal that is evaluated outside the context of previous or following signals. Each signal is initially defined as corresponding to a certain concept, in our case the numbers one to five. The test subjects were asked first to remember the sound and later to recognize it again—making a connection back to the numeric definition given earlier.

It is well known that the human hearing system is quite capable of discriminating even small changes in audio signals in terms of pitch or tempo [7]. However, the absolute recognition of signals is considered much more difficult and usually dependent on musical abilities and training (the "perfect pitch" phenomena is an extreme example for the absolute recognition of sound). Because absolute recognition is acknowledged to be a difficult problem, our study concentrated on a fairly simple setup. In our study, we defined an audio signal by three simple parameters. These parameters, pitch (frequency), instrument (timbre) and tempo (tone repeat rate) are used to generate (musical) tones via a MIDI (Musical Instrument Digital Interface) synthesizer.

The goal of our study was to determine if the subjects could be trained to reliably recognize a certain small set of audio signals, and connect them back to an equal number of numeric values. We chose a small number of integer values to represent a partition of the overall data set in a number of "bins". This so-called "ballpark setup" connects, in our case, five logically progressive musical notes to a sequence such as "very low, low, medium, high and very high" in order to give the user a rough idea about local data values rather then their actual, precise values. This technique was used in GDIS in addition to the more traditional relative sonification, where a change sound symbolizes a general, relative change in the sonified data set; for example a rise in pitch could be used to sonify a rise in temperature.

3.1 Setup of the experiment

The idea of the ballpark scenario raised several interesting questions: How should the audio signals (using pitch, tempo and instrument) be composed to allow for a short training and easy recognition? Should we use only single parameters (e.g., only pitch or only instrument) or some kind of combination? The study aimed to answer these questions by reducing the ballpark scenario to a simple "game" in which the test subject was guided through seven conditions and asked to recognize a series of audio signals from a set of five different sounds. We postulated, that, not only should the subjects be able to learn how to differentiate between audio signals, but also, that the subjects would be more successful in recognizing conditions where more than one parameter was changing (conditions 4 to 7). The composition of the five audio signals for each of the seven conditions is shown in the table below.

Condition	Setup of the five audio signals
1	*Pitch* only: single note C in five different octaves: C2, C3, C4, C5, C6, played by a grand piano.
2	*Instrument* only: single note C4 played by an church organ, a grand piano, a pan flute, a muted trumped and a cembalo.
3	*Tempo:* 3 clearly separate notes (C4 played by grand piano) with speed increasing in five stages from overall 1.5 to 0.3 seconds.
4	Combination of pitch and instrument (single notes).
5	Combination of pitch and tempo (played by grand piano).
6	Combination of tempo and instrument (playing a C4).
7	Combination of pitch, instrument and tempo.

The study was conducted on a standard SGI workstation (Octane or O2) equipped with small PC speakers and a software MIDI synthesizer (midisynth), which is part of the SGI digital media software package. At the beginning of each experiment, the subject was asked to complete a questionnaire about his or her age, gender, musical background, listening habits and professional background. The subject was guided through seven conditions. Each condition began with a *training phase* (Figure 1a), in which the subject was familiarized with the current meaning of the five audio signals. This was followed by a *recognition phase,* in which the subject was asked to identify a random series of 15 audio signals (which was different for each condition but the same for each subject), and indicate his or her confidence in the choice.

Figure 1b depicts the recognition phase, where the trained subject plays the audio signal (PLAY), chooses a number (1–5) that corresponds to the audio signal, and indicates his or her confidence (0–100). We also recorded the time spent for the recognition of each condition. The total time spent to complete the experiment varied between 20 and 40 minutes.

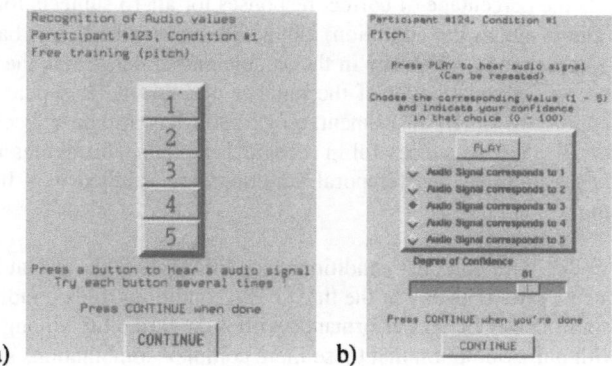

Figure 1. Example snapshots of the training phase (a) and of the recognition phase (b).

3.2 Results of the study

A total of 13 subjects (11 male, 2 female) were tested in this study. Their ages ranged from 21 to 55 (average 33.15). From the background questionnaire, it appears that the subjects are fairly evenly mixed in most categories. Three subjects have an under-graduate degree, five a Master's degree and five a Ph.D. degree. Four of the subjects are students, four are researchers and four are professionals. None of the subjects were music majors or professional musicians, although some like classical music and some are amateur musicians. However, the subjects' musicality had no significant impact on their performance. The results of three dependent variables were analyzed: success rate, time to recognize, and subjective confidence. To ensure statistical significance, Analysis of Variances tests (ANOVA) and planned comparison tests were conducted for all dependent variables.

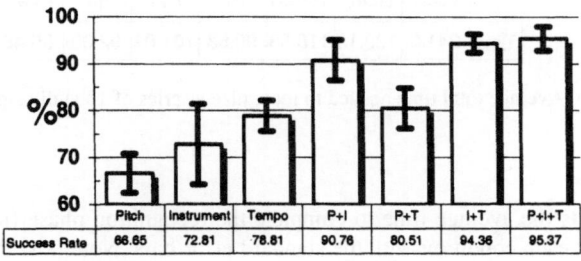

Figure 2. Average success rate across the seven conditions.

Figure 2 depicts the percentage of correct responses for all 13 subjects for each condition (with 15 audio signals per condition) along with the standard error bars. The standard error is a measure of variability in the sample and is defined as the standard deviation divided by the square root of the number of subjects. It appears, that of the three simple conditions (pitch, instrument, tempo) subjects are most successful in recognizing tempo, while least successful in recognizing pitch. This agrees with the generally accepted assumption that temporal variations are handled well by the human hearing system.

Although all of the more complex conditions (conditions 4–7) showed at least marginally higher success rates than any of the three basic conditions, only conditions 6 and 7 resulted in a significantly better performance with little variability among the subjects. This agrees with our assumption that these more complex combinations should be easier to recognize and therefore result in higher success rates. Although subjects had lower success rates in condition 2 (instrument only) than in condition 3 (tempo only), the instrument property seems to positively affect performance when used within a combination. Of the conditions that combine two properties, the change in the instrument property seems to make the recognition more successful. For example, condition 4 (Pitch + Instrument) and condition 6 (Instrument + Tempo) are more successful than condition 5 (Pitch + Tempo). Although condition 7 (Pitch + Instrument + Tempo) seems to result in higher success rates, it is not much different from condition 6 (Instrument + Tempo). This may indicate that pitch does not contribute much to the success rate within a combined condition.

Time to Recognize (13 subjects)

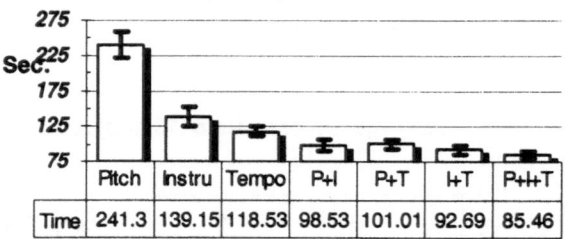

Figure 3. Average total time needed to recognize a series of 15 audio signals.

Figure 3 depicts the average time to complete the recognition phase (in seconds) for all subjects for each condition with the standard error bars. Note how much faster the subjects are able to recognize the signals in conditions after the first one (nearly twice as fast!). This time still improves in conditions 2 and 3 and reaches a plateau for the more complex conditions (4 to 7). This improvement may indicate a very strong learning curve, were learning is essentially complete after the subject has become familiar with three different sound properties (after condition 3).

Confidence (13 subjects)

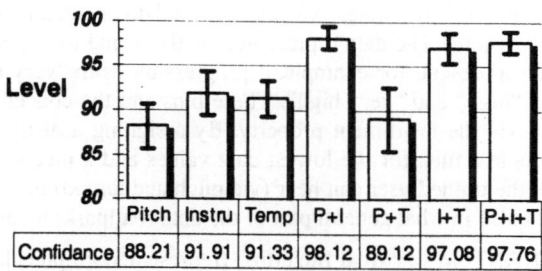

	Pitch	Instru	Temp	P+I	P+T	I+T	P+I+T
Confidance	88.21	91.91	91.33	98.12	89.12	97.08	97.76

Figure 4. Average subjective confidence levels.

A measure of the subject's confidence about his or her choice was recorded on a scale of 0 to 100. Figure 4 depicts the average confidence level for all 13 subjects for each condition; along with the standard error bars.Of the four more complex conditions, subjects reported the least confidence in condition 5 (Pitch + Tempo), with only slight differences among the other three conditions (4, 6 and 7). This unexpected finding is interesting because condition 5 is the only complex condition that lacks a change in the instrument. A similar trend was also observed in the success rate. This suggests that the combination of changing instrument and either pitch or tempo may enhance both success in recognizing the audio signal as well as confidence in the ability to correctly recognize the audio signal.

The study demonstrated that any subject, musical or not, can be trained to differentiate between five different audio signals and connect those audio signals with numbers 1 to 5 (a so-called ballpark setup). After running though several different conditions (scenarios), each divided into a short training phase and a recognition phase, the subjects achieved a 95% success rate for a recognition scheme that simultaneously varies pitch, tempo and instrument. This result indicates that it is possible to use the auditory channel for applications that need to convey a secondary data stream and training can be achieved in a short period of time (20 to 40 minutes).

4. Multi-sensory investigation of geoscientific data with GDIS

We conducted a case study of GDIS's multi-sensory abilities by investigating surface-mesh based geoscientific data. A high-resolution elevation model of a seafloor (derived from multi-beam bathymetry data) was texture-mapped with the residual mantle-boguer-anomaly (RMBA gravity map, see Figure 6). The gravity is visually mapped with a Blue-Green-Red color map, a map of the age of the oceanic crust (calculated from magnetics) was mapped into the audio domain (sound map) and the change of slope of the bathymetry was expressed in a "friction map".

As a result of this study, the sonification in GDIS uses both a relative and an absolute form of sonification. Only a single tone, defined by pitch, tempo and instrument are played. The data value of the sound map is used to synthesize the current tone with a

MIDI synthesizer. The pitch and the tempo use a simple linear mapping: low data values map to a low pitched, slow sequence of notes whereas high data values map to a series of high pitched, fast notes. Most users of GDIS seemed to intuitively understand this form of mapping. The data represented by the sound map can be divided into five "bins", which represent, for example, a progression from "very low" and "low" over "medium" to "high" and "very high". These bins and the concepts connected to them can be heard via the instrument property. By assigning a distinct instrument to each bin (for example a tuba for the lowest data values and a piccolo flute for the highest data values), the trained user can hear (via pitch and tempo) not only how the data are changing, but also in what general part of the data (ballpark) it currently falls.

When the PHANToM's "virtual fingertip" (represented graphically by a cone) touches the surface, the force feedback allows the user to feel the sometimes-delicate surface features and get an idea of the age of the surface at this point by listening to the pitch, instrument and duration of the currently played notes. The user is also able to perceive inflection points by feeling an increase in friction. The user navigates by "grabbing" the surface with the stylus. Holding down the stylus button attaches the surface to the stylus. With the surface attached to the user's hand, the user can easily look at the surface from different angles and distances and, in general, investigate the change of surface curvature. The application offers a special surface coloring, which shows not only the magnitude of the slope, but also its direction as a change of color (Figure 7). By changing the position of the global light source (virtual sun) and by using a virtual flashlight attached to the end of the virtual fingertip, the user can enhance the appearance of geologically interesting surface structures.

Besides the aforementioned settings, GDIS can be configured to freely explore other combinations of visual (texture) maps, sound maps and friction maps by allowing the user to designate any surface variable as either a visual, sound or friction map. Although the pure exploration of a data set by multi-sensory means is one important aspect of our research, we also augmented the commonly employed, interactive task of digitizing line segments onto a surface and extracting surface data from digitized polygons. We were particularly interested in the PHANToM's ability to interact with surface morphology while simultaneously receiving input about several surface attributes via visual, acoustic and haptic means. Hitting the space key on the keyboard digitizes line segments–a new point will be dropped exactly where the virtual fingertip touches the surface. The system then drapes a new three-dimensional line segment on the surface. Line segments can be closed to form polygons from which internal points can be extracted. Figure 6 depicts the seafloor data set after digitizing an important geological feature, a major fault line (bright line left).

Using GDIS, we have investigated a tectonically interesting area on the Mid-Atlantic Ridge, where a new oceanic crust is created in water depth of 3000 m to 6000 m. By precisely digitizing fault structures while simultaneously accessing other surface properties, we are working to improve models of recently discovered dome-like structures called mega-mullions, which promise to grant insight into the deeper structures of the ocean floor. Figure 7 depicts a structural model of fault planes directly the on Mid-Ocean ridge. Note the "virtual fingertip", represented as a cone and its flashlight-effect (lower left corner). In the future, we plan to apply this system to other surface data with multiple overlapping properties. For example, in the remote sensing domain,

GDIS may offer a way to aid in the interactive planning of a pipeline by giving access to multiple constraints at once.

5. User-feedback and Discussion

GDIS was presented on several occasions to a number of geoscientists at the Annual Conference of the Society of Exploration Geophysics (SEG 2000) in Calgary, Canada in August 2000. At the conclusion of the twenty minute demonstration of GDIS' functionality, the audience was invited to experience the multi-sensory system first-hand by "test-riding" GDIS. About 50 geoscientists with different backgrounds used the system for different periods of time (5 min to 60 min), during which they were brought into contact with all major points of the system: three-dimensional visualization, haptic feedback, and sonification. All users were geoscientists, from a variety of geoscientific backgrounds including seismic interpretation, visualization, remote-sensing/GIS, structural geology, seismic acquisition, and geophysical data processing. Some of the users (27) recorded their impressions on a short questionnaire. This section reports on the results of these impressions. The results may be indicative of general trends and provide feedback with regard to a deployment in other parts of the geosciences. The users spent between 5 and 60 minutes with the demonstrator (13.15 minutes on average). Of the 27 geoscientists, 18 use some form of digitizing as part of their work. The user's ratings for the system's three components and the overall system (Figure 9) were very high: on the average the graphics part was rated 8.52; haptics was rated 7.59; sound was rated 6.52 and the overall system 7.93.

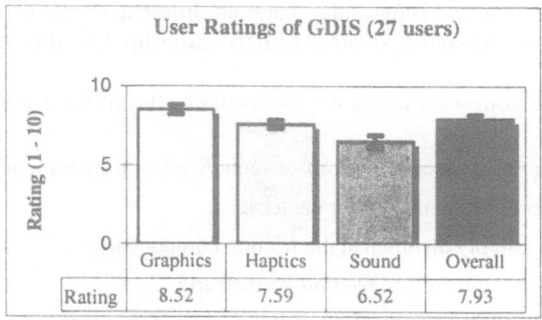

Figure 9. GDIS User Ratings with standard error bars.

Among the users, 16 could see a system like GDIS as useful in their current job and that it would add value to it. When asked about what they liked most about GDIS, the users mentioned the ability to perceive information through more than the visual channel, the ability to navigate and manipulate in three-dimensional with the PHANToM and the ability to switch between the surface's attributes. The following list contains some of the users responses:

- The ability to realize unseen information
- Haptics provides additional sensory input for investigation surfaces
- Ability to follow faults in the "inverted" surface
- PHANToM is a natural tool for navigation and manipulating three-dimensional objects
- Feeling the surface to "visualize" the structural framework
- Interesting application of sound
- Tactile feedback
- Navigation and the spotlight
- Liked the sound idea (hearing your data)
- Graphics (rotating, illumination, texture)
- The way to hear "steps" between instruments
- Overall system integration
- Interactive feedback
- Multiple parameters sensed simultaneously
- The uniqueness of the touch/hear system
- The option, ease to change between attributes
- Integration of senses - ability to query multiple data sets at once

When asked what they liked least, the users mentioned the lack of available customization, problem in distinguishing between different levels of friction and the limitation to perceive the volume of areas with the stylus tip. Specifically:

- Sound is somewhat distracting but could be useful for listening to additional data-sets
- Needs user interface to customize sound/haptics for parameters
- Needs to be personalized for each user
- Hard to reliably distinguish the friction component
- Did not feel the surface friction very clearly
- Hard to scan large surfaces or volume with single point

When asked, if the users could imagine an area for a "killer application" they mentioned seismic interpretation with multiple attributes, aiding GIS-related planning tasks and help for the visually challenged:

- Digitizing faults on coherence time slices
- Planning of logistics and routes
- Shaping the surface to volume data
- Medical (surgical) application

- Adding 3D graphics to GIS
- Color blind, blind interpreters, interactive visualization, learning tool
- Attribute interpretation based on engineering and geophysical data
- Assisting seismic interpretation by tracking amplitude and hearing other attributes simultaneously

GDIS was generally very well received and praised for its novel approach. Most users were exploration geophysicists with a seismic interpretation or visualization background from the oil and gas or minerals domain. Although the demonstrated task, the structural exploration of high-resolution bathymetry data with gravity data and age data mapped onto it, was not familiar to many users, most of them could see a deployment of the system for a similar task closer to their own work. Users specifically liked the ability to receive information about "invisible" data via friction or sound while exploring the surface and the ability to explore and to interact with the surface in true 3D.

The demonstrations were expected to stimulate the users about the potential deployment of multi-sensory systems in their specific line of work. Several users voiced suggestions for applications in other areas, for example seismic interpretation and planning tasks based on remote sensing/GIS type data. In the seismic interpretation area, it was suggested to be used for modeling of faults or other structural elements on subsurface horizons with multi-sensory access to several different seismic attributes such as amplitude or coherence or geostatistically derived properties such as rock facies, permeability, or porosity. For the remote sensing area, users suggested possible applications related to planning of structures such as pipelines or roads on topographic or bathymetric surfaces within a three-dimensional GIS system, with haptics conveying the slope and sound warning about unseen problems.

6. Summary

In this paper, we have presented a system that integrates a novel sonification method and haptic force feedback with three-dimensional visualization to form a multi-sensory user interface for exploring geoscientific data. The system has been developed for the specific geoscientific task of investigating geological structures on the seafloor. However, it is generic enough to be used in the exploration and modeling of other geoscientific surfaces. The system has been demonstrated to geoscientists and was very well received. Many geoscientists acknowledged the system's potential for a future deployment in other parts of the geosciences and provided valuable feedback.

14

Using a combination of graphics, haptics and sound to solve a particular geoscientific task is still a very new field that needs to be explored in a wider context. The fields of haptics and sonification are still at an early stage of their development—perhaps comparable to computer graphics 10 or 15 years ago. Only time will tell if these fields will experience a similar explosion of technological possibilities over the next years and become an integral part of our life as well. The way that our system makes use of the three senses is clearly just a first step and a glimpse of what could be possible in the future.

References

1. Aviles, W.A. and J.F. Ranta. Haptic Interaction with Geoscientific Data. In: Fourth PHANToMS User Group (PUG) Meeting. 1999. Cambridge, MA.

2. Barrass, S. and B. Zehner. Responsive Sonification of Well-logs. In ICAD 2000. 2000. Atlanta, GA: ICAD.

3. Fröhlich, B., et al. Exploring Geo-Scientific Data in Virtual Environments. In: IEEE Visualization. 1999.

4. Harding, C., Multi-Sensory Investigation of Geoscientific Data: Adding Touch and Sound to 3D Visualization and Modeling, PhD Thesis, Department of Geosciences. 2000, University of Houston: Houston. p. 170.

5. Harding, C., R.B. Loftin, and A. Anderson, Visualization and modeling of geoscientific data on the Interactive Workbench. The Leading Edge, 2000. 19(5): p. 506-511.

6. Kramer, G., An Introduction to Auditory Display, in Auditory Display: Sonification, Audification and Auditory Interfaces, G. Kramer, Editor. 1994, Addison Wesley: Reading, MA. p. 1-58.

7. Moore, B.C.J., An introduction to the psychology of hearing. 4th ed. 1997, Orlando, FL: Academic Press.

Editors' Note: see Appendix, p. 333 for colored figures of this paper

Acquisition and Display of Real-Time Atmospheric Data on Terrain

Tian-yue Jiang, William Ribarsky, Tony Wasilewski, Nickolas Faust, Brendan Hannigan, and Mitchell Parry

GVU Center, Georgia Institute of Technology

{jiangf, ribarsky, brendan, parry}@cc.gatech.edu

{tony.wasilewski, nickolas.faust}@gtri.gatech.edu

Abstract. This paper investigates the integrated acquisition, organization, and display of data from disparate sources, including the display of data acquired in real-time. In this case real-time acquisition and display refers to the capture and visualization of data as they are being produced. The particular application investigated is 3D dynamic atmospheric data on terrain, but key elements presented here are applicable more generally to other types of real-time data. 3D Doppler radar data are acquired and visualized with global, high resolution terrain. This is the first time such data have been displayed together in a real-time environment and provides the potential for new vistas in forecasting and analysis. Associated data such as buildings and maps are displayed along with the weather data and the terrain. A global hierarchical structure makes these disparate data available for integrated visualization in real-time. Requirements for effective 3D visualization for decision-making are identified, and it is shown that the applications presented meet most of these requirements.

Keywords: weather, visualization, atmosphere, decision-support, terrain, large scale data, real-time

1 Introduction

The advent of nationwide meteorological networks such as 3D Doppler radar, high resolution weather satellites, and automated surface sensors has lately given weather forecasters and researchers who develop improved analysis and predictive tools access to much more observational data for making decisions in severe weather situations than ever before. These data sources provide higher spatial and temporal resolution than was previously available, but processing this vast amount of information in order to extract and display what is useful to the forecaster or analyst presents a formidable challenge. For the weather forecaster the challenges and benefits are especially acute. Even as the forecasting community struggles to take advantage of abundant observational data, the general public is expecting data that are both more precise and more highly customized to their particular geography and lifestyle. To meet this expectation, it is necessary to integrate very detailed thematic data (e.g., terrain, roadways, rivers/streams, political boundaries, landmarks) with the weather data. There are significant advantages to combining these data into a new type of universal dataset with significant limits on times of access, display, and analysis.

In this paper we present initial results that concentrate on the integrated acquisition, organization, and visualization of atmospheric data with high resolution terrain. Among the atmospheric results that can be included are 3D Doppler radar, satellite imagery, chemical plumes, weather simulations, and other data. With respect to the Doppler radar measurements, this is the first time such data have been acquired and displayed together in a real-time environment. Heretofore the Doppler radar features have only been displayed, after analysis, as moving 2D icons on 2D maps with no terrain elevation information and with no direct display of radar feature heights or volumes (though that information is available). The real-time environment is crucial for decision-making in severe storm situations where, for example, weather forecasters may have only a few minutes to issue warnings to affected areas [Eil95]. The integrated data permits observation of geospatial features that may affect weather patterns and the combining, for example, of flood and terrain models for accurate flood prediction. In addition the results present an extension of current real-time terrain data organizations and visualization systems since these are not designed to handle the 4D (time + space or dynamical) aspects of the integrated data sets.

2 Related Work

Weather visualization has been a fairly popular topic over the years. We will not give here an exhaustive review of this work but will rather concentrate on representative work related to interactive visualization of weather data or simulations. In many cases these visualizations are intended for analysis of weather patterns and in a few cases for decision-making.

There are visualization/analysis tools developed on top of toolkits. For example, the Vis5D display and analysis tool [Hib96] and Display 3D (D3D94) developed by the National Oceanic and Atmospheric Administration (NOAA) Forecast Systems Laboratory have 3D display capability. However, these tools have been designed not as general 3D visualization interfaces, but rather to focus on numerical model output. In addition, these tools do not have an operational decision support focus (attained by supplying the most pertinent information for decision-makers). Finally they do not address the issues of scalably large data and integration of geospatial data. There are also tools developed on platforms such as AVS Express [Che96]. These tend to be for analysis of fixed data rather than real-time visualization of potentially large and constantly updated datasets.

There is also research that has combined visualization with the value-added of storm tracking analysis. Cheng et. al. [Che98] use fuzzy logic to develop a storm tracking algorithm that presents results that closely match an expert meteorologist's perception. This work is relevant to our approach of displaying results of automatic analysis to aid in decision-making. However, it does not address the issues of real-time 3D visualization and does not present weather along with other relevant data, such as terrain. So far, the major real-time acquisition and display system [NWS98] is the Advanced Weather Interactive Processing System (AWIPS). However, it only provides 2D visualization. The present paper provides the first example of real-time acquisition and interactive visualization of 3D weather data from multiple sources.

Treinish [Tre98] has considered the general issues of visualization design for weather forecasting, including 2D and 3D visualizations. The latter are classified as either 3D browsing or analysis tools. The perceptual issues in using color and the representations for quantities such as wind vectors are also considered. Our approach combines browsing and analysis in a coherent 3D setting. We also build on some of the design issues presented in [Tre98].

Djurcilov and Pang [Dju99] address the issue of missing data in Doppler radar and other gridded data. Because of the curvilinear nature of the radar scan (Fig. 1), the data are rather sparse when usual visualization techniques (i.e., isosurfaces and volumetric rendering), which use regular grids, are applied. To the extent that the pre-analysis tools in our approach take into account the data non-uniformity (and the possibility of false readings), some of this uncertainty is accounted for. However, it is still useful to show the locations and scan patterns (and location-dependent uncertainties) for the radars, especially since in the future overlapping radars will be used. In the future we also expect to apply more direct and detailed rendering of data; here the considerations in [Dju99] will be important.

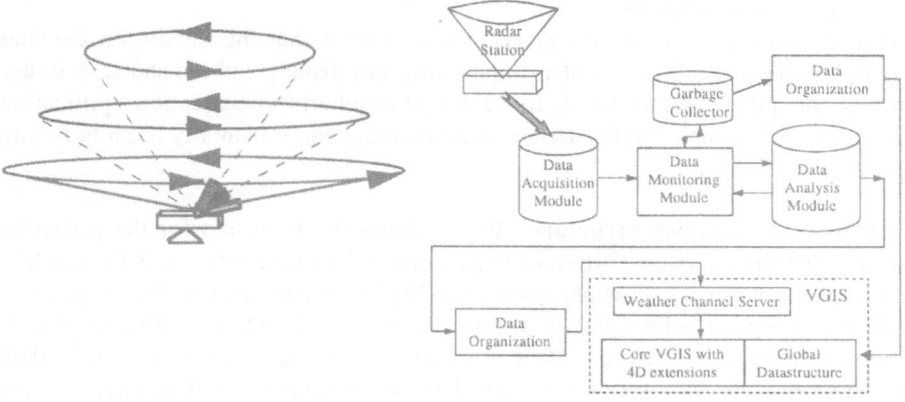

Fig. 1 Sparse scanning geometry for
3D Doppler radar.

Fig. 2 Flow chart of acquisition and display of
Real-time weather data on terrain.

In this work we present time-dependent weather and other atmospheric patterns in a global terrain visualization environment [Fau00]. We have built a highly interactive exploratory visual interface [War99a, War99b] for this environment that will be useful for decision-makers. The global environment is effective for out-of-core visualization, which is necessary for the large-scale data considered here. A principle reason for the effectiveness of our approach is that it uses application-controlled demand-paging [Cox97, Dav98], where the system knows something about what data are needed and when. We have shown that the same concept can be extended beyond terrain to other geospatial data, such as buildings and other static objects [Dav99]. We are confident that the global hierarchy can be extended to moving objects and other geo-located features.

3 Data Acquisition and Visualization

Requirements for Decision-Makers. We list first some requirements for real-time acquisition and effective visualization for decision-making (including weather and chemical/biological situation forecasting).

- Real-time data acquisition and data communication. This requires that real-time acquired data should be organized and inserted in data structures, then transferred to data analysis modules for analysis, and finally displayed in an interactive environment.

- End-to-end real-time capability. This means not only real-time acquisition but also visualization and on-the-spot analysis in appropriate time budgets.
- Capabilities for both browsing (or exploring) and analyzing time-dependent 3D data, including historical data.
- Easy-to-use 3D navigation and manipulation.
- Details on demand, to eliminate clutter presenting the important details initially.
- Integration of relevant data in a combined scene (e.g., weather and terrain).
- Easy-to-use quantitative tools and information-retrieval tools to augment the qualitative visualizations.

Schneiderman and co-workers [Tan97] have shown that the details on demand strategy can be very effective when coupled with fast display updates and easy-to-use controls for adjusting the detail. Since the atmospheric visualization application presented here is coupled to our terrain navigation system, it naturally has a browsing capability.

Acquisition and Display Structure. Fig. 2 shows the flowchart for the particular case of 3D Doppler radar. Different acquisition points would be used for satellite imagery and other types of data. As shown in Fig. 2, remote radar stations (operating as shown in Fig. 1) collect sets of volumetric data. Each set of volumetric data is composed of multiple sweeps. Once a sweep is done, data are passed to a data acquisition module via a set of T-1 lines. The data acquisition module organizes the raw volumetric data into a form appropriate for the hierarchical global data structure [Fau00]. The data analysis module then applies a set of pre-analysis and modeling tools, using methods developed by the National Severe Storms Lab (NSSL) [Eil95, Joh98, Mit98]. The tools are embedded in the Warning Decision Support System (WDSS), used by weather forecasters to make severe storm and tornado warning decisions. The pre-analysis is described further in the results section below. Since it is made for operational weather forecasting, the WDSS can analyze 3D Doppler radar on-the-fly. An extension of the structure of our terrain visualization system permits immediate insertion of these data and the accompanying raw volumetric data as a time-stamped stream of objects for real-time display.

Typically it takes about 5-7 minutes to collect a set of volumetric data. In order to prevent the constant stream of networked data from backing up, there is an incoming data monitoring module (Fig. 2). If the data analysis module can not analyze the incoming data faster than the data acquisition module delivers it, the data monitoring module will dump obsolete data to a collector for later organization and recording. After volumetric data is analyzed, the organized data (both raw and analyzed data) is sent in an efficient, compact form to the real-time server channel of VGIS. VGIS, which supports the interactive display part of our visual interface, then provides an updated visualization.

Visual Interface. Any interface for extended 3D data presents challenges for the development of effective and intuitive tools for navigation and selection, especially when using a mouse. There are two navigation modes, orbital and fly in our interface

(C.P. 1 and C.P. 3—see the color plates). Since we use a variety of interaction tools, including the mouse, joystick, and devices with 6DOF trackers, we must consider how to best operate these modes for each tool. The orbital mode presents a god's eye view and has navigation characteristics similar to those for a 2D interface (including panning, zooming, and rotation). This mode is straightforward to operate with all interaction tools. For the fly mode (similar to flying along the Earth in an airplane) using the mouse and to some extent the joystick, we have found after significant testing and evaluation of alternatives that it is best to constrain degrees of freedom. In the present work we map pitch to vertical mouse movement and yaw to horizontal mouse movement. The attitude of the viewpoint is fixed on a platform parallel to the Earth's surface, so there is no roll. One navigates forward by pushing the left mouse button and backward by pushing the middle mouse button. One changes speed through a menu option. This interface is reasonably straightforward to use and makes operations such as taking a 360° look about a location easy to do.

We have implemented a more intuitive 3D interface for the 6DOF device on the virtual workbench (C.P. 4). Here one switches from orbital to fly mode simply by turning a tracked button stick (shown in C.P. 4) to vertical position. The interface automatically switches to a mode where the stick acts like an airplane joystick. Turning it to left or right causes yaw to left or right; pushing forward or backwards causes flight in that direction; lowering or raising the stick causes descending or ascending flight. This interface is reasonably intuitive to use. In the following discussion we will concentrate on the mouse-based interface, though the capabilities described can also be applied to the workbench interface.

For the remainder of this section, we will discuss the visual interface in terms of the weather application. However, applications using other types of atmospheric data can be invoked, as discussed in the next section. One can individually turn on or off the time-flows of raw and analyzed weather data, or view them simultaneously, as in C.Ps. 1 and 2. This permits forecasters to quickly isolate details or view correlated phenomena. The latter can be useful since the detailed development of the raw data may reveal behavior not shown in the analyzed data. The forecaster can fly in for closer inspection, including correlation with terrain or map features, or back out for an overview. The bright cylinders in C.Ps. 1 and 2) reveal analyzed mesocyclone cells with possible tornadic action. These correlate with the underlying patterns in the raw data.

In addition to on-the-fly results, the forecaster must often analyze previously archived histories. For efficient use of this mode, we have provided a control panel. This control shows where the displayed frame is in a given time series and permits the frame rate of weather steps to be slowed down or speeded up. The user can also specify a time range or a time step and immediately see it. Thus the forecaster can look more carefully and more slowly at particular sequences of steps.

A user will want to quickly obtain detailed 3D views from any location and angle. To do this we have provided a selection and jump mode. Selection is straightforward in orbital mode but more challenging in fly mode. For the latter the system casts a ray from the eyepoint through the selected point and finds the terrain intersection point. If the ray misses the terrain, no selection is made. This ray and the intersection point are displayed, which helps the user perceive the location of the intersection point and also select an altitude value. The latter is done by moving the mouse up or down, with the ray end following the moving cursor. The altitude value is displayed at the top of the window (Fig. 3). Upon completion of the selection, the ray is replaced by a ray perpendicular to the Earth ellipsoid [War9a] at the intersection point and of length equal to the chosen altitude (Fig. 3). Upon selection the user's view now jumps to the selected 3D location and switches to fly mode. The user is then free to look or move around. A jump-back option on the menu permits the user to return to the original viewpoint position and navigation mode. Not only does the jump mode permit the user to quickly move in for detailed views, it reduces the possibility of getting lost during navigation. Selected positions remain displayed for future use but can be turned off via a menu option.

4 Results

Real-Time Weather Volumes. We have used our system on sequences of analyzed data from a Doppler radar located at the National Weather Service (NWS) facility in Peachtree City, GA. We visualize both raw data and analyzed results, using simple shapes for the latter. Our display meets the real-time requirement. The process of data acquisition, analysis, preparation for display, and display takes about 1 minute (Fig. 2). Most of this time is spent in the analysis step, which could be speeded up. This should be compared with the 5-7 minutes it takes to collect the 3D Doppler radar volume, which involves a series of 2D scans at progressively larger angle with respect to the ground (Fig. 1). For a series of time steps, display is typically in the range of 5-15 updates per second in a stereoscopic environment and about double that in a monoscopic environment. (The update rate depends on the amount of data displayed. It is 15 fps if only the analyzed shapes are shown.)

To show the capabilities of our methods in detail, we use some previously captured severe storm data. Forecasters will often look at histories as well as at current data. The data presented are from a series of severe storms that occurred over Georgia from 2 am till 10 pm on March 19, 1996. There are 72 time steps in these results, but the system could handle a much larger group of time steps. The analyzed data were in the form of mesocyclones and tornadic vorticity signatures. The mesocyclones are areas of large coherent rotation and are possible precursors to tornadoes. The tornadic signatures are obtained by looking for compactness, intensity, and shear in adjacent radar bins. At least some of these signatures were actual tornadoes. The heights and elevations off the ground (which can be seen clearly in fly mode) of the mesocyclones are indicators of their power and potential for damage. Both types of signatures were built from the stacked 2D scans using simple spatial correlation. The 3D view of the arrangement and time evolution of these structures can provide the forecaster with useful additional information.

We used semi-transparent cylinders and cones to represent the mesocyclones and tornadic signatures respectively. The rotation intensity was mapped to the color with the range of color chosen to be similar to the range employed in 2D visualizations used by forecasters. Icons similar to the ones used in the 2D visualizations were attached to the tops of the mesocyclones so that in orbital mode the 3D scene looks similar to the 2D scene. However, these features are also shown in correlation with the raw 3D radar scans (C.Ps. 1 and 2). In addition, a lot more information is displayed showing correlation with terrain features, urban areas, roads, and rivers. (C.Ps. 2 and 3, and Fig. 4). Details from the full time-dependent Doppler radar set, retrieved on-the-fly from the hierarchical data structure, are shown in C.P. 3. Here we have mapped the 3D data onto cones representing the Doppler scans. We can even display weather features over maps (Fig. 4), which can be interactively blended in or removed. All this additional information has never been visualized in detail simultaneously with time-dependent 3D weather data. Although the correlation between analyzed storm cell features and underlying Doppler radar features is good in C.P. 2, these features can diverge over time—some of the divergence may be due to terrain features or even human activity. In addition, the integrated visualizations give ample, immediate information about what the storm cells are hitting or about to hit. We have added a grid that can be turned on and off through a menu option (C.P. 3) to show the coverage of the radar.

In overview, one can easily see the sweep of the storms as they progress from the Alabama border across North Georgia. (See, for example, C.P. 1.) Since the weather visualization is embedded in a global terrain framework, one does not have discontinuous or truncated views but can move smoothly to any view with higher resolution terrain and feature data coming in wherever they are available. For example, the visualization does not stop abruptly at the Georgia border. In the next generation of weather forecasting tools, data will be collected from overlapping Doppler radars and other sensors over a much larger area. The global terrain capability will be even more useful when this occurs.

The orbital and fly modes permit a continuous movement between browsing and more detailed analysis. For example, Fig. 4 shows an overview of storm cells as they increase in size and intensity over Atlanta. The forecaster can use the jump mode (and then can fly around) to see the relation of these features to urban areas and can also see the heights of these features and their relation to each other and to the ground. The jump mode also permits one to quickly get to the tornadic signatures, which show clearly as bright blue objects with crosses on top (Fig. 3).

We have found that terrain features, such as mountains or buildings, plus the 3D shapes of the storm signatures are especially prominent on the virtual workbench (C.P. 4). Here the stereoscopic display makes the 3D structure "stand out" automatically, even without changes of view by the user. Further, the workbench provides a large work surface for analysis and collaboration among users. The navigation modes are also fast and intuitive.

Chemical/Biological Clouds. Aside from real-time weather data on terrain, our techniques and data modules can also be applied to acquisition and display of chemical/biological clouds in real-time. Through the Center for Emergency Response Technology, Instrumentation, and Policy (CERTIP), we recently worked on urban emergency response to a terrorist attack. This culminated in an exercise on the Georgia Tech campus. The visual interface and global data structure described here were used for situation assessment. The exercise began with the Atlanta Fire Department arriving on the scene. As soon as it was determined that a toxic chemical (Sarin gas) was being released from the roof of a campus building (C.Ps. 5, 6 in orbital and fly modes, respectively), a special team was called in and set up an Emergency Operations Center (EOC). The EOC team used a set of simulation codes to generate time series plumes every 5 minutes. The large colored area emanating from the top of the building in C.Ps. 5, 6 is one of those simulated plumes displayed in real-time. Large red human icons (C.P. 6) are EOC first responders with GPS units and wireless LAN transceivers. Their positions are also updated in real-time. Letters "A, B, C", and "D" (C.P. 5) are labels for operations areas at different locations around the attacked building.

The first responders, wearing level A suits (special garb to protect again toxic chemicals), entered the building, using a chemical sensor to detect Sarin and a radar flashlight to detect whether people were behind closed doors. They used high resolution imagery and 3D buildings for the Georgia Tech campus (C.Ps. 5, 6) to plan their movements. Query capability was installed in a web-server version of the visual interface and the personnel could click on the building where the attack took place and retrieve floor plans for each floor. They could also retrieve overview maps for the nearby downtown Atlanta area. The visualizations were dynamically updated with simulation and other movement information every 40 seconds. The mobile, wireless LAN sent the products from the EOC (maps and plumes) and retrieved medical info from the first responders (blood pressure, pulse, heart rate, and EKG along with a web-cam pictures of the victims after removal from the building). This "Reachback" was intended to allow doctors at a nearby medical evacuation location to advise the first responders on diagnosis and treatment. In the future, we will augment the urban database employed here and use it with entirely mobile computers (e.g., laptops and wearables) with wireless communication. We will also develop new interfaces for the mobile applications.

5 Conclusions and Future Work

We have presented methods and results for real-time acquisition, organization, and visualization of atmospheric data in a geospatial environment, emphasizing capabilities that will be of use to forecasters and other decision-makers. We expect these capabilities will also be useful to others concerned with the analysis of weather or other atmospheric data, such as researchers or planners. Our atmospheric visualization application meets most of the requirements listed at the beginning of Sec. 3. In particular the application provides real-time acquisition, end-to-end real-time capability, integrated browsing and analysis, details on demand, and integration of relevant data in one visualization. This last capability can help researchers develop better models of storm development, which will yield rules for how storms behave in the presence of hills or mountains and other features.

The results show how time-dependent atmospheric data in a geospatial environment can be effectively explored visually using appropriate interactive tools. These include direct manipulation tools for navigation and manipulation, and interface elements for controlling animation and scale. With these tools and with multiresolution global visualization the user is able to quickly get to features of interest and to gather more information than was available before for decision-making.

We have several avenues of future work. We expect to redesign our prototype application and bring more elements from the menus into an always visible control panel for faster access. We will also produce a finder window that will give a simultaneous wide area orbital view. As the user moves around the main window, an icon showing current position and direction will be updated in the finder window. The moving storm cells will also be displayed in the finder window, and the user can execute jumps from either window. Once these features are in place, we will give the application to colleagues at the NSSL and to NWS forecasters for evaluation. In the longer term we will develop new methods for detailed rendering of the 3D radar data. These data will be in the form of "hierarchical geospatial volumes" so that they have levels of detail that fit into our overall global data structure [Fau00, Dav99]. This will be a major step towards fully implementing the most detailed and accurate views of the evolving 3D weather patterns.

Acknowledgments

This work was performed in part under a grant from the NSF Large Scientific and Software Data Visualization program, under a MURI grant through ARO, and under a grant from ONR in conjunction with the Naval Research Lab.. The global terrain visualization system was developed under a contract from the U.S. Army Research Laboratory. We would like to thank Kurt Hondl, V. Lakshmanan, Tom Vaughan, and their group at the NSSL for supplying 3D Doppler radar data and the WDSS analysis tools.

References

Che96   Chen, P.C. Climate and weather simulations and data visualization using a supercomputer, workstations and microcomputers. *Proceedings of the SPIE*, Vol.2656, pp. 254-264 (1996).

Che98   D. Cheng, R. Mercer, J. Barron, and P. Joe. Tracking severe weather storms in Doppler radar images. *Int. Journal of Imaging Systems and Technology*, Vol.9, pp. 201-213 (1998).

Cox97   M. Cox and D. Ellsworth. Application-Controlled Demand Paging for Out-of-Core Visualization. *IEEE Visualization '97*, pp. 235-244 (1997).

D3D94   For a description, see www-sdd.fsl.noaa.gov/~jwake/WFO-A-intro.html.

Dav98   D. Davis, T.Y Jiang, W. Ribarsky, and N. Faust. Intent, Perception, and Out-of-Core Visualization Applied to Terrain. Rep. GIT-GVU-98-12, pp. 455-458, *IEEE Vis. '98*.

Dav99   D. Davis, W. Ribarsky, T.Y. Jiang, N. Faust, and Sean Ho. Real-Time Visualization of Scalably Large Collections of Heterogeneous Objects. *IEEE Visualization '99*, pp. 437-440.

Dju99   S. Djurcilov and A. Pang. Visualizing gridded datasets with large number of missing values. *Proceedings IEEE Visualization '99*, pp. 405-408.

Eil95   M.D. Eilts, J.T. Johnson, E.D. Mitchell, S. Sanger, G. Stumpf, A. Witt, K. Hondl, and K. Thomas. Warning Decision Support System. *11th Inter. Conf. on Interactive Information and Processing Systems (IIPS) for Meteorology*, Oceanography, & Hydrology, pp. 62-67 (1995).

Fau00   N. Faust, W. Ribarsky, T.Y. Jiang, and T. Wasilewski. Real-Time Global Data Model for the Digital Earth. *Proceedings of the INTERNATIONAL CONFERENCE ON DISCRETE GLOBAL GRIDS (2000)*. An earlier version is in Report GIT-GVU-97-07

Hib96   W.L Hibbard, J. Anderson, I. Foster, B.E. Paul, R. Jacob, and C. Schafer. Exploring coupled atmosphere-ocean models using Vis5D. *International Journal of Supercomputer Applications and High Performance Computing*, vol. 10, no. 2-3, pp 211-222 (1996).

24

Joh98   J.T. Johnson, Pamela MacKeen, ArthurWitt, E. DeWayne Mitchell, Greg Stumpf, Michael D. Eilts, and Kevin Thomas. The Storm Cell Identification and Tracking Algorithm: An Enhanced WSR-88D Algorithm. <u>Weather and Forecasting</u>, vol . 13, pp. 263-276 (1998).

Mit98   E.D. Mitchell, S. Vasiloff, G. Stumpf, M.D. Eilts, A, Witt, J. T. Johnson, and K. Thomas. The National Severe Storms Laboratory Tornado Detection Algorithm. *Weather and Forecasting*, vol. 13, no. 2, pp. 352-36 6 (1998).

NWS98 National   Weather   Service.   A WIPS   Program   Information. http://www.nws.noaa.gov/msm/awips/awipsmsm.html, March 1998

Tan97   E Tanin, R Beigel, and B Schneiderman. Design and Evaluation of Incremental Data Structures and Algorithms for Dynamic Query Interfaces. *Proc. InfoVis '97*, pp. 81-86 (1997)

Tre98   Treinish, L.A. Task-specific visualization design: a case study in operational weather forecasting. *Proceedings IEEE Visualization '98*, pp. 405-409.

War99a Z. Wartell, W. Ribarsky, and L. Hodges. Efficient Ray Intersection for Visualization and Navigation of Global Terrain. Eurographics-IEEE Visualization Symposium 99, <u>Data Visualization '99</u>, pp. 213-224 (Springer-Verlag, Vienna, 1999).

War99b Z. Wartell, W. Ribarsky, and L. Hodges. Third Person Navigation of Whole-Planet Terrain in a Head-tracked Stereoscopic Environment. Report GIT-GVU-98-31, *IEEE Virtual Reality 99*, pp. 141-149.

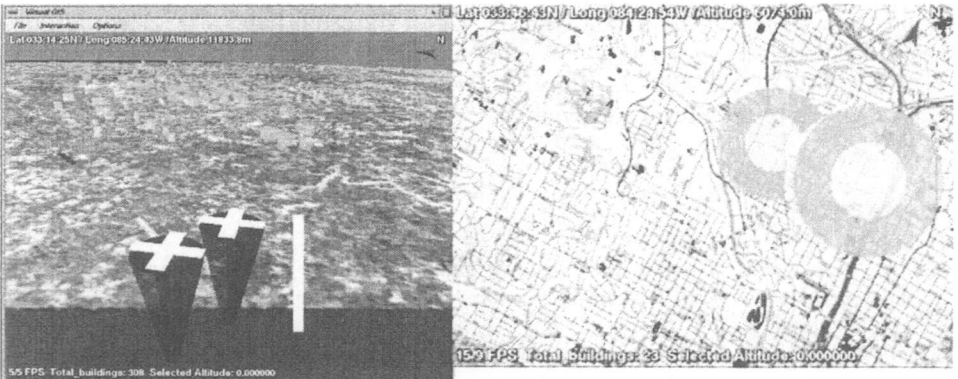

Fig. 3 Tornadic signatures with altitude measure (right line).

Fig 4. Detailed Atlanta map turned on interactively to correlate with mesocyclone features.

Editors' Note: see Appendix, p. 334 for colored figures of this paper

Extraction of Crack-free Isosurfaces from Adaptive Mesh Refinement Data

Gunther H. Weber[1,2,3], Oliver Kreylos[1,3], Terry J. Ligocki[3], John M. Shalf[3,4], Hans Hagen[2], Bernd Hamann[1,3], and Kenneth I. Joy[1]

[1] Center for Image Processing and Integrated Computing (CIPIC), Department of Computer Science, University of California, Davis
[2] Department of Computer Science, University of Kaiserslautern, Germany
[3] National Energy Research Scientific Computing Center (NERSC), Lawrence Berkeley National Laboratory, Berkeley
[4] National Center for Supercomputing Applications (NCSA), University of Illinois, Urbana-Champaign

Abstract. Adaptive mesh refinement (AMR) is a numerical simulation technique used in computational fluid dynamics (CFD). It permits the efficient simulation of phenomena characterized by substantially varying scales in complexity of local behavior of certain variables. By using a set of nested grids at different resolutions, AMR combines the simplicity of structured rectilinear grids with the possibility to adapt to local changes in complexity and spatial resolution. Hierarchical representations of scientific data pose challenges when isosurfaces are extracted. Cracks can arise at the boundaries between regions represented at different resolutions. We present a method for the extraction of isosurfaces from AMR data that avoids cracks at the boundaries between levels of different resolution.

1 Introduction

AMR was introduced to computational physics by Berger and Oliger [3] in 1984. A modified version of their algorithm was published by Berger and Colella [2]. AMR has become increasingly popular in the computational physics community, and it is used in a variety of applications. For example, Bryan et al. [4] use a hybrid approach of AMR and particle simulations for simulation of astrophysical phenomena.

Fig. 1 shows a simple two-dimensional (2D) AMR hierarchy produced by the Berger–Colella method. The basic building block of a d–dimensional Berger-Colella AMR hierarchy is an axis-aligned structured rectilinear grid. Each grid g consists of hexahedral cells. Each grid can be positioned by specifying its origin o_g. The underlying simulation method is a finite-difference method. Typically, a *cell-centered* data format is used, i.e., dependent function values are associated with the centers of the cells. We denote the region covered by the grid by Γ_g. Each grid contains a pointer to an array containing the dependent data values. These are stored in a simple array

An AMR hierarchy consists of several levels Λ_l comprising one or multiple grids. All grids in the same level have the same resolution, i.e., all grids in a level share the same cell size δ_{Γ_l}. The region covered by a level Γ_{Λ_l} is the union of regions covered by the grids of that level.

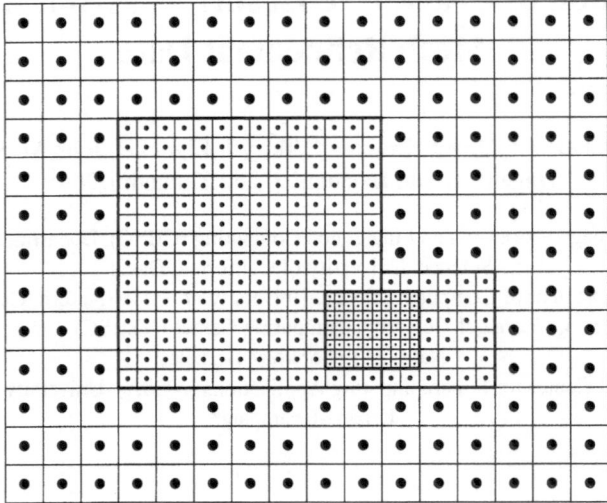

Fig. 1. AMR hierarchy consisting of four grids in three levels. The root level consists of one grid. This grid is refined by a second level consisting of two grids. A fourth grid refines the second level. It overlaps both grids of the second level. Boundaries of the grids are drawn as bold lines. Locations at which dependent variables are given are indicated by solid discs

The hierarchy starts with the *root level* Λ_0, the coarsest level. Each level Λ_l may be refined by a finer level Λ_{l+1}. A grid of the refined level is commonly referred to as a *coarse grid* and a grid of the refining level as a *fine grid*. The *refinement ratio* r specifies how many fine grid cells fit into a coarse grid cell, considering all axial directions. The value of r is always a positive integer. A refining grid refines an entire level Λ_l, i.e., it is completely contained in Γ_{Λ_l} but not necessarily in the region covered by a single grid of that level. Each refining grid can only refine complete grid cells of the parent level, i.e., it must start and end at the boundaries of grid cells of the parent level. Furthermore, there is always a layer with a width of at least one grid-cell between a refining grid and the boundary of the refined level. Due to the hierarchical nature of AMR simulations, the resulting data lend themselves to hierarchical visualization. We discuss a new method for the direct extraction of isosurfaces from AMR data sets.

2 Related Work

Little research has been published regarding the visualization of AMR data. Norman et al. [12] convert an AMR hierarchy into finite-element hexahedral cells with cell-centered data that can take advantage of standard visualization tools (like AVS [1], IDL [7], or VTK [13]), while preserving the hierarchical nature of the data. Ma [9] describes a parallel rendering approach for AMR data. Even though he re-samples the data to vertex-centered data, he still uses the hierarchical nature of AMR data and contrasts it to re-sampling it to the highest resolution-level available. Max [10] describes a sorting scheme for cells for volume rendering, and uses AMR data as one application of his method.

Isosurface extraction is a commonly used technique for the visual exploration of scalar fields. Our work is based on the marching-cubes (MC) method, introduced by Lorensen and Cline [8]. A volume is traversed cell-by-cell, and the part of the iso-surface within each cell is constructed using a look-up table (LUT). The LUT of the original article by Lorensen and Cline contained a minor error that could lead to cracks in the extracted isosurface. This is due to ambiguous cases where different isosurface triangulations in a cell are possible. Nielson and Hamann [11], among others, addressed this problem and proposed a solution to it. Van Gelder and Wilhelms [5] have provided a survey of solutions to this problem. They show that, in order to extract a topologically correct isosurface, more than one cell must be considered at a time. If topological correctness of the isosurface is not required, it is possible to avoid cracks without looking at surrounding cells. In our implementation, we use the LUT from VTK [13] that avoids cracks by taking special care during LUT generation.

Octree-based methods are among the methods used to speed up the extraction of isosurfaces. Shekhar et al. [14] use an octree as a hierarchical representation of the data. By adaptively traversing the octree and merging cells that satisfy certain criteria, they reduce the amount of triangles generated for an isosurface. Their scheme removes the cracks in the resulting isosurface. Westermann et al. [15] modified this approach by adjusting the traversal criteria and improving the crack-removal strategy. Gross et al. [6] used a combination of wavelets and quadtrees to approximate surfaces, e.g., from terrain data. Using an estimate based on a wavelet transform their approach chooses a level in the quadtree structure to represent a given region. Handling transitions between quadtree levels is similar to handling those between levels in an AMR hierarchy.

3 Dual Grids

The MC method assumes that data values are associated with the cell vertices, but the AMR method produces values at cell centers. To deal with this incompatibility problem one can, for example, re-sample the data set to a vertex-centered format. However, re-sampling causes "dangling nodes" in the fine level. Even if the re-sampling scheme assigns the same values to the dangling nodes as the interpolation scheme assigns to them in the coarse level, dangling nodes can cause cracks when using the MC method (see [15]). We solve these problems by using a *dual grid* for isosurface extraction. This dual grid is defined by the function values at the cell centers. The cell centers become the vertices of the vertex-centered dual grid.

The dual grids for the first two levels of the AMR hierarchy shown in Fig. 1 are shown in Fig. 2. We note that the dual grids have "shrunk" by one cell in each axial direction with respect to the original grid. The result is a gap between the coarse grid and the embedded fine grids. Due to the existence of this gap, there are no dangling nodes that could cause discontinuities in an extracted isosurface.

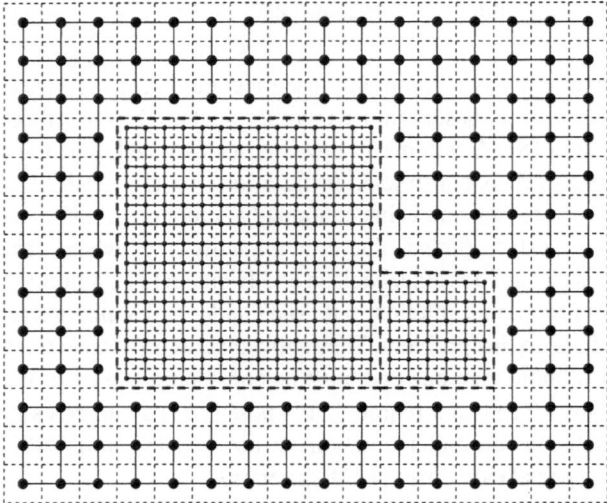

Fig. 2. Dual grids for the three AMR grids comprising the first two hierarchy levels shown in Figure 1. The original AMR grids are drawn in dashed lines and the dual grids in solid lines

4 Stitching 2D Grids

To avoid cracks in extracted isosurfaces as a result of gaps between grids, a tessellation scheme is needed that "stitches" grids of two different hierarchy levels. The resulting *stitch mesh* is constrained by the boundaries of the coarse and the fine grids and can be used to merge levels seamlessly. The stitch mesh must not subdivide any boundary elements of the existing grids. In the 2D case, this is achieved by requiring that only existing vertices are used and no new vertices generated. Since one of the reasons for using the dual grids is to avoid the insertion of new vertices, whenever possible, this poses no problems.

In the 2D case, a constrained Delaunay triangulation can be used to fill the gaps between grids. For two reasons, we chose not to do this. While in the 2D case only edges must be shared between the stitching grid and the dual grids, entire faces must be shared in the 3D case. The boundary faces of rectilinear grids are quadrilaterals and cannot be shared by tetrahedra without being subdivided, thus causing cracks when used in an MC-based isosurface extraction scheme. Furthermore, an index-based approach is more efficient, since it takes advantage of the regular structure of the boundaries while avoiding problems that might be caused by this regular structure when using a Delaunay-based approach.

The stitching process for a refinement ratio of two is shown in Fig. 3. Stitch cells must be generated for edges along the boundary and for the vertices of the fine grid. The stitch cells generated for the edges are shown in dark grey, while the stitch cells generated for the vertices are drawn in light grey. For the transition between one fine and one coarse grid, each edge of the fine grid is connected alternatingly to either a vertex or an edge of the coarse grid. This yields triangles and deformed quadrilaterals

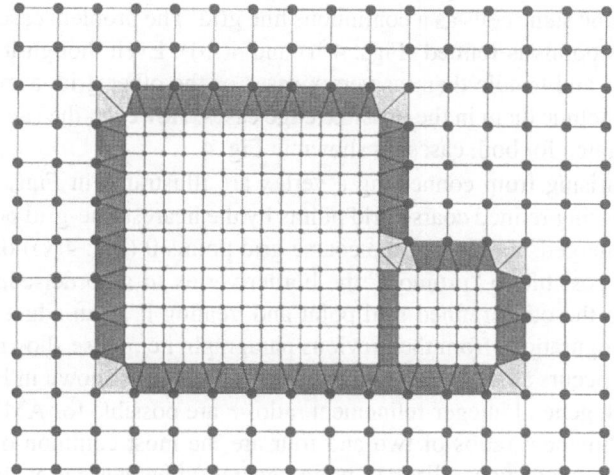

Fig. 3. "Stitch cells" for first two levels of AMR data set shown in Fig. 2

as additional cells. The quadrilaterals are not subdivided, since such a subdivision is not unique. (This in turn would cause problems in the 3D case when these quadrilaterals become boundary faces shared between cells.) The vertices are connected to the coarse grid via two triangles. Here, a consistent partition of the deformed quadrilateral is possible. The obvious choice is to connect each edge to the two coarse edges that are "visible" from it.

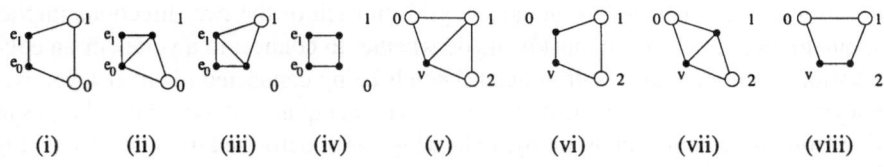

(i)	(ii)	(iii)	(iv)	(v)	(vi)	(vii)	(viii)

Fig. 4. Possible cases for connecting a boundary edge $\overline{e_0 e_1}$ ((i)–(iv)) or a boundary vertex v ((v)–(viii)) to a coarse grid. If cells of the coarse grid are refined, the coarse grid points (circles) are replaced by the corresponding refining point (solid black discs)

In the case of multiple grids, a check must be performed: Are the grid points in the coarse grid refined or not? If a fine edge is connected to a coarse point, this check is simple. If the coarse point is refined, the fine edge must be connected to another fine edge; this yields a rectilinear instead of a triangular cell. The case of connecting to a coarse edge is more complicated and illustrated in Figs. 4 (i)–(iv). If both points are refined (Fig. 4(iv)), the fine edge is connected to another fine edge. As a result, adjacent fine grids yield the same cells as a continuous fine grid. The problem cases occur where only one of the points is refined (Figs. 4(ii) and 4(iii)). Even though it is possible to skip these cases and handle them as vertex cases of the other grid, a more consistent approach is to include them in the possible edge cases. However, the same tessellations should be generated for both cases, as shown in Fig. 4.

The cases arising from connecting a vertex are illustrated in Figs. 4 (v)-(vii). In addition to replacing refined coarse grid points by the nearest fine-grid point, adjoining grids must be merged. If either of the coarse grid points 0 (Fig. 4(v)) or 2 (Fig. 4(v)) is refined, it is possible to "promote" the border vertex to a border-edge segment by connecting it to the other refined grid point and treating it as an edge, and using the connection configurations from the previous paragraph, i.e., those shown in Fig. 4 (i)–(iv). (This case occurs along the bottom edge of the fine grids shown in Fig. 3.)

Even though general integer-refinement ratios r are possible for AMR grids, in 2D simulations, refinement ratios of two and four are the most common ones used. The stitching process can be generalized to more general refinement ratios. Instead of connecting edge segments of the refining grid alternatingly to a coarse-grid edge segment and point, $(r-1)$ consecutive edge segments must be connected to one common coarse-grid point. Every r-th fine edge must be connected to a coarse edge. Even though the valence of the grid points of the coarse grid is increased, this is not a problem with the commonly used refinement ratios. Furthermore, general refinement ratios do not add more refinement configurations, since the fundamental connection strategies remain the same.

5 Stitching 3D Grids

Our index-based approach can be generalized to 3D AMR grids. In the simple case of one fine grid embedded in a coarse grid, quadrilaterals, edges and vertices of the fine grid must be connected to the coarse grid. In each of the two directions implied by a quadrilateral, a decision must be made whether to connect to a vertex or an edge. The various combinations result in quadrilaterals being connected to either a vertex, a line segment (in the two possible directions) or another quadrilateral. The cell types resulting from these connections are pyramids (Fig. 5(i)), deformed triangle prisms (Fig. 5(ii), and deformed hexahedral cells (Fig. 5(iii)).

The edge case can be viewed as a combination of the vertex and edge cases of the 2D case. If the viewing direction is parallel to the edge (such that it appears to the viewer as a point), it must always be connected to two perpendicular edges of the coarse grid. In the direction along the edge, one connects it to a point or a parallel edge. The combination results in the edge to be connected to either two perpendicular edges or two quadrilaterals of the coarse grid. This results in two tetrahedra, shown in Fig. 5(iv), or two deformed triangle prisms, shown in Fig. 5(v), as connecting cells. The vertex case is the combination of two 2D vertex cases. This results in each vertex being connected to three quadrilaterals of the coarse grid via pyramid cells, as shown in Figure 5(vi).

When the coarse grid is refined by more than one fine grid, one must check each coarse-grid point for refinement. Edges might be "upgraded" to the quadrilateral case

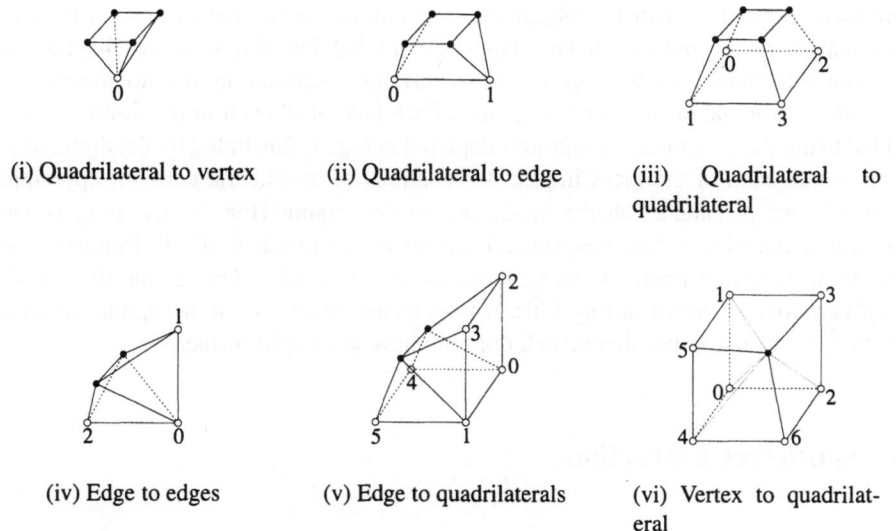

(i) Quadrilateral to vertex   (ii) Quadrilateral to edge   (iii) Quadrilateral   to quadrilateral

(iv) Edge to edges   (v) Edge to quadrilaterals   (vi) Vertex to quadrilateral

Fig. 5. Possible connection types for quadrilateral, edge and vertex in 3D case

(for two adjacent edges). This occurs for the hexahedral cell (Fig. 5(iii)) when either grid points 2 and 3 or grid points 4 and 5 are refined. Vertices can be promoted to edges, or even quadrilaterals, when more than two grids meet at a given location. The fine vertex shown in Fig. 5(vi) can be promoted to an edge, if any of the coarse grid points 3, 5, or 6 is refined.

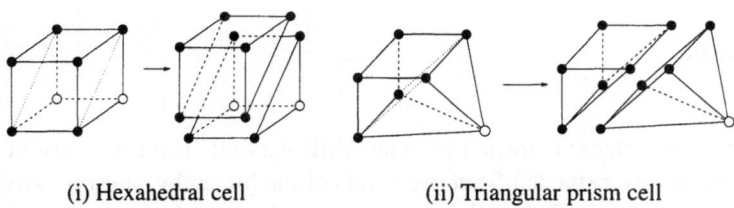

(i) Hexahedral cell   (ii) Triangular prism cell

Fig. 6. Tessellations for 3D cells

The possible refinement configurations result in a large number of cases to be considered. In situations where a fine quadrilateral, edge, or vertex is connected to coarse quadrilaterals, eight points are considered. These points form a deformed hexahedral cell. Each of the faces of the cell corresponds to a possible 2D refinement configuration shown in Fig. 4. It is important to note that the 2D refinement configurations that produce subdivided quadrilaterals are the same configurations that yield non-planar cell boundaries, i.e., boundaries that need to be subdivided. Fig. 6(i) illustrates that this subdivision information alone is sufficient to determine a tessellation. It is not necessary to consider the actual positions of the points. Each face of the cell in the figure is subdivided using the canonical tessellations depicted in Fig. 4, illustrated by the dotted lines. This subdivision of the faces implies tessellations of hexahedral cells into pyramids, triangular prisms, and tetrahedra. In the case of the pyramid (Fig. 5(i)), a refined coarse point is replaced by a fine quadrilateral; the result is a hexahedral cell. Refined coarse points in triangular prisms must be replaced by a fine edge. One of the the possible configurations is shown in Fig. 6(ii). If both coarse points of the triangular prism are refined, the resulting hexahedral cell does not have to be split further.

6 Isosurface Extraction

Within the individual grids, we apply a slightly modified MC approach. Instead of considering all cells of a grid for isosurface generation, we consider only those cells that are not refined by a finer grid. We do this by pre-computing a map with refinement information for each grid. For each grid cell, this map contains an index of a refining grid or an entry that the cell is unrefined. This enables us to quickly skip refined portions of the grid. For the generation of isosurface within the stitch cells, the MC method must be extended to handle the cell types generated during the stitching process. This is a straightforward extension achieved by generating case tables for each of the new cell types. These new case tables must be compatible with the one used in the MC approach, i.e., the ambiguous cases mentioned in Section 2 must be handled in exact the same way as for the hexahedral cells.

7 Results

Fig. 7 shows isosurfaces extracted from an AMR data set. The isosurface in Fig. 7(i) shows an isosurface extracted from two levels of the hierarchy, and Fig. 7(ii) one extracted from three levels. To highlight the transitions between levels, parts of the isosurface extracted from different levels of the hierarchy are colored differently. Isosurface parts extracted from the root, the first and the second level are colored red, orange and light blue, respectively. Portions extracted from the stitch meshes between the root and the first level are colored in green, and portions extracted from the stitch mesh between the first and second level are colored in yellow. The root level and the first level of the used AMR hierarchy each consist of one $32 \times 32 \times 32$ grid. The second level consists of 12 grids with dimensions $6 \times 12 \times 6$, 6×4, $8 \times 12 \times 10$, $6 \times 4 \times 4$, $14 \times 4 \times 10$, $6 \times 6 \times 12$, $12 \times 10 \times 12$, $10 \times 4 \times 8$, $6 \times 6 \times 2$, $16 \times 26 \times 52$, $14 \times 16 \times 12$, and $36 \times 52 \times 36$. All measurements were performed on an standard PC with a 700Mhz Pentium III processor.

8 Future Work

One possible extension of our method is to use a generic triangulation scheme ensuring crack-free isosurface extraction. This would allow the use of our method for other, more general AMR data, where grids might not necessarily be axis-aligned, e.g., data sets produced by the AMR method of Berger and Oliger [3]. Furthermore, it is possible to use the computed tessellation for other purposes, not just for extraction of isosurfaces. One other possible application is to use the tessellation for direct volume rendering to obtain high-quality volume-rendered images, see Max [10].

9 Acknowledgments

This work was supported by the Directory, Office of Science, Office of Basic Energy Sciences, of the U.S. Department of Energy under Contract No. DE-AC03-76SF00098; the Lawrence Berkeley National Laboratory; the National Science Foundation under contracts ACI 9624034 (CAREER Award), through the Large Scientific and Software Data Set Visualization (LSSDSV) program under contract ACI 9982251, and through the National Partnership for Advanced Computational Infrastructure (NPACI); the Office of Naval Research under contract N00014-97-1-0222; the Army Research Office under contract ARO 36598-MA-RIP; the NASA Ames Research Center through an NRA award under contract NAG2-1216; the Lawrence Livermore National Laboratory under ASCI ASAP Level-2 Memorandum Agreement B347878 and under Memorandum Agreement B503159; the Los Alamos National Laboratory; and the North Atlantic Treaty Organization (NATO) under contract CRG.971628.

We also acknowledge the support of ALSTOM Schilling Robotics and SGI. We thank the members of the NERSC/LBNL Visualization Group; the LBNL Applied Numerical Algorithms Group; the Visualization and Graphics Research Group at the Center for Image Processing and Integrated Computing (CIPIC) at the University of California, Davis, and the AG Graphische Datenverarbeitung und Computergeometrie at the University of Kaiserslautern, Germany.

References

[1] AVS5. Product of Advanced Visual Systems, further details can be found at http://www.avs.com/products/AVS5/avs5.htm.

[2] Marsha Berger and Phillip Colella. Local adaptive mesh refinement for shock hydrodynamics. *Journal of Computational Physics*, 82:64–84, May 1989. Lawrence Livermore National Laboratory, Technical Report No. UCRL-97196.

[3] Marsha Berger and Joseph Oliger. Adaptive mesh refinement for hyperbolic partial differential equations. *Journal of Computational Physics*, 53:484–512, March 1984.

[4] Greg L. Bryan. Fluids in the universe: Adaptive mesh refinement in cosmology. *Computing in Science and Engineering*, 1(2):46–53, March/April 1999.

[5] Allen Van Gelder and Jane Wilhelms. Topological considerations in isosurface generation. *ACM Transactions on Graphics*, 13(4):337–375, October 1994.

[6] Markus H. Gross, Oliver G. Staadt, and Roger Gatti. Efficient triangular surface approximations using wavelets and quadtree data structures. *IEEE Transactions on Visualization and Computer Graphics*, 2(2):130–143, June 1996.

34

[7] Interactive Data Language (IDL). Product of Research Systems, Inc., see http://www. rsinc.com/idl/index.cfm for details.

[8] William E. Lorensen and Harvey E. Cline. Marching cubes: A high resolution 3D surface construction algorithm. *Computer Graphics (SIGGRAPH '87 Proceedings)*, 21(4):163–169, July 1987.

[9] Kwan-Liu Ma. Parallel rendering of 3D AMR data on the SGI/Cray T3E. In: *Proceedings of Frontiers '99 the Seventh Symposium on the Frontiers of Massively Parallel Computation*, pages 138–145, IEEE Computer Society Press, Los Alamitos, California, February 1999.

[10] Nelson L. Max. Sorting for polyhedron compositing. In: Hans Hagen, Heinrich Müller, and Gregory M. Nielson, editors, *Focus on Scientific Visualization*, pages 259–268. Springer-Verlag, New York, New York, 1993.

[11] Gregory M. Nielson and Bernd Hamann. The asymptotic decider: Removing the ambiguity in marching cubes. In: Gregory M. Nielson and Larry J. Rosenblum, editors, *IEEE Visualization '91*, pages 83–91, IEEE Computer Society Press, Los Alamitos, California, 1991.

[12] Michael L. Norman, John M. Shalf, Stuart Levy, and Greg Daues. Diving deep: Data management and visualization strategies for adaptive mesh refinement simulations. *Computing in Science and Engineering*, 1(4):36–47, July/August 1999.

[13] William J. Schroeder, Kenneth M. Martin, and William E. Lorensen. *The Visualization Toolkit*, second edition, 1998. Prentice-Hall, Upper Saddle River, New Jersey.

[14] Raj Shekhar, Elias Fayyad, Roni Yagel, and J. Fredrick Cornhill. Octree-based decimation of marching cubes surface. In: Roni Yagel and Gregory M. Nielson, editors, *IEEE Visualization '96*, pages 335–342, 499, IEEE Computer Society Press, Los Alamitos, California, October 1998.

[15] Rüdiger Westermann, Leif Kobbelt, and Thomas Ertl. Real-time exploration of regular volume data by adaptive reconstruction of iso-surfaces. *The Visual Computer*, 15(2):100–111, 1999.

Editors' Note: see Appendix, p. 335 for colored figure of this paper

Fast Multiresolution Extraction of Multiple Transparent Isosurfaces

Thomas Gerstner

Department for Applied Mathematics, University of Bonn
Wegelerstr. 6, 53115 Bonn, Germany
gerstner@iam.uni-bonn.de

Abstract. In this paper, we present a multiresolution algorithm which is capable to render multiple transparent isosurfaces under real–time constraints. To this end, the underlying 3D data set is covered with a hierarchical tetrahedral grid. The multiresolution extraction algorithm is then based on an adaptive traversal of the tetrahedral grid with the help of error indicators. The display of transparent isosurfaces using alpha blending requires a back–to–front rendering of the isosurface triangles. This is achieved by a hierarchical sorting procedure of the tetrahedra and the hierarchical computation of data gradients. We will also comment on the automated selection of suitable isovalues for visualization applications.

1 Introduction

Interactive rendering of large volumetric data sets is a hard task. Besides direct volume rendering methods, such as ray casting, splatting, or 3D texture mapping, indirect volume rendering techniques, such as isosurface extraction, are frequently applied. In both settings, hardware acceleration and multiresolution techniques are often required in order to achieve real–time visualization performance.

Isosurface extraction algorithms rely on the ability of current graphics processors to render large amounts of triangles very quickly. Still, the total rendering time is often dominated by the extraction time of the isosurface, that is the computation of the isosurface triangle vertices. Here, multiresolution methods allow significant reductions of the number of isosurface triangles through suitable approximations of the volume, thereby speeding up both extraction and rendering time.

The display of multiple isosurfaces can be used as a surrogate for direct volume rendering techniques, especially when spiky transfer functions are used. Thereby the selection of isovalues can be done statically, automatically adapted to the data set, or defined by the user. Since the number of displayed isosurfaces will not be very large in interactive applications, the huge number of degrees of freedom in transfer function design is also drastically reduced. Isosurfaces with different isovalues are completely nested and therefore the inner isosurfaces are completely obscured by the outer isosurface except at the boundary of the data set. Thus, multiple isosurfaces have to be rendered transparently which is usually also supported by the graphics hardware. However, then the triangles have to be processed in a strict back–to–front fashion. This requires a view–dependent sorting of all the isosurface triangles. Once the user changes the viewpoint, the sorting time will dominate the total rendering time.

The goal of this paper is to show that in volumetric multiresolution methods this sorting step to be done hierarchically in constant time. Thereby we will focus on a specific well–known multiresolution method based on recursive bisection of tetrahedra. The hierarchical sorting is done in three phases: sorting of the initial tetrahedra, recursive sorting of child tetrahedra during the tree traversal, and sorting of the isosurface components inside each tetrahedron. View–dependent sorting also requires the computation of data gradients inside each tetrahedron. Although these gradients can be precomputed, they require large amounts of memory. We will therefore show how gradients can quickly be computed hierarchically on–the–fly. Finally, we will comment on the automated selection of suitable isovalues.

This paper is organized as follows. Section 2 reviews related work. Section 3 shortly discusses the construction of multiresolution isosurfaces based on tetrahedral bisection. Sorting is done in Section 4. Section 5 explains hierarchical gradient computation. Visualization examples are shown in Section 6. Section 7 describes a technique for automated isovalue selection. The final remarks of Section 8 conclude the paper.

2 Related Work

Multiresolution techniques have been successfully applied to the four most popular direct volume rendering algorithms such as 3D texture mapping [11, 27], ray casting [4, 18, 28], splatting [9, 12, 13], and the shear–warp transformation [3, 31]. For a detailed (non–multiresolution) comparison of these methods see [17].

Isosurface extraction can be very slow when marching algorithms [15, 21] which scan the complete data set are used. Therefore, a variety of methods have been designed which try to avoid to search through regions where no intersection with the isosurface occurs. To this end, hierarchical partitions of either the geometric [30] or the span space [14] are constructed. For a survey and comparison of available methods see [1, 26].

Multiresolution isosurface extraction methods are characterized by a hierachical decomposition of the underlying geometric space. Through suitable approximations of the data, they are also able to extract approximate isosurfaces with varying complexity. The various methods mainly differ in the type of hierarchy and interpolation, such as octrees [24], red tetrahedral refinement [8, 10, 29], tetrahedral bisection [5–7, 20, 32], hierarchical Delaunay triangulations [2], or wavelet techniques [25]. With the help of bounds for the minimum and maximum data value inside each subdomain, the scanning of empty regions is also avoided.

If the data domain is refined adaptively, it can happen that the extracted isosurface contains cracks at transition zones where the mesh resolution changes. For this problem, different solutions have been devised such as remeshing [8], point insertion [24], projection [19], blending [10], and saturation [5–7, 32].

3 Multiresolution Isosurface Extraction

In this section, we will explain the construction of multiresolution isosurfaces based on tetrahedral bisection and error indicators. The algorithms have already been described in detail in previous works but for clarity we shortly repeat the basic steps here.

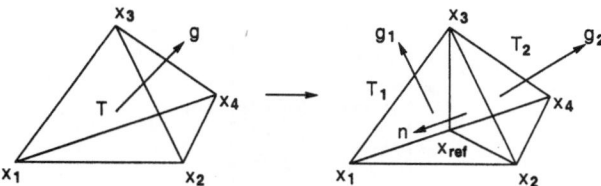

Fig. 1. Bisection of a tetrahedron T into two child tetrahedra T_1 and T_2.

Let us consider a nested hierarchy of tetrahedral grids where the tetrahedra are refined by recursive bisection [16, 23]. For a tetrahedron T the midpoint of a predetermined (in our case the longest) edge $e_{ref}(T)$ is chosen as a new vertex $x_{ref}(T)$. Then, the tetrahedron is split at the face spanned by $x_{ref}(T)$ and the two vertices of T opposite to $e_{ref}(T)$ into two child tetrahedra $T_1(T)$ and $T_2(T)$ (Figure 1). Through recursive application of the refinement rule a binary tree hierarchy is inferred on the tetrahedra.

The adaptive multiresolution isosurface algorithm is based on a depth first traversal of the binary tree. On every tetrahedron for a stopping criterion is checked. If it is true, the algorithms stops and renders the local isosurfaces using the look–up table of the marching tetrahedra algorithm [21]. Otherwise, the two children are visited recursively.

If the algorithms stops on a specific tetrahedron T and refines another tetrahedron which shares the refinement edge, an inconsistency occurs at the hanging node $x_{ref}(T)$. This leads to cracks in the isosurface. Therefore, we ensure that whenever a tetrahedron is refined, all tetrahedra sharing its refinement edge are refined as well. This can be achieved by definition of error indicators η on the refinement vertices, i.e. $\eta(T) = \eta(x_{ref}(T))$, and choosing $\eta(T) < \varepsilon$ as a stopping criterion for some user specified threshold value ε. If the error indicator values are saturated [5–7, 32], no hanging nodes can occur for all possible values of ε. This way, the extraction algorithm is completely local and information from neighboring tetrahedra is never required.

Furthermore, the traversal of the binary tree is also stopped if the tetrahedron is not a candidate for an intersection with one of the isosurfaces. In our case, it is checked whether any of the isovalues is contained in the interval consisting of all the data values inside the tetrahedron. This information can either be explicitly computed in a bottom–up traversal of the tree [30] or be obtained from already available error indicator values [7]. This way, the complexity of the extraction algorithm is of the order of the output (the number of drawn triangles), independent of the size of the input in practice.

4 Transparency Sorting

The most efficient way to display transparent surfaces is through alpha blending, which is supported by basically all manufacturers of graphics cards. Alpha blending requires a back–to–front sorting of the rendered primitives, though. In principle, the isosurface meshes could first be extracted and then the triangles be sorted, but it turns out that the sorting time then dominates the total rendering time. This is especially bad since rotation of the isosurface is the predominant user action in visualization and thus sorting has to be done for almost every frame.

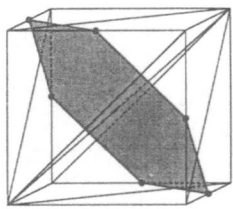

Fig. 2. Sorting points and normal plane of the initial tetrahedral mesh.

On the other hand, multiresolution methods allow a back–to–front sorting during the extraction phase which eliminates above problem. We will now show how this sorting can be done at virtually no extra cost for recursive bisection tetrahedral meshes. Thereby three different sorting problems arise: sorting of the initial tetrahedra, recursive sorting of the child tetrahedra during the adaptive tree traversal, and sorting of the isosurface components inside a tetrahedron during extraction. Let us emphasize here that only those tetrahedra that are visited during the adaptive tree traversal are sorted.

4.1 Initial Tetrahedra

We assume that the input volume data is arranged in a uniform grid with n^3 nodes, $n = 2^k + 1$. The initial tetrahedral mesh consists of the six tetrahedra whose vertices are adjacent corners of the cube and which all share the same diagonal of that cube (Figure 2). Applying then the refinement scheme of the previous section, all refinement vertices $x_{ref}(T)$ will fall onto grid points of the original data set.

These tetrahedra have to be sorted starting with the most distant tetrahedron and ending with the closest one. Let v be the viewing vector (from the eye to the object) and n be the normal of a separating plane of a pair of non–intersecting tetrahedra T_i and T_j. Let us assume that T_i lies in direction of the normal n and T_j is in opposite direction. Then, T_i is behind T_j if $v \cdot n > 0$ and before T_j if $v \cdot n < 0$. If $v \cdot n = 0$ those tetrahedra could be processed in parallel.

For the six initial tetrahedra, 15 normals of separating planes between any two tetrahedra are possible. These normals can be computed as the vector between two distinguished points each located on the boundary of one tetrahedron. These six points are the midpoints of those edges whose endpoints are not endpoints of the diagonal of the cube thereby spanning a normal plane (as shown in Figure 2). With this information, any sorting algorithm such as quicksort can immediately be applied to sort the tetrahedra.

4.2 Child Tetrahedra

During the tree traversal, a tetrahedron is split into two child tetrahedra. The separating plane between those tetrahedra is spanned by the points x_2, x_3 and x_{ref} (Figure 1). Let the normal n of the separating plane point towards T_1. Then, similarly to the previous section, T_1 has to be processed before T_2 if $v \cdot n > 0$ and vice versa otherwise.

The normal n could be computed by $\overline{x_2 x_{ref}} \times \overline{x_3 x_{ref}}$ on–the–fly, but it is more efficient to precompute and store this information for each type of reference tetrahedron. In our

N	1	2	3	4	5	6	7	8	9	10	11	12	13	14	15	16	17	18	19	20	21	22	23	24
N_1	36	43	42	47	48	26	44	35	33	46	45	37	25	27	29	31	33	35	34	32	39	41	42	43
N_2	26	28	30	32	34	36	37	38	40	27	25	44	45	46	31	29	40	38	48	47	41	39	30	28
N	25	26	27	28	29	30	31	32	33	34	35	36	37	38	39	40	41	42	43	44	45	46	47	48
N_1	49	51	53	55	57	59	61	63	65	67	69	71	52	60	62	56	58	66	70	72	68	64	50	54
N_2	50	52	54	56	58	60	62	64	66	68	70	72	71	55	57	59	61	69	65	51	63	67	53	49
N	49	50	51	52	53	54	55	56	57	58	59	60	61	62	63	64	65	66	67	68	69	70	71	72
N_1	11	16	18	7	2	19	2	9	16	21	23	18	4	9	4	14	21	11	7	23	6	14	6	19
N_2	17	1	12	1	20	10	8	24	22	3	17	3	10	15	13	8	12	5	22	5	15	20	24	13

Table 1. Mapping table for the reference tetrahedra numbers.

case, there are 72 reference tetrahedra. Of the three basic types of tetrahedra which cycle all three refinement levels (see [5]) there are 24 instances from the respective rotation and mirror symmetry classes. For example, the six tetrahedra of the first type all share the same diagonal of the cube and there are four different diagonals possible.

Not surprisingly, the numbers of the reference tetrahedra can be determined hierarchically. So given the reference number N of a tetrahedron, the reference numbers of the two child tetrahedra N_1 and N_2 need to be determined. It turns out that this mapping appears to be quite erratic and no simple formula can be given for $N_1(N)$ and $N_2(N)$ (or, at least, we found none). Since the complete mapping is fairly difficult to obtain due to its cyclic structure, we state it in table 1. The six initial tetrahedra are numbered from 1 to 6.

4.3 Isosurface Components inside a Tetrahedron

Now that we've finally sorted all the tetrahedra in the adaptive tetrahedral mesh the only thing left is to sort the isosurface components within each tetrahedron. This case rarely arises for fine resolution datasets and large differences in between isovalues. But our multiresolution algorithm tries to extract and render isosurfaces on coarse tetrahedra and thereby often several isosurfaces will intersect a given tetrahedron making this sorting necessary.

Let $\{i_1, \ldots, i_m\}$ the ordered set of isovalues starting with i_1 being the smallest and i_m being the largest. Let us assume that for the ordered subset $\{i_j, \ldots, i_k\}$ of all isovalues the corresponding isosurfaces intersect the current tetrahedron. Then, the isosurfaces have to be rendered either starting with the lowest isovalue i_j and ending with the highest isovalue i_k, or starting with i_k and ending with i_j. Since linear interpolation is used inside each tetrahedron, all isosurface components are parallel to each other. The order of the isosurfaces is therefore determined by the inner product of the normal of the isosurfaces with the viewing vector. The normal of the isosurface triangles is just the gradient g of the linear function spanned by the data values at the vertices of the tetrahedron. We have therefore the sorting test: if $v \cdot g > 0$, then the highest isovalue has to be processed first, if $v \cdot g < 0$, the lowest one. Note that for $v \cdot g = 0$ the isosurface triangles are parallel to the viewing vector and therefore nearly invisible. How these gradients can be computed efficiently will be shown in the next section.

5 Hierarchical Gradient Computation

As we have seen in the previous section, transparency sorting requires the data gradients. These gradients can in principle be precomputed but they will then require a lot of memory. Since the number of tetrahedra in the mesh is about six times the number of vertices, the required memory would be roughly $6 \cdot 3n^3$ floating point numbers in addition to the n^3 data values. It is therefore advisable to compute these gradients on–the–fly, but in straightforward implementation, gradient computation would significantly decrease the performance of the multiresolution algorithm. We will now show how gradients can be computed very efficiently hierarchically.

Let us recall that the gradient g on the tetrahedron T for a linear function $f(x, y, z) = ax + by + cz + d$ is given by $g = \nabla f = (a\ b\ c)^T$. The coefficients a, b, c can be computed for data values $f_1, \ldots, f_4$ at the respective vertices $(x_1, y_1, z_1), \ldots, (x_4, y_4, z_4)$ by the solution of the linear system

$$
\begin{pmatrix}
x_1 & y_1 & z_1 & 1 \\
x_2 & y_2 & z_2 & 1 \\
x_3 & y_3 & z_3 & 1 \\
x_4 & y_4 & z_4 & 1
\end{pmatrix}
\begin{pmatrix}
a \\ b \\ c \\ d
\end{pmatrix}
=
\begin{pmatrix}
f_1 \\ f_2 \\ f_3 \\ f_4
\end{pmatrix}.
$$

Given the solution of this system, we now want to compute the gradients g_1, g_2 of the two child tetrahedra. Let us first look at the first child T_1 (see Figure 1). The coefficients a_1, b_1, c_1 of g_1 are given by the following system

$$
\begin{pmatrix}
x_1 & y_1 & z_1 & 1 \\
x_2 & y_2 & z_2 & 1 \\
x_3 & y_3 & z_3 & 1 \\
x_{\text{ref}} & y_{\text{ref}} & z_{\text{ref}} & 1
\end{pmatrix}
\begin{pmatrix}
a_1 \\ b_1 \\ c_1 \\ d_1
\end{pmatrix}
=
\begin{pmatrix}
f_1 \\ f_2 \\ f_3 \\ f_{\text{ref}}
\end{pmatrix},
$$

where f_{ref} is the data value at the refinement vertex. Let us now rewrite the first system through replacement of the fourth row with the sum of the first and fourth rows divided by two,

$$
\begin{pmatrix}
x_1 & y_1 & z_1 & 1 \\
x_2 & y_2 & z_2 & 1 \\
x_3 & y_3 & z_3 & 1 \\
\frac{x_1+x_4}{2} & \frac{y_1+y_4}{2} & \frac{z_1+z_4}{2} & 1
\end{pmatrix}
\begin{pmatrix}
a \\ b \\ c \\ d
\end{pmatrix}
=
\begin{pmatrix}
f_1 \\ f_2 \\ f_3 \\ \frac{f_1+f_4}{2}
\end{pmatrix}.
$$

Since $x_{\text{ref}} = (x_1 + x_4)/2$ (and similarly for y and z), we see that the system for the child differs from the system for the parent only by the fourth component of the right hand side. Let the inverse of above matrix be given by (w_{ij}). Then we have for a

$$
a = w_{11}f_1 + w_{12}f_2 + w_{13}f_3 + w_{14}(f_1 + f_4)/2
$$

and for a_1

$$
a_1 = w_{11}f_1 + w_{12}f_2 + w_{13}f_3 + w_{14}f_{\text{ref}}.
$$

Differencing yields

$$
a_1 = a + w_{14}(f_{\text{ref}} - (f_1 + f_4)/2).
$$

Repeating this step for b and c we end up with

$$\begin{pmatrix} a_1 \\ b_1 \\ c_1 \end{pmatrix} = \begin{pmatrix} a \\ b \\ c \end{pmatrix} + (f_{\text{ref}} - (f_1 + f_4)/2) \begin{pmatrix} w_{14} \\ w_{24} \\ w_{34} \end{pmatrix}.$$

Completely analogously, the gradient g_2 of T_2 can be computed from g by replacement of w_{i4} with w_{i1}. Now, only the corresponding w_{ij} for all 72 reference tetrahedra have to be computed and stored in advance. The numbers of the reference tetrahedra can be determined during the traversal by the numbering scheme of the previous section. In comparison to direct computation (matrix inversion) or the plain usage of reference elements (matrix–vector multiplication), the hierarchical method requires just scalar multiplication and vector addition. Note that the factor $f_{\text{ref}} - (f_1 + f_4)/2$ is exactly the wavelet coefficient in the piecewise linear lazy wavelet representation. Of course the gradients can be reused for flat illumination shading of the isosurface triangles, although we use Goraud shading in our examples.

6 Visualization Examples

As a first example serves the well–known buckyball data set (courtesy of AVS). Figure 5 shows three isosurfaces for isovalues of 0.05, 0.15 and 0.25. The colors of the isosurfaces are blue, red and green (from outer to inner) with opacities of 0.3, 0.5, and 0.7. The error thresholds ε of the six images are 0.0, 0.01, 0.02, 0.04, 0.08, and 0.16. The number of triangles are 1775762, 979730, 438042, 277698, 171465, and 92309.

The second example in Figure 6 shows the electron density around the cap of a nanotube (courtesy of A. Caglar, Univ. of Bonn). In addition to the isosurfaces, the different atoms of the molecule are shown (hydrogen in blue, boron in red, and nitrite in green). The isovalues are 0.05 (yellow), 0.2 (green), and 0.35 (blue) with opacities of 0.3. The ε–values are identical to the buckyball example resulting in 230452, 175054, 143126, 96327, 55933, and 42630 triangles. The atoms are rendered as small textured balls and are inserted during the tree traversal at the appropriate position. In this example, it was not necessary to split the textures across tetrahedra since they were small enough and placed in the centers of the tetrahedra. In other cases, it may be necessary to split such textures, though.

The combined rendering and extraction time of the algorithm is about 270000 triangles/sec on an SGI Onyx^2 (R10000, 195 MHz). The same algorithm for opaque isosurfaces achieves about 300000 triangles/sec, so the time required for sorting is moderate (about 10%) and does not degrade the total performance significantly.

7 Isovalue Selection

In many applications there is a close corresponence between data values and useful isovalues. For example, in medical imaging, bone, tissue and blood vessels have certain known reflectivities returned from medical scanners. In many other applications this correspondence may be unknown or changing in between data sets. Certainly, it is often

42

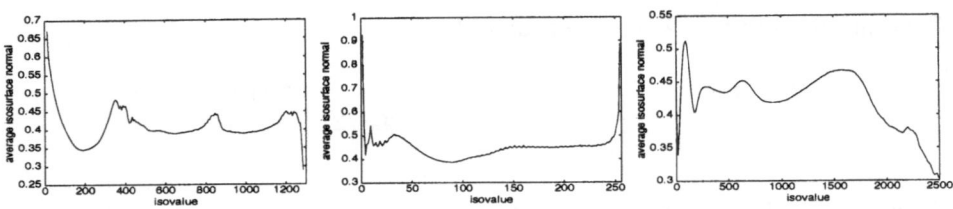

Fig. 3. Average isosurface normal sizes $\bar{g}$ for the tooth, sheep and knee data sets.

the best way to determine suitable isovalues by trial and error. However, in some cases, it may also be helpful to offer the user first guesses of suitable isosurfaces and let him or her decide whose are useful and whose not.

Probably the most straightforward way to determine suitable isovalues is by looking at the gradient field of the data set. Let us define the (discrete) average size of the normals of an isosurface as

$$\bar{g}(i) = \frac{\displaystyle\sum_{T_j : s(i) \cap T_j \neq \emptyset} area(s(i) \cap T_j) \cdot |g(T_j)|}{\displaystyle\sum_{T_j : s(i) \cap T_j \neq \emptyset} area(s(i) \cap T_j)}$$

where i is the isovalue, $s(i)$ the triangulated isosurface, and $g(T_j)$ the data gradient on T_j (which is the normal of the isosurface). The local minima and maxima of $\bar{g}(i)$ then characterize possible isovalues. The maxima will separate homogeneous areas and the minima are the centers of these areas.

Let us take a look at some concrete graphs $\bar{g}(i)$ for the three different data sets (courtesy of B. Lorensen, General Electric) which have been used for the transfer function bake–off at Visualization 2000 [22]. Those data sets, CT respectively MRI scans of a tooth, a sheep's heart, and a knee serve as benchmarks in transfer function design. The computed $\bar{g}(i)$ for the three data sets are shown in Figure 3.

In all cases, we found the minima more useful. For the tooth data set there are three clearly visible minima. The leftmost minimum corresponds to the cylindrical outer shell of the medium in which the tooth was set prior to scanning. The other two minima give the surface and the enamel of the tooth as shown in Figure 4, left. The sheep data set shows only one distinguished minimum indicating the heart's surface and inner blood vessels (Figure 4 middle). Manual scanning of other isovalues revealed no further useful isosurfaces. The knee data sets shows three clear minima despite the high amount of noise in the data corresponding to skin, muscular tissue, and bone (Figure 4 right). Although we did not do so, such noisy data should be smoothed before isosurfacing.

8 Concluding Remarks

In this paper, we have shown how multiple transparent isosurfaces can be extracted interactively using the tetrahedral bisection hierarchy. This was achieved by an adaptive tree traversal, a hierarchical sorting procedure of the tetrahedra and isosurface triangles,

Fig. 4. The tooth, sheep and knee data sets rendered with multiple transparent isosurfaces.

and the hierarchical computation of data gradients. Furthermore, we have shown how isovalues can be selected (semi-)automatically based on the local minima of the average isosurface normal graphs.

Let us remark that in comparison to direct volume rendering methods isosurface extraction requires no special purpose hardware and gives images with sharp boundaries. Also, besides the isovalues, colors and opacities, which can be obtained quickly, no further design parameters are necessary. On the downside, details in between isosurfaces are lost and cannot be displayed with this methodology. Of course, the sorting and numbering algorithms for the tetrahedra can be used in direct volume rendering algorithms based on tetrahedral splats or ray casting (with inverted sorting order).

References

1. C. Bajaj, V. Pascucci, and D. Schikore. Accelerated Isocontouring of Scalar Fields. In C. Bajaj, editor, *Data Visualization Techniques*. John Wiley and Sons, 1998.
2. P. Cignoni, L. De Floriani, C. Montani, E. Puppo, and R. Scopigno. Multiresolution Representation and Visualization of Volume Data. *IEEE Transactions on Visualization and Computer Graphics*, 3(4):352–369, 1997.
3. F. Dong, M. Krokos, and G. Clapworthy. Fast Volume Rendering and Data Classification using Multiresolution Min–Max Octrees. *Computer Graphics Forum*, 19(3):359–367, 2000.
4. T. Ertl, R. Westermann, and R. Grosso. Multiresolution and Hierarchical Methods for the Visualization of Volume Data. *Future Generation Computer Systems*, 15(1):31–42, 1999.
5. T. Gerstner and R. Pajarola. Topology Preserving and Controlled Topology Simplifying Multiresolution Isosurface Extraction. In *Proc. IEEE Visualization 2000*, pages 259–266. IEEE Computer Society Press, 2000.
6. T. Gerstner and M. Rumpf. Multiresolutional Parallel Isosurface Extraction based on Tetrahedral Bisection. In M. Chen, A. Kaufman, and R. Yagel, editors, *Volume Graphics*, pages 267–278. Springer, 2000.
7. T. Gerstner, M. Rumpf, and U. Weikard. Error Indicators for Multilevel Visualization and Computing on Nested Grids. *Computers & Graphics*, 24(3):363–373, 2000.
8. R. Grosso, C. Lürig, and T. Ertl. The Multilevel Finite Element Method for Adaptive Mesh Optimization and Visualization of Volume Data. In *Proc. IEEE Visualization '97*, pages 387–394. IEEE Computer Society Press, 1997.

44

9. B. Guo. A Multiscale Model for Structure–based Volume Rendering. *IEEE Transactions on Visualization and Computer Graphics*, 1(4):291–301, 1995.
10. D. Holliday and G. Nielson. Progressive Volume Model for Rectilinear Data using Tetrahedral Coons Volumes. In W. de Leeuw and R. van Liere, editors, *Data Visualization 2000*, pages 83–92. Springer, 2000.
11. E. LaMar, B. Hamann, and K. Joy. Multiresolution Techniques for Interactive Texture–based Volume Visualization. In *Proc. IEEE Visualization '99*, pages 355–362. IEEE Press, 1999.
12. D. Laur and P. Hanrahan. Hierarchical Splatting: A Progressive Refinement Algorithm for Volume Rendering. *Computer Graphics (SIGGRAPH '91 Proc.)*, pages 285–288, 1991.
13. L. Lippert and M. Gross. Fast Wavelet based Volume Rendering by Accumulation of Transparent Texture Maps. *Computer Graphics Forum*, 14(3):431–444, 1995.
14. Y. Livnat, H. Shen, and C. Johnson. A Near Optimal Isosurface Extraction Algorithm using the Span Space. *IEEE Trans. on Visualization and Computer Graphics*, 2(1):73–83, 1996.
15. W. Lorensen and H. Cline. Marching Cubes: A High Resolution 3D Surface Construction Algorithm. *Computer Graphics*, 21(4):163–169, 1987.
16. J. Maubach. Local Bisection Refinement for n-simplicial Grids generated by Reflection. *SIAM J. Sci. Comp.*, 16:210–227, 1995.
17. M. Meissner, J. Huang, D. Bartz, K. Mueller, and R. Crawfis. A Practical Evaluation of Popular Volume Rendering Algorithms. In *Proc. Volume Visualization 2000*, pages 81–91. ACM Press, 2000.
18. S. Muraki. Approximation and Rendering of Volume Data using Wavelet Transforms. *Computer Graphics and Applications*, 13(4):50–56, 1993.
19. M. Ohlberger and M. Rumpf. Adaptive Projection Methods in Multiresolutional Scientific Visualization. *IEEE Trans. on Visualization and Computer Graphics*, 4(4):74–94, 1998.
20. V. Pascucci and C. Bajaj. Time Critical Isosurface Refinement and Smoothing. In *Proc. Volume Visualization 2000*, pages 33–42. ACM Press, 2000.
21. B. Payne and A. Toga. Surface Mapping Brain Function on 3D Models. *IEEE Computer Graphics and Applications*, 10(5):33–41, 1990.
22. H. Pfister (org.), B. Lorensen, C. Bajaj, G. Kindlmann, and W. Schroeder. The Transfer Function Bake–Off. Panel session at IEEE Visualization '00, 2000.
23. M. Rivara and C. Levin. A 3D Refinement Algorithm suitable for Adaptive and Multi-Grid Techniques. *Comm. Appl. Num. Meth.*, 8:281–290, 1992.
24. R. Shekhar, E. Fayyad, R. Yagel, and J. Cornhill. Octree–based Decimation of Marching Cubes Surfaces. In *Proc. IEEE Visualization '96*, pages 335–344. IEEE Press, 1996.
25. O. Staadt, M. Gross, and R. Weber. Multiresolution Compression and Reconstruction. In *Proc. IEEE Visualization '97*, pages 337–364. IEEE Computer Society Press, 1997.
26. P. Sutton, C. Hansen, H.-W. Shen, and D. Schikore. A Case Study of Isosurface Extraction Algorithm Performance. In W. de Leeuw and R. van Liere, editors, *Data Visualization 2000*, pages 259–268. Springer, 2000.
27. M. Weiler, R. Westermann, C. Hansen, K. Zimmerman, and T. Ertl. Level–of–Detail Volume Rendering via 3D Textures. In *Proc. Volume Vis. 2000*, pages 7–13. ACM Press, 2000.
28. R. Westermann. A Multiresolution Framework for Volume Rendering. In *Proc. Volume Visualization 94*, pages 51–57. ACM Press, 1994.
29. R. Westermann, L. Kobbelt, and T. Ertl. Real–Time Exploration of Regular Volume Data by Adaptive Reconstruction of Isosurfaces. *The Visual Computer*, 15:100–111, 1999.
30. J. Wilhelms and A. Van Gelder. Octrees for Faster Isosurface Generation. *ACM Transactions on Graphics*, 11(3):201–227, 1992.
31. Y. Yang, F. Lin, and H. Seah. Fast Multi–Resolution Volume Rendering. In M. Chen, A. Kaufman, and R. Yagel, editors, *Volume Graphics*, pages 185–197. Springer, 2000.
32. Y. Zhou, B. Chen, and A. Kaufman. Multiresolution Tetrahedral Framework for Visualizing Volume Data. In *Proc. IEEE Visualization '97*, pages 135–142. IEEE Press, 1997.

Editors' Note: see Appendix, p. 336 for colored figures of this paper

Multiresolution Maximum Intensity Volume Rendering by Morphological Pyramids

Jos B.T.M. Roerdink

Institute for Mathematics and Computing Science
University of Groningen
P.O. Box 800, 9700 AV Groningen, The Netherlands
roe@cs.rug.nl

Abstract We propose a multiresolution representation for maximum intensity projection (MIP) volume rendering, based on morphological pyramids which allow progressive refinement and have the property of perfect reconstruction. The pyramidal analysis and synthesis operators are composed of morphological erosion and dilation, combined with dyadic downsampling for analysis and dyadic upsampling for synthesis. The structure of the multiresolution MIP representation is very similar to wavelet splatting, the main differences being that (i) linear summation of voxel values is replaced by maximum computation, and (ii) linear wavelet filters are replaced by (nonlinear) morphological filters.

1 Introduction

Interactive rendering and transfer of volume data is still a demanding problem due to the sizes of the data sets. For this purpose multiresolution models are developed, which can be used to visualize data incrementally ('progressive refinement'). An extensively studied class of such multiresolution models is based on wavelets [4, 12, 18]. Recent methods for X-ray rendering include wavelet splatting [7, 8], which extends splatting [19] by using wavelets as reconstruction filters, and Fourier-wavelet volume rendering [14, 17], which extends standard Fourier volume rendering [10], and uses a frequency domain implementation of the wavelet transform.

The goal of this paper is to propose a multiresolution representation for Maximum Intensity Projection (MIP) volume rendering, where one computes not the (opacity-weighted) integral, but the *maximum* along the line of sight. Because of its computational simplicity, this algorithm is widely used in the display of magnetic resonance angiography (MRA) and ultrasound data. Our approach makes use of the concept of *morphological pyramids*, following recent work of Goutsias and Heijmans [3, 6], who present a general framework for multiresolution signal decomposition, which includes linear wavelet analysis as a special case. Even though the morphological operators are nonlinear and non-invertible, the pyramid scheme does allow perfect reconstruction as well as progressive refinement, just as in the linear wavelet case. We restrict ourselves here to the so-called *flat* pyramids, where minima and maxima are computed in a local neighbourhood of each voxel, requiring only integer computations. Flatness in particular means that no new grey values are introduced in the analysis of a signal. Also, flat pyramids allow global error control, since they have the property that the approximation error decreases monotonically as we add detail signals. Note also that the morphological pyramids used in this paper are not auxiliary data, but an exact representation of the initial data. After the pyramid has been constructed, the original volume data can be discarded, since the pyramid allows perfect reconstruction of the data.

Morphological methods have a well-established mathematical basis and are widely used in image processing for filtering, segmentation, and shape analysis [5, 15]. Applications of morphological methods in visualization have so far mostly been restricted to preprocessing of volume data, but this is beginning to change. For example, Lürig and Ertl [9] used multiscale morphological operators as an alternative to transfer functions in traditional colour-opacity volume rendering. Visualization of solids defined by morphological operators was considered in [13].

Morphological pyramids are useful in the context of MIP for several reasons. First, from a mathematical point of view, the morphological operations of erosion and dilation (involving minimum and maximum computation) are exactly the right ones for the case of MIP, which involves maximum computation, just as linear wavelet representations are the right tool for the case of linear X-ray rendering. Second, the feature extraction capabilities of morphological operators can be incorporated within the volume rendering process. This allows processing based on geometric information, not just on grey value properties, as usually is the case. For example, when processing angiographic data, the multiresolution scheme will systematically remove small veins when going higher up in the pyramid, while keeping larger ones. Whether or not this is a desired property can only be answered in the context of the concrete medical application. Finally, pyramids are one of many possibilities for accelerating MIP. Many methods already exist for that purpose, including distance encoding [20], splatting in sheared object space [2], or MIP at warp speed [11], which preprocesses the data to remove non-contributing voxels from the volume.

We stress that in this paper the main issue is the presentation of a new multiresolution MIP representation. Computational efficiency is a separate issue: any existing fast MIP implementation can in principle be used for computing the maximum projections which are required to render different levels of the pyramid, as long as such an implementation can work directly on the data structures used to represent the pyramid. In the examples below we will use the voxel projection method of Mroz *et al.* [11]. A detailed study of computational aspects will be presented in future work.

The organization of this paper is as follows. Section 2 gives a few preliminaries on morphological operators, and summarizes the work of Goutsias and Heijmans [3, 6] on morphological pyramids. Section 3 contains the new material, i.e. the derivation of a multiresolution MIP rendering algorithm (MMIP) allowing progressive refinement. An example is given in section 4. Section 5 contains a discussion of future work.

2 Morphological pyramids

Before we consider multiresolution signal decomposition, first some elementary morphological operators are introduced.

Morphological operators Morphological operations for grey value images have been defined in analogy with the binary case [16]. For a mathematical treatment, see e.g. [5]. We consider signals or functions, defined on a subset of the discrete grid $\mathbb{Z}^d$, where $d = 2$ or $d = 3$ (image and volume data).

Let f be a signal with domain $F \subseteq \mathbb{Z}^d$, and A a subset of $\mathbb{Z}^d$ called the structuring element. Then the *dilation* $\delta_A(f)$ and *erosion* $\varepsilon_A(f)$ of f by A are defined by

$$\delta_A(f)(x) = \max_{y \in A, x-y \in F} f(x - y), \quad \varepsilon_A(f)(x) = \min_{y \in A, x+y \in F} f(x + y). \quad (1)$$

So dilation and erosion simply replace each value by the maximum or minimum in a neighbourhood defined by the structuring element A. By taking products of dilation and erosion we can construct *openings* and *closings*. The opening $\alpha_A(f)$ and closing $\phi_A(f)$ of f by A are defined by

$$\alpha_A(f)(x) = \delta_A(\varepsilon_A(f))(x), \quad \phi_A(f)(x) = \varepsilon_A(\delta_A(f))(x). \quad (2)$$

The opening has the property that it is increasing ($f \leq g$ implies that $\alpha_A(f) \leq \alpha_A(g)$), anti-extensive ($\alpha_A(f) \leq f$) and idempotent ($\alpha_A(\alpha_A(f)) = \alpha_A(f)$). Similar properties hold for the closing, with the difference that closing is extensive ($\phi_A(f) \geq f$). The opening eliminates peaks, the closing valleys.

Pyramids We outline here the multiresolution signal decomposition scheme as recently introduced by Goutsias and Heijmans [3, 6], which encompasses linear (e.g. laplacian) and nonlinear pyramid schemes.

Consider signals in a d-dimensional signal space V_0, which is assumed to be the set of functions on (a subset of) the discrete grid $\mathbb{Z}^d$ that take values in a finite set of nonnegative integers. The goal is to decompose the original signal $f \in V_0$ into a number of coarser signals $f_j, j = 0, 1, 2, \ldots$. Here j is called the level of the decomposition. It is assumed that the signals f_j are elements of associated signal spaces V_j, which have the same structure as V_0.

Signal decomposition or *analysis* proceeds by analysis operators $\psi_j^{\uparrow} : V_j \to V_{j+1}$, which map a signal to a level higher in the pyramid, thereby reducing information. Signal reconstruction or *synthesis* proceeds by synthesis operators $\psi_j^{\downarrow} : V_{j+1} \to V_j$, which map a signal to a level lower in the pyramid. To guarantee that information lost during analysis can be recovered in the synthesis phase in a non-redundant way, one needs the so-called *pyramid condition*:

$$\psi_j^{\uparrow} \psi_j^{\downarrow}(f) = f \text{ for all } f \text{ on } V_{j+1}. \quad (3)$$

Decomposition of a signal $f \in V_0$ proceeds by

$$f_0 = f$$
$$f_{j+1} = \psi_j^{\uparrow}(f_j), \quad j \geq 0$$
$$d_j = f_j \dot{-} \psi_j^{\downarrow}(f_{j+1}).$$

In a decomposition of L levels, this results in a sequence $d_0, d_1, \ldots, d_{L-1}, f_L$, where $\{d_j\}$ are detail signals and f_L an approximation signal at the coarsest level. Here $\dot{-}$

is a generalized subtraction operator. Assuming there exists an associated generalized addition operator $\dot{+}$ such that

$$\hat{f} \dot{+} (f \dot{-} \hat{f}) = f, \quad \text{if } f \in V_j \text{ and } \hat{f} = \psi_j^{\downarrow}\psi_j^{\uparrow}(f),$$

we have perfect reconstruction, that is, $f \in V_0$ can be *exactly* reconstructed from the sequence $d_0, d_1, \ldots, d_{L-1}, f_L$ by the recursion

$$f_j = \psi_j^{\downarrow}(f_{j+1}) \dot{+} d_j. \tag{4}$$

The operators $\dot{+}$ and $\dot{-}$ can be ordinary addition and subtraction, but other choices are possible, as we will see below.

By *approximations* of $f \in V_0$ we will mean signals which are reconstructed from higher levels by omitting some of the detail signals. To make this notion precise, we introduce the multilevel analysis operator $\psi_{i,j}^{\uparrow} = \psi_{j-1}^{\uparrow}\psi_{j-2}^{\uparrow} \cdots \psi_i^{\uparrow}, j > i$, which maps an element of V_i to an element of V_j. Similarly, the multilevel synthesis operator $\psi_{i,j}^{\downarrow} = \psi_i^{\downarrow}\psi_{i+1}^{\downarrow} \cdots \psi_{j-1}^{\downarrow}, j > i$, maps an element of V_j back to an element of V_i. The operator $\hat{\psi}_{i,j} = \psi_{i,j}^{\downarrow}\psi_{i,j}^{\uparrow}$ can be regarded as an *approximation operator* that maps the information obtained at level j by the analysis operator $\psi_{i,j}^{\uparrow}$ back to level i by the synthesis operator $\psi_{i,j}^{\downarrow}$. Now we define a level-j approximation $\hat{f}_{0,j}$ of $f \in V_0$ as

$$\hat{f}_{0,j} = \hat{\psi}_{0,j}(f) = \psi_{j,0}^{\downarrow}\psi_{0,j}^{\uparrow}(f) = \psi_{j,0}^{\downarrow}(f_j).$$

Adjunction pyramids We now introduce the class of so-called *morphological adjunction pyramids* [3], for which (i) the analysis and synthesis operators are independent of level ($\psi_j^{\uparrow} = \psi^{\uparrow}, \psi_j^{\downarrow} = \psi^{\downarrow}$), and (ii) $\psi^{\uparrow} : V_0 \to V_1$ and $\psi^{\downarrow} : V_1 \to V_0$ form a so-called *adjunction* between V_0 and V_1, implying that $\psi^{\uparrow}$ is an *erosion*, i.e. commutes with minima, and $\psi^{\downarrow}$ is a *dilation*, i.e. commutes with maxima[1]. In this case, the analysis and synthesis operators acting on a d-dimensional signal f have the form

$$\psi_A^{\uparrow}(f)(n) = \sigma^{\uparrow} \varepsilon_A(f), \qquad \psi_A^{\downarrow}(f)(k) = \delta_A \sigma^{\downarrow}(f). \tag{5}$$

Here $\delta_A(f)$ and $\varepsilon_A(f)$ are the dilation and erosion defined in (1), whereas $\sigma^{\uparrow}$ and $\sigma^{\downarrow}$ denote dyadic downsampling and dyadic upsampling in each spatial dimension:

$$\sigma^{\uparrow}(f)(n) = f(2n)$$

$$\sigma^{\downarrow}(f)(m) = \begin{cases} f(n), & \text{if } m = 2n \\ 0, & \text{otherwise} \end{cases}$$

So in the analysis phase we first compute an erosion, and then downsample; in the synthesis phase we first upsample and then dilate. Note that the notation is somewhat confusing: the arrow on σ for downsampling points upwards, and vice versa for upsampling. This is because downsampling is related to going to coarser levels in the pyramid,

---

[1] In a more general setting, 'maxima' and 'minima' should be replaced by 'suprema' and 'infima', respectively.

which traditionally are the higher levels. We could have inverted the arrows, so that the pyramid in upside-down, but decided to adhere to the notation of [3]. The pyramid condition (3) is satisfied, if there exists an $a \in A$ such that the translates of a over an even number of grid steps are never contained in the structuring element A; see [3] for more details.

In an adjunction pyramid, the product $\psi_A^{\downarrow} \psi_A^{\uparrow}$ is an *opening*, i.e. an operator which is increasing, anti-extensive and idempotent. The anti-extensivity property means that $\psi_A^{\downarrow} \psi_A^{\uparrow}(f) \leq f$. Therefore, we can define the generalized addition and subtraction operators by (cf. [3]):

$$t \dotplus s = t \vee s = \max(t, s) \tag{6}$$

$$t \dotminus s = \begin{cases} t, & \text{if } t > s \\ 0, & \text{if } t = s \end{cases} \tag{7}$$

where 0 is the smallest element, that is, the smallest image or voxel value. As a consequence, the detail signals are non-negative:

$$d_j(n) = f_j(n) \dotminus \psi_A^{\downarrow}(f_{j+1})(n) = f_j(n) \dotminus \psi_A^{\downarrow} \psi_A^{\uparrow}(f_j)(n) \geq 0. \tag{8}$$

Note that (7) implies that the detail signal $d_j(n)$ equals $f_j(n)$, except at points n for which $f_j(n) = \psi_A^{\downarrow} \psi_A^{\uparrow}(f_j)(n)$, where $d_j(n) = 0$. So, detail signals are not 'small' in regions where the structuring element does not fit well to the data. As long as we only look at the approximation signals, this is not a problem, but for compression purposes other choices of addition and subtraction operators are more suitable.

For an adjunction pyramid with the addition operator defined by (6), the reconstruction takes a special form. Making use of the fact that $\psi_A^{\downarrow}$ is a dilation, hence commutes with maxima, we derive from (4) and (6):

$$f = {\psi_A^{\downarrow}}^{L}(f_L) \vee \bigvee_{k=0}^{L-1} {\psi_A^{\downarrow}}^{k}(d_k), \tag{9}$$

where L is the decomposition depth and ${\psi_A^{\downarrow}}^{k}$ denotes k-fold composition of $\psi_A^{\downarrow}$ with itself. This representation is quite similar to the (linear) laplacian pyramid representation [1]. The main difference is that sums have been replaced by maxima.

3   Multiresolution maximum intensity projection

Now we come to the new part of this paper, which is the derivation of a multiresolution MIP volume rendering algorithm with progressive refinement based on morphological pyramids. To emphasize the main ideas, we first consider projections[2] along one of the coordinate axes, and then briefly indicate how the method can be extended to arbitrary viewing directions.

---

[2] By 'projection', we mean maximum intensity projection in what follows.

50

Axial projections Consider a 3-D volume data set f, and project parallel to the z-axis by computing the maximum value. The result is denoted by $\mathcal{M}(f)$:

$$\mathcal{M}(f)(x,y) = \max_z f(x,y,z), \qquad (x,y,z) \in \mathbb{Z}^3.$$

Applying the pyramid representation (9), and the fact that $\mathcal{M}$ evidently distributes over maxima, we get

$$\mathcal{M}(f) = \left(\mathcal{M}(\psi_A^{\downarrow L}(f_L))\right) \vee \bigvee_{k=0}^{L-1} \left(\mathcal{M}(\psi_A^{\downarrow k}(d_k))\right) \tag{10}$$

In principle, this formula allows us to do multiresolution MIP. Computationally, however, this expression is inefficient, because to compute the projections at a certain level k, we have to reconstruct first to full resolution by $\psi_A^{\downarrow k}$ and then apply the maximum operator $\mathcal{M}$. It would be desirable to first compute the maxima along the line of sight on a coarse level, where the size of the data is reduced, before applying a synthesis operator to perform reconstruction to a finer resolution level. This is possible, as is shown next.

Computing the maxima before synthesis As (5) shows, the synthesis operator $\psi_A^{\downarrow}$ is composed of upsampling, followed by a dilation. Therefore, our problem is to rewrite $\mathcal{M}\,\psi_A^{\downarrow}(f) = \mathcal{M}\delta_A\,\sigma^{\downarrow}(f)$ such that the projection operator $\mathcal{M}$ is 'moved to the right'. The problem can be split in two parts. First we consider the projection of a dilated function, then the projection of an upsampled function, and finally combine the two results.

Now, both $\mathcal{M}$ and δ_A involve the computation of maxima. Therefore, it is easy to see that to compute $\mathcal{M}\delta_A(f)$, we can first project f along the z-axis, and then dilate the resulting 2-D function by a structuring element $\tilde{A}$, which is the projection of A. So,

$$\mathcal{M}\,\delta_A(f) = \delta_{\tilde{A}}\,\mathcal{M}(f), \tag{11}$$

with $\tilde{A} := \{(x,y) \in \mathbb{Z}^2 | (x,y,z) \in A \text{ for some } z \in \mathbb{Z}\}$. Note that δ_A is a 3-D dilation, while $\delta_{\tilde{A}}$ is a 2-D dilation (both defined by the formula (1) which holds for any dimension).

Next, consider projection of an upsampled function: $\mathcal{M}\,\sigma^{\downarrow}(f)$. Upsampling has the effect of inserting zeroes between neighbouring voxels in all three spatial dimensions. If we project the upsampled function, then for those (x,y) which are in the projection of the support of the original function f the outcome will be unaffected, since the inserted zero values never contribute to the maximum, zero being the minimum data value possible. On the other hand, for those (x,y) which are not in the projection of the support of the original function f, projection means computing the maximum of a vertical line of zeroes, which results in a zero at (x,y). Therefore,

$$\mathcal{M}\,\sigma^{\downarrow}(f) = \sigma^{\downarrow}\,\mathcal{M}(f), \tag{12}$$

where $\sigma^{\downarrow}$ on the right-hand side is a 2-D upsampling operator (the dimension of $\sigma^{\downarrow}$ is clear from the dimension of the functions on which it acts).

Now we can take the final step, which is to combine (11) and (12). We find,

$$\mathcal{M}\,\psi_A^\downarrow\,(f) = \mathcal{M}\delta_A\,\sigma^\downarrow(f) = \delta_{\bar{A}}\,\mathcal{M}\,\sigma^\downarrow(f) = \delta_{\bar{A}}\,\sigma^\downarrow\mathcal{M}(f) = \psi_{\bar{A}}^\downarrow\,\mathcal{M}(f),$$

where $\psi_{\bar{A}}^\downarrow = \delta_{\bar{A}}\,\sigma^\downarrow$ is a 2-D synthesis operator of the same form as $\psi_A^\downarrow$ (the 3-D structuring element A has only been replaced by a 2-D structuring element $\bar{A}$). It is evident that a similar formula holds for iterated versions of $\psi_A^\downarrow$.

As a result of the above analysis, we have proved the main result of this paper, which is a multiresolution representation of the maximum intensity projection $\mathcal{M}(f)$ of a 3-D voxel array f:

$$\mathcal{M}(f) = \left(\psi_{\bar{A}}^{\downarrow^L} \mathcal{M}(f_L)\right) \vee \bigvee_{k=0}^{L-1} \left(\psi_{\bar{A}}^{\downarrow^k} \mathcal{M}(d_k)\right) \tag{13}$$

As long as a user is interacting with the data (preview mode), only a coarse approximation $\hat{\mathcal{M}}_j(f)$ may be used, which can be refined to full resolution for close inspection.

The MMIP algorithm The multiresolution MIP algorithm can be summarized as follows.

- *Preprocessing.* Compute an L-level 3-D morphological pyramid of the volume data, resulting in a sequence $d_0, d_1, \dots, d_{L-1}, f_L$
- *Actual MIP volume rendering.*

  1. Compute a low resolution approximation $\hat{\mathcal{M}}_L(f)$ by projecting f_L followed by applying the 2-D synthesis operator $\psi_{\bar{A}}^{\downarrow^L}$.
  2. Refine the image progressively by taking the detail signals d_k, $k = L-1, \dots, 0$ into account. From a level j approximation $\hat{\mathcal{M}}_j(f)$, compute a level $j-1$ approximation $\hat{\mathcal{M}}_{j-1}(f)$ by projecting d_{j-1}, applying the 2-D pyramid synthesis operator $\psi_{\bar{A}}^\downarrow$ to the projection, and finally taking the maximum of the 2-D signal so obtained with the previous approximation:

$$\hat{\mathcal{M}}_{j-1}(f) = \psi_{\bar{A}}^{\downarrow^{j-1}}(\mathcal{M}(d_{j-1})) \vee \hat{\mathcal{M}}_j(f). \tag{14}$$

  3. The recursion terminates with $\hat{\mathcal{M}}_0(f) = \mathcal{M}(f)$, the exact MIP of f.

The structure of this algorithm is very similar to that of wavelet splatting [7, 8, 17], with the difference that (i) linear summation of voxel values is replaced by maximum computation, and (ii) linear wavelet filters have been replaced by morphological filters (dilation and erosion).

From (14) we immediately deduce that $\hat{\mathcal{M}}_j(f) \leq \hat{\mathcal{M}}_{j-1}(f)$. So if we define a global error measure by the maximum (or L_∞) norm, then approximation error decreases monotonically as we go down the pyramid.

Implementation We implemented the MIP projections $\mathcal{M}$ required in the MMIP algorithm by means of the object order voxel projection method of Mroz *et al.* [11], that treats voxels as cells which a constant data value, which are simply projected on the viewing plane, each voxel contributing to exactly one pixel. The method also uses an efficient volume data storage scheme, by histogram-based sorting of interesting voxels according to grey value, and storing these in a value-sorted array of voxel positions. An additional array contains the cumulative histogram values. For our case of multiresolution data, all levels of the pyramid were created and stored as value-sorted arrays. To prevent holes forming in the projection image for non-axial viewing directions, we postprocessed the projection images by performing a closing with a flat structuring element of size 2×2. By $\mathcal{M}_{\text{discrete}}$ we will denote the operator consisting of voxel projection followed by the final closing. In the experiments to be discussed in Section 4, we define interesting voxels simply as those with a non-zero grey value (i.e., we did not use the preprocessing scheme of [11] to identify and remove other types of non-contributing voxels). In practice, especially for angiographic data, a substantial reduction (up to 99%) in the amount of voxels to be processed is possible when only nonempty voxels are stored.

Arbitrary view directions To deal with general view directions a continuous function has first to be reconstructed from the discrete samples of the data set f. The MIP operator $\mathcal{M}$ now computes the maximum of the reconstructed function along a given direction vector. The reconstruction operator is assumed to be a dilation (one choice is the voxel model, which treats voxels as cells which a constant data value). Then formulas (9) and (10) still hold for the reconstructed functions. Formulas (11) and (12) have to be slightly adapted to deal with discretization effects. In the experiments below, we have simply replaced $\mathcal{M}$ by the discrete MIP operator $\mathcal{M}_{\text{discrete}}$ of the previous subsection. The structuring element $\tilde{A}$ in (11) is replaced by the result of applying $\mathcal{M}_{\text{discrete}}$ to the 3-D structuring element A. In the experiments, we found this approximation to be quite accurate. A careful study of the associated discretization error will be presented elsewhere.

4 Examples

Experiments were carried out on a PC with a 500 MHz Pentium III processor and 128 Mb memory. We performed MMIP rendering of a CT head data set and an MR angiography data set, both of size 256^3, using a 2-level pyramid. Dilations and erosions with a $2 \times 2 \times 2$ structuring element (3-D morphological Haar pyramid) were used. The sampling distance in the view plane was taken equal to the sampling distance of the original volume data. For the CT data, about 26% of the data consisted of nonzero voxels; for the angiography data, this was 1.25%. Creation of the pyramid took about 25 seconds in both cases. Rendering times were found to be almost independent of view angle. Sizes in value-sorted array format and rendering times of the successive levels of the pyramid are given in Table 1. For comparison, the numbers for direct MIP rendering of the full-size volume data are given as well. All times are excluding I/O. The timings show that computing a level-2 or level-1 approximation takes considerably less time than a full-size MIP, especially for data sets with a relatively large number of nonzero voxels. Figure 1 shows successive approximations for the CT data. MMIP approximations quickly remove details of the data, due to the fact that the approximations essentially are morphological openings by a structuring element whose size increases with level. Note in particular in Fig. 1 that small details such as the tube from the mouth almost disappear in the level 1 approximation. To be useful for angiographic data, the method has to be adapted so that small details are better preserved in higher levels of the pyramid (see discussion).

Table1. Data sizes (value-sorted array format) and rendering times of MIP (full image) and MMIP (progressive renderings of approximation and detail data).

MRA data 256 × 256 × 256	size (kbytes)	time (s)	CT data 256 × 256 × 256	size (kbytes)	time (s)
full image	838.5	0.423	full image	17433	6.92
level 2 approximation	0.812	0.110	level 2 approximation	253	0.20
add detail level 1	30.2	0.129	add detail level 1	1861	0.87
add detail level 0	801.6	0.417	add detail level 0	15171	6.04

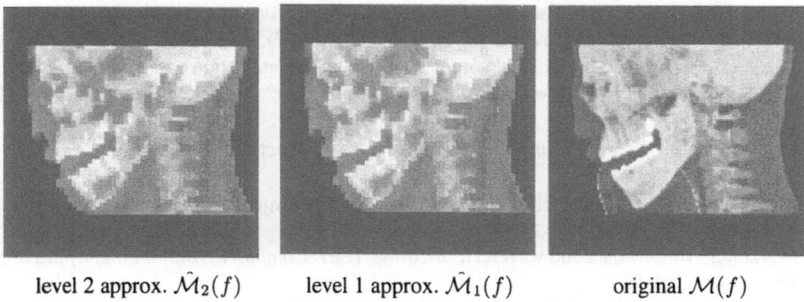

level 2 approx. $\hat{\mathcal{M}}_2(f)$ level 1 approx. $\hat{\mathcal{M}}_1(f)$ original $\mathcal{M}(f)$

Figure1. MMIP reconstruction from a 2-level morphological adjunction pyramid using a $2 \times 2 \times 2$ structuring element.

5 Discussion

Several extensions to the proposed multiresolution extension of MIP volume rendering are possible. Adjunction pyramids with non-flat structuring functions can be used. The shape of this structuring function can be adapted in such a way that smaller details of the data at higher approximation levels are retained. For the same purpose, other operators than erosions, such as openings, can be used for the analysis phase. This implies however, that the representation formula (9) no longer holds. To maintain an acceptable level of efficiency, we still require that the synthesis operator $\psi^{\downarrow}$ is a dilation, so that it commutes with maxima. For compression purposes, other choices of addition and subtraction operators can be considered, and morphological wavelets can be used [6],

54

which have the advantage that they provide a non-redundant multiresolution representation.

References

1. Burt, P. J., and Adelson, E. H. The Laplacian pyramid as a compact image code. *IEEE Trans. Commun. 31* (1983), 532–540.
2. Cai, W., and Sakas, G. Maximum intensity projection using splatting in sheared object space. *Computer Graphics Forum (Proc. Proc. Eurographics'98) 17*, 3 (1998), C113–124.
3. Goutsias, J., and Heijmans, H. J. A. M. Multiresolution signal decomposition schemes. Part 1: Linear and morphological pyramids. Tech. Rep. PNA-R9810, Centre for Mathematics and Computer Science, Amsterdam, Oct. 1998.
4. Grosso, R., and Ertl, T. Biorthogonal wavelet filters for frequency domain volume rendering. In *Proceedings of Visualization in Scientific Computing '95* (1995), J. van Wijk, R. Scateni, and P. Zanarini, Eds.
5. Heijmans, H. J. A. M. *Morphological Image Operators*, vol. 25 of *Advances in Electronics and Electron Physics, Supplement.* Academic Press, New York, 1994.
6. Heijmans, H. J. A. M., and Goutsias, J. Multiresolution signal decomposition schemes. Part 2: morphological wavelets. Tech. Rep. PNA-R9905, Centre for Mathematics and Computer Science, Amsterdam, June 1999.
7. Lippert, L., and Gross, M. H. Fast wavelet based volume rendering by accumulation of transparent texture maps. *Computer Graphics Forum 14*, 3 (1995), 431–443.
8. Lippert, L., Gross, M. H., and Kurmann, C. Compression domain volume rendering for distributed environments. In *Proc. Eurographics'97* (1997), pp. 95–107.
9. Lürig, C., and Ertl, T. Hierarchical volume analysis and visualization based on morphological operators. In *Proc. IEEE Visualization '98* (1998), IEEE Computer Society Press, pp. 335–341.
10. Malzbender, T. Fourier volume rendering. *ACM Transactions on Graphics 12*, 3 (1993), 233–250.
11. Mroz, L., König, A., and Gröller, E. Maximum intensity projection at warp speed. *Computers & Graphics 24* (2000), 343–352.
12. Muraki, S. Volume data and wavelet transforms. *IEEE Computer Graphics and Applications 13*, 4 (1993), 50–56.
13. Roerdink, J. B. T. M., and Blaauwgeers, G. S. M. Visualization of Minkowski operations by computer graphics techniques. In *Mathematical Morphology and its Applications to Image Processing*, J. Serra and P. Soille, Eds. Kluwer Acad. Publ., Dordrecht, 1994, pp. 289–296.
14. Roerdink, J. B. T. M., and Westenberg, M. A. Wavelet-based volume visualization. *Nieuw Archief voor Wiskunde 17 (Fourth Series)*, 2 (July 1999), 149–158.
15. Serra, J. *Image Analysis and Mathematical Morphology.* Academic Press, New York, 1982.
16. Sternberg, S. R. Grayscale morphology. *Comp. Vis. Graph. Im. Proc. 35* (1986), 333–355.
17. Westenberg, M. A., and Roerdink, J. B. T. M. Frequency domain volume rendering by the wavelet X-ray transform. *IEEE Trans. Image Processing 9*, 7 (2000), 1249–1261.
18. Westermann, R., and Ertl, T. A multiscale approach to integrated volume segmentation and rendering. In *Proc. Eurographics'97, Vienna*, D. Fellner and L. Szirmay-Kalos, Eds., vol. 16. 1997, pp. C–117–C–127.
19. Westover, L. A. Footprint evaluation for volume rendering. *Computer Graphics 24*, 4 (1990), 367–376.
20. Zuiderveld, K. J., Koning, A. H. J., and Viergever, M. A. Techniques for speeding up high-quality perspective Maximum Intensity Projection. *Pattern Recognition Letters 15* (1994), 507–517.

Subdivision Surfaces for Scattered-data Approximation

Martin Bertram and Hans Hagen

University of Kaiserslautern
Department of Computer Science
P.O. Box 3049
D-67653 Kaiserslautern
Germany
{bertram,hagen}@informatik.uni-kl.de

Keywords: multiresolution methods, scattered data, subdivision surfaces, terrain modeling, triangulation.

Abstract. We propose a modified Loop subdivision surface scheme for the approximation of scattered data in the plane. Starting with a triangulated set of scattered data with associated function values, our scheme applies linear, stationary subdivision rules resulting in a hierarchy of triangulations that converge rapidly to a smooth limit surface. The novelty of our scheme is that it applies subdivision only to the ordinates of control points, whereas the triangulated mesh in the plane is fixed. Our subdivision scheme defines locally supported, bivariate basis functions and provides multiple levels of approximation with triangles. We use our subdivision scheme for terrain modeling.

1 Introduction

Subdivision surfaces [4, 8] are widely used for modeling surfaces of arbitrary topological genus. They are defined by polygonal control meshes that are recursively refined analogously to knot insertion for B-spline surfaces [15]. This refinement process converges to smooth limit surfaces that are in many cases piecewise polynomials. The subdivision schemes by Catmull/Clark [7] and Doo/Sabin [9], for example, reproduce uniform B-Splines on regular, rectilinear meshes.

The strength of subdivision surfaces is their ability to deal with irregular meshes defining arbitrary two-manifolds. Extraordinary points, *i.e.*, surface points corresponding to vertices with other than four adjacent edges in a control mesh, are typically surrounded by an infinite number of smaller and smaller polynomial patches satisfying certain continuity constraints. Eigen-analysis of local subdivision matrices can be used to compute surface normals and to evaluate a limit surface at arbitrary parameter values [25, 23, 14].

Multiresolution modeling techniques, like wavelet transforms [26], are required for real-time visualization of large-scale data sets. Subdivision surfaces and wavelet transforms can be combined to a single, highly efficient multiresolution modeling tool [20, 24, 18, 3]. Subdivision-surface wavelets with finite filters have been constructed for compression and multiresolution representation of functions defined on planar tessellations and surfaces of arbitrary topology like isosurfaces [1–3]. Related multiresolution methods [12, 17] have been designed for completely irregular triangulated surfaces without subdivision connectivity. Wavelets and subdivision techniques are also successfully being applied to computational fluid dynamics (CFD) and flow visualization problems [27, 21, 6].

Despite of their simplicity and flexibility for modeling surfaces of arbitrary topology, subdivision surfaces have never been used for an apparently simpler problem—representing graph surfaces, *i.e.*, modeling smooth functions defined on planar domains. Instead, piecewise polynomial constructions like the Clough-Tocher and Powell-Sabin interpolants [15] are frequently used, splitting every triangle into multiple macro-triangles that may have bad aspect-ratios. Additionally, interpolation constraints may cause unwanted variations that are not present in surfaces defined by control points without interpolation, like B-Splines. Other approaches are based on multiquadrics [10, 11]. Some aspects of subdivision surfaces, however, have been exploited by a scattered-data fitting method based on triangular B-splines [22].

Classical visualization problems, like terrain modeling, have not taken advantage of subdivision techniques, in the past. In this paper we propose a simple and efficient variant of Loop's subdivision scheme [19, 26] for modeling scattered data in the plane. We expect that our subdivision technique will successfully be used for applications like terrain modeling and that trivariate constructions can be developed for volume modeling data defined on tetrahedral grids, as well.

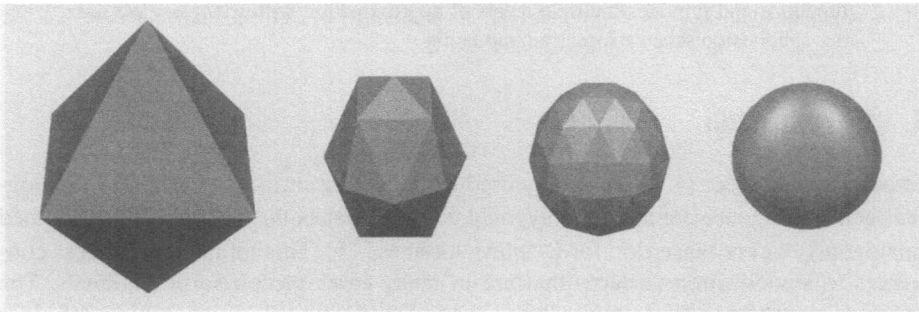

Fig. 1. Loop subdivision process. Starting with a triangulated control mesh (octahedron, left), a hierarchy of triangulations converges to a smooth limit surface (right).

2   Loop's Subdivision

We now review Loop's subdivision scheme [19] generalizng quartic box splines [15] to arbitrary triangular control meshes. The big deal about subdivision schemes is that they generate smooth surfaces from irregular control meshes with *extraordinary* points. Extraordinary points correspond to vertices that have other than four incident edges in a locally rectilinear mesh and vertices that have other than six incident edges in a triangle mesh. Thus, it is possible to define smooth surfaces of arbitrary topology by simple control meshes.

Starting with a triangulated control mesh, Loop's scheme splits every triangle into four by inserting a new vertex on every edge. This subdivision step is recursively repeated, resulting in a mesh hierarchy that converges to a smooth limit surface, see Figure 1. The coordinates of the control points are updated by linear, stationary subdivision rules, $i.e.$, the new points depend linearly on a local stencil of old control points and the masks for the updates are the same in every subdivision step.

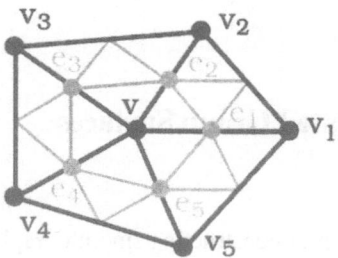

Fig. 2. Local indexing for the subdivision masks.

To explain the subdivision masks, we consider a particular vertex v with valence n (number of incident edges) and use a local indexing for the adjacent vertices v_i in some order ($i = 1, 2, \ldots, n$), see Figure 2. All indices are applied modulo n. The new vertices located on triangle edges are denoted by e_i. The first subdivision mask defines these vertices as

$$e_i = \frac{1}{8}(3v + 3v_i + v_{i-1} + v_{i+1}). \tag{2.1}$$

The second mask then updates the old vertices

$$v' = \alpha_n v + (1 - \alpha_n)\bar{v}, \quad \text{where}$$

$$\bar{v} = \frac{1}{n} \sum_{i=1}^{n} v_i \quad \text{and} \tag{2.2}$$

$$\alpha_n = \frac{3}{8} + \left(\frac{3}{8} + \frac{1}{4} cos\frac{2\pi}{n}\right)^2.$$

Splitting every triangle into four and applying these two masks results in a finer mesh that is further subdivided using the same rules again and again.

Alternatively to the second mask we can update the old vertices v without duplicating their coordinates. Therefore, we replace equation (2.2) by the equivalent mask

$$v' = (1 - \beta_n)v + \beta_n \bar{e}, \quad \text{where}$$

$$\bar{e} = \frac{1}{n} \sum_{i=1}^{n} e_i \quad \text{and} \tag{2.3}$$

$$\beta_n = \frac{8}{5}(1 - \alpha_n).$$

58

In the case of a regular triangulation (without extraordinary points) using $\alpha_6 = \frac{5}{8}$, Loop's scheme reproduces quartic box splines. The choice of weights α_n ($n \geq 3$) implies C^1-continuity of the limit surface at extraordinary points [19]. The limit behavior of subdivision surfaces at extraordinary points can be computed from the eigen-structure of a local subdivision matrix, which first was done by Doo/Sabin [9]. This eigen-structure can be used to compute local control points of polynomial patches, resulting in an efficient algorithm for computing a subdivision surface and its partial derivatives at arbitrary parameter values [25].

3 Recursively Generated Graph Surfaces

We now consider the problem of constructing smooth graph surfaces $f(x,y)$ for triangulated scattered data in the plane. We start with a set of planar points $v_i = (x_i, y_i)$ ($i = 1, 2, \ldots, m$) with associated function values v_i^f and a triangulation Δ of these points. When applying Loop's subdivision to the triangulated mesh with control points (v_i, v_i^f), the planar triangulation is deformed by piecewise quartic polynomials, see Figure 3. Evaluating the surface at an arbitrary point (x,y) would be difficult since we would have to estimate corresponding local parameters in a certain triangle. Hence, we need to subdivide the triangles linearly and use Loop's scheme only for computing the function values. This, however, results in creases at the edges of Δ, see Figure 4. Only in the case of a regular triangulation with congruent triangles Loop's subdivision coincides with linear subdivision in the plane, due to linear precision, and a smooth graph surface is generated.

We want to modify the masks for Loop's subdivision in equations (2.1) and (2.3) such that they smooth out creases caused by the parametrization. Therefore, these masks must take into account the shapes of adjacent triangles. Only if these triangles are congruent, then the new masks should coincide with Loop's.

We use the same indexing as in equations (2.1) and (2.3), see Figure 5. First, we define the mask for the ordinates e_i^f at the new points e_i that are now midpoints of the corresponding edges. Therefore, we compute a parametric least-squares plane $\pi(x,y)$ [5] satisfying

$$3\left(\pi(v) - v^f\right)^2 + 3\left(\pi(v_i) - v_i^f\right)^2 + \left(\pi(v_{i-1}) - v_{i-1}^f\right)^2 + \left(\pi(v_{i+1}) - v_{i+1}^f\right)^2$$
$$\longrightarrow min.$$

$$(3.1)$$

This plane, evaluated at e_i, provides the new ordinate e_i^f.

As we will show in the following, this mask is linear, stationary, and reproduces equation (2.1) if both triangles are congruent, i.e., if the vectors $v_{i+1} - v$ and $v_i - v_{i-1}$

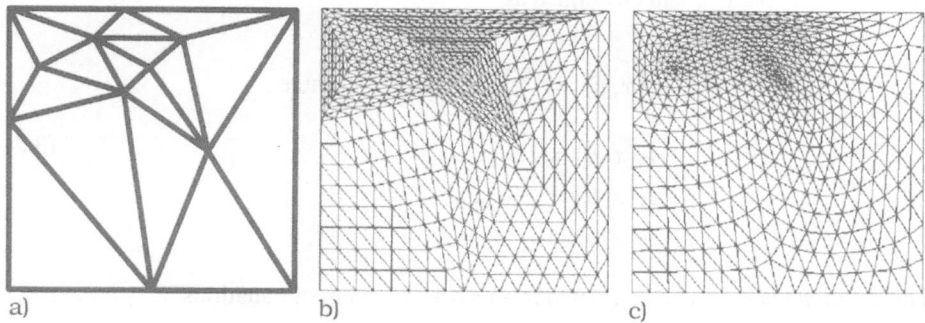

Fig. 3. Parametrizations for subdivision process. a) initial triangulation; b) linear subdivision; c) Loop subdivision deforming the triangles (boundary points are fixed).

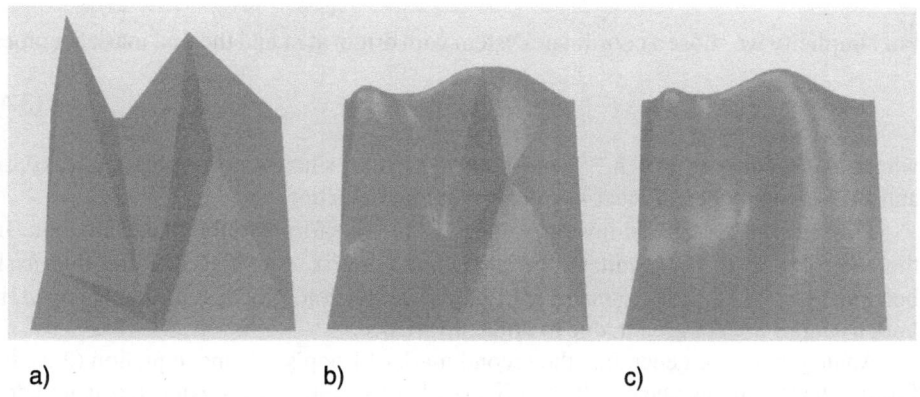

Fig. 4. Effect of parametrizations. a) initial control mesh based on the triangulation shown in Figure 3.; b) Loop's subdivision applied to ordinates causing creases at triangle edges; c) Loop's subdivision applied to all coordinates generating a smooth graph surface with deformed domain.

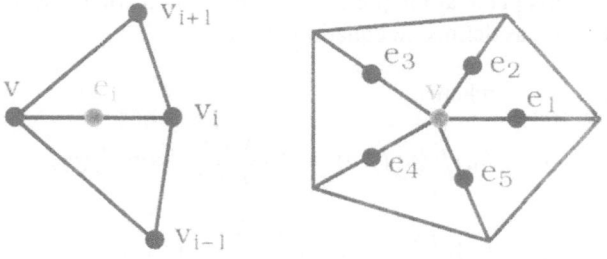

Fig. 5. Defining subdivision masks for the ordinates of e_i and v.

are equal. The plane π can be defined as

$$\pi(x,y) = \sum_{i=1}^{3} c_i f_i(x,y), \quad \text{where}$$

$$f_0(x,y) = 1,$$
$$f_1(x,y) = x, \quad \text{and} \tag{3.2}$$
$$f_2(x,y) = y.$$

The coefficients c_i can be computed from a 3×3-system of equations

$$\mathbf{Ac} = \mathbf{b}, \quad \text{where}$$
$$a_{kj} = 3f_k(v)f_j(v) + 3f_k(v_i)f_j(v_i)$$
$$\quad + f_k(v_{i-1})f_j(v_{i-1}) + f_k(v_{i+1})f_j(v_{i+1}), \quad \text{and} \tag{3.3}$$
$$b_k = 3v^f f_k(v) + 3v_i^f f_k(v_i) + v_{i-1}^f f_k(v_{i-1}) + v_{i+1}^f f_k(v_{i+1}).$$

For simplicity we chose a coordinate system with origin at e_i and the first mask becomes

$$e_i^f = c_0 = \tilde{\mathbf{a}} \cdot \mathbf{b} \tag{3.4}$$

where $\tilde{\mathbf{a}}$ is the first row of $\mathbf{A}^{-1}$, which always exists, since the four points v, v_i, v_{i-1} and v_{i+1} can only be collinear in a degenerate triangulation.

This mask is linear and invariant under affine transforms in the planar domain. In the special case of two equilateral triangles, the matrix A is diagonal and the mask becomes $\frac{1}{8}b_0$ which is the same as for Loop's subdivision. This result remains valid, if both triangles are congruent, due to affine invariance.

Analogously, we generalize the second mask of Loop's scheme, equation (2.3), by fitting a least-squares plane π. The ordinate v^f for a point v with valence n is updated by computing π, such that

$$(1 - \beta_n)\left(\pi(v) - v^f\right)^2 + \beta_n \frac{1}{n} \sum_{i=1}^{n} \left(\pi(e_i) - e_i^f\right)^2 \longrightarrow min, \tag{3.5}$$

and by evaluating this plane at the point v. Again, we construct a system of equations for the coefficients c_i, as defined in equation (3.2):

$$\mathbf{Ac} = \mathbf{b}, \quad \text{where}$$

$$a_{kj} = (1 - \beta_n)f_k(v)f_j(v) + \beta_n \frac{1}{n} \sum_{i=1}^{n} f_k(e_i)f_j(e_i), \quad \text{and} \tag{3.6}$$

$$b_k = (1 - \beta_n)v^f f_k(v) + \beta_n \frac{1}{n} \sum_{i=1}^{n} e_i^f f_k(e_i).$$

This system is solved analogously to the system for the first mask. The same arguments hold for linearity, affine invariance, and existence of the solution. This second mask reproduces Loop's scheme only for a vertex v surrounded by six congruent triangles, *i.e.*, by triangles resulting from equilaterals when applying an affine map.

We note that these masks can be pre-computed. Due to affine invariance, there is at most one different mask for every vertex and for every edge of the initial triangulation Δ. Thus, our scheme is stationary. All vertices generated on the same edge of Δ use the same mask. For the majority of newly generated vertices that are not located on an edge or vertex of the initial triangulation Δ, we can apply Loop's subdivision rules directly to compute the corresponding ordinates.

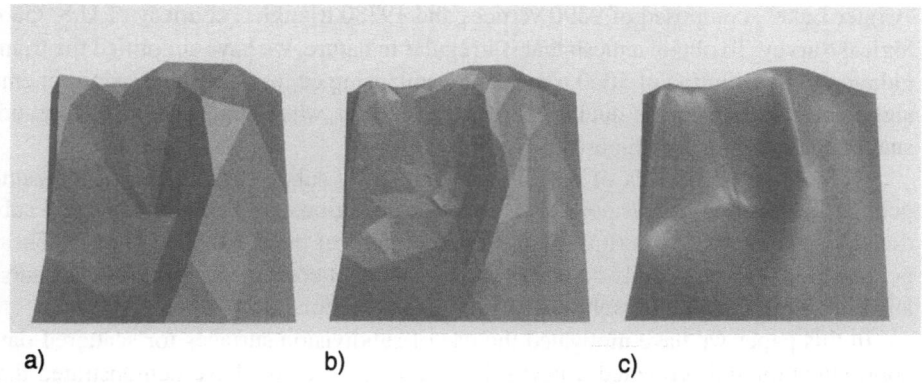

a)   b)   c)

Fig. 6. Subdivision process and limit surface for our subdivision scheme. a) first subdivision level; b) second level; c) resulting surface (sixth level).

4 Results and Future Work

Our linear, stationary subdivision scheme provides the following properties:

- It is invariant under affine transforms in the plane.
- It has linear precision.
- Our scheme generates piecewise polynomial surfaces.
- It coincides with Loop's subdivision and with quartic box-splines on a regular grid composed of congruent triangles.

Surface regions that are strictly located inside the triangles of Δ are eventually determined by Loop's subdivision rules, after applying a finite number of subdivision steps. This implies that our subdivision surfaces are composed of quartic patches and are C^1-continuous in all points strictly inside the individual triangles of Δ. For the remaining points a rigorous mathematical analysis of the limit-behavior considering the eigen-structure of local subdivision matrices needs still to be done.

The subdivision process for a surface generated with our scheme using a control mesh (Figure 4a.) composed of 13 vertices and 17 triangles is shown in Figure 6. This surface appears to be much smoother than the one generated with Loop's subdivision

62

masks applied to the ordinates, shown in Figure 4b. However, it does not look as fair as the surface generated by Loop's subdivision applied to all coordinates, see Figure 4c. This is due to the constraints for the parametrization that do not allow the surface to relax parallel to the plane. These constraints are useful, however, since they lead to limit surfaces that better preserve the shape of the initial control mesh. Additionally, these constraints imply that the vertices of the initial triangulation Δ are fixed in the plane. Thus, we can easily interpolate certain ordinates at these vertices by solving for the corresponding control points. Interpolating normals and surface fairing can also be accomplished with subdivision surfaces [14, 16].

Figure 7. shows our subdivision scheme applied to the triangulated terrain model "Crater Lake", composed of 9890 vertices and 19380 triangles, courtesy of U.S. Geological Survey. To obtain a mesh that is irregular in nature, we have simplified the triangulation to a resolution of 5000 triangles by collapsing edges. Our subdivision scheme significantly increases the quality of rendered images when compared to a Gouraud-shaded triangulation, see Figure 7.

We note that all levels of resolution obtained by subdivision represent the same geometric information. It is possible to introduce additional surface detail at every subdivision level by locally perturbing the control points of the subdivided meshes. These perturbations can compactly be represented by wavelet coefficients, providing a sparse and highly efficient multiresolution surface representation [20].

In this paper we have motivated the use of subdivision surfaces for scattered data approximation and presented a new subdivision scheme. We have demonstrated that our method is well suited for terrain modeling. It is also possible to use our subdivision scheme for approximating bivariate functions with multi-dimensional ranges, like planar tensor fields and color images. Future work will be directed at the construction of trivariate subdivision schemes for volume modeling. Additionally, we want to improve the fairness of our functional subdivision surfaces by incorporating variational principles [13, 16] into our subdivision scheme.

References

1. M. Bertram, *Multiresolution Modeling for Scientific Visualization*, Ph.D. Thesis, University of California at Davis, July 2000.
   http://graphics.cs.ucdavis.edu/~bertram
2. M. Bertram, M.A. Duchaineau, B. Hamann, and K.I. Joy, *Wavelets on planar tesselations*, Proceedings of the International Conference on Imaging Science, Systems, and Technology, CSREA Press, 2000, pp. 619–625.
3. M. Bertram, M.A. Duchaineau, B. Hamann, and K.I. Joy, *Bicubic subdivision-surface wavelets for large-scale isosurface representation and visualization*, Proceedings of IEEE Visualization 2000, pp. 389–396 & 579.
4. H. Biermann, A. Levin, and D. Zorin, *Piecewise smooth division surfaces with normal control*, Proceedings of ACM Siggraph 2000, pp. 113–120.
5. W. Boehm and H. Prautzsch, *Geometric Concepts for Geometric Design*, A.K. Peters, Ltd., Wellesley, Massachusetts, 1994.
6. G.-P. Bonneau, *Optimal triangular Haar bases for spherical data*, Proceedings of IEEE Visualization 1999, pp. 279–284 & 534.

7. E. Catmull and J. Clark, *Recursively generated B-spline surfaces on arbitrary topological meshes*, Computer-Aided Design, Vol. 10, No. 6, 1978, pp. 350–355.

8. T. DeRose, M. Kass, and T. Truong, *Subdivision surfaces in character animation*, Proceedings of ACM Siggraph 1998, pp. 85–94.

9. D. Doo and M. Sabin, *Behaviour of recursive division surfaces near extraordinary points*, Computer-Aided Design, Vol. 10, No. 6, 1978, pp. 356–360.

10. R. Franke and H. Hagen, *Least squares surface approximation using multiquadrics and parametric domain distortion*, Computer-Aided Geometric Design, Vol. 16, No. 3, Elsevier, 1999, pp.177–196.

11. R. Franke, H. Hagen, and G. Nielson, *Repeated knots in least squares multiquadric functions*, Computing Supplement 10, Geometric Modelling, Springer, 1995, pp. 177–187.

12. I. Guskov, W. Sweldens, and P. Schröder, *Multiresolution signal processing for meshes*, Proceedings of ACM Siggraph 1999, pp. 325–334.

13. H. Hagen, G. Brunnett, and P. Santarelli, *Variational principles in curve and surface design*, Surveys on Mathematics for Industry, Vol. 3, No. 1, 1993, pp. 1–27.

14. M. Halstead, M. Kass, and T. DeRose, *Efficient, fair interpolation using Catmull-Clark surfaces*, Proceedings of ACM Siggraph 1993, pp. 35–44.

15. J. Hoscheck and D. Lasser, *Fundamentals of Computer Aided Geometric Design*, A.K. Peters, 1993.

16. L. Kobbelt, *Discrete fairing and variational subdivision for freeform surface design*, Visual Computer, Vol. 16, No. 3–4, Springer, 2000, pp. 142–158.

17. L. Kobbelt, J. Vorsatz, and H.-P. Seidel, *Multiresolution hierarchies on unstructured triangle meshes*, Computational Geometry: Theory and Applications, Vol. 14, No. 1-3, Elsevier, 1999, pp. 5–24.

18. L. Kobbelt and P. Schröder, *A multiresolution framework for variational subdivision*, ACM Transactions on Graphics, Vol. 17, No. 4, 1998. pp. 209–237.

19. C.T. Loop, *Smooth subdivision surfaces based on triangles*, M.S. Thesis, Department of Mathematics, University of Utah, 1987.

20. M. Lounsbery, T. DeRose, and J. Warren, *Multiresolution analysis for surfaces of arbitrary topological type*, ACM Transactions on Graphics, Vol. 16, No. 1, 1997, pp. 34–73.

21. G.M. Nielson, I.-H. Jung, and J. Sung, *Haar wavelets over triangular domains with applications to multiresolution models for flow over a sphere*, Proceedings of IEEE Visualization 1997, pp. 143–150.

22. R. Pfeifle, H.-P. Seidel, *Fitting triangular B-splines to functional scattered data*, Computer Graphics Forum, Vol. 15, No. 1, Backwell Publishers, 1996, pp. 15–23.

23. U. Reif, *A unified approach to subdivision algorithms near extraordinary vertices*, Computer-Aided Geometric Design, Vol. 12, No. 2, 1995, pp.153–174.

24. P. Schröder and W. Sweldens, *Spherical wavelets: efficiently representing functions on the sphere*, Proceedings of ACM Siggraph 1995, pp. 161–172.

25. J. Stam, *Exact evaluation of Catmull-Clark subdivision surfaces at arbitrary parameter values*, Proceedings of ACM Siggraph 1998, pp. 395–404.

26. E.J. Stollnitz, T.D. DeRose, D.H. Salesin, *Wavelets for Computer Graphics—Theory and Applications*, Morgan Kaufmann, 1996.

27. H. Weimer and J. Warren, *Subdivision schemes for fluid flow*, Proceedings of ACM Siggraph 1999, pp. 111–120.

Editors' Note: see Appendix, p. 337 for colored figure of this paper

The Rendering of Unstructured Grids Revisited

Rüdiger Westermann

Scientific Visualization and Imaging Group
University of Technology Aachen

Abstract. In this paper we propose a technique for resampling scalar fields given on unstructured tetrahedral grids. This technique takes advantage of hardware accelerated polygon rendering and 2D texture mapping and thus avoids any sorting of the tetrahedral elements. Using this technique, we have built a visualization tool that enables us to either resample the data onto arbitrarily sized Cartesian grids, or to directly render the data on a slice-by-slice basis. Since our approach does not rely on any pre-processing of the data, it can be utilized efficiently for the display of time-dependent unstructured grids where geometry as well as topology change over time.

1 Introduction and related work

Rendering unstructured volume data is still challenging because no existing algorithm allows for the accurate display of reasonably sized data sets at interactive frame rates. Although considerable efforts have been made during the last couple of years, frame rates are still not competitive with those that can be reached for high resolution Cartesian grids using specialized software solutions or dedicated graphics hardware. In addition, many of the proposed techniques fail in practical applications, where memory issues play a major concern and dynamic changes of geometry as well as topology happen frequently.

In particular two basic problems have to be addressed when developing rendering algorithms for unstructured grids. The first one is to determine the correct visibility ordering of elements. The second one is to implement an appropriate algorithm that allows one to render each element in the ascertained order.

Two different classes of algorithms exist to solve for the latter problem as illustrated in Figure 1. Image space techniques compute for each view ray the entry and the exit point for every element that is hit by that ray. At both points the data is interpolated between the values given at the elements vertices. Finally, taking into account an optical model the integration along the ray is performed. If pre-shaded samples are used for interpolation it can be exploited that the resampled signal along the ray depends linearly on the values at the entry and the exit point. In post-shading, however, the scalar field has to be reconstructed along the ray by taking an appropriate step size with respect to the selected transfer function.

Object space techniques, on the other hand, accomplish the rendering by projecting each element onto the viewing plane such as to approximate the visual stimulus of viewing the element with regard to the chosen optical model. Two principal methods have been shown to be very effective in performing this task: *slicing* and *cell projection*.

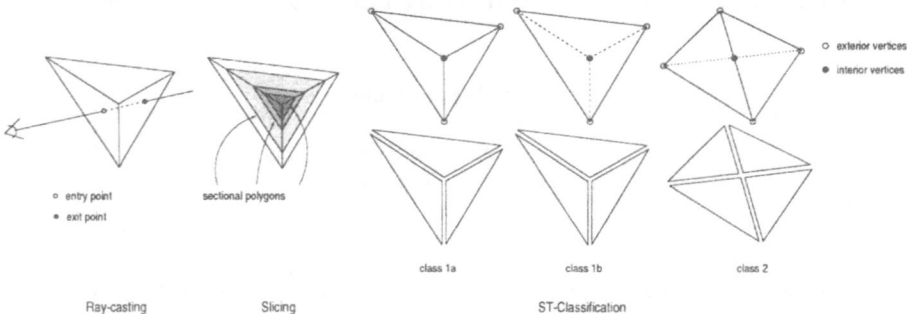

Fig. 1. *Different rendering techniques for tetrahedral elements are illustrated.*

Using the former method, each element gets projected on a slice-by-slice basis. Therefore, the cross sections of each element and the current slicing plane are computed. At each intersection point with one of the element edges the color value is interpolated from the pre-shaded samples at cell vertices. The cross sections are then rendered by letting the graphics hardware resample the color values onto the discrete pixel buffer during rasterization. The computation of the sectional polygons can either be done explicitly [19, 15], or implicitly on a per-pixel basis by taking advantage of dedicated graphics hardware providing per-fragment tests and operations [14].

A different object space approach is cell-projection [8]. Based on the current viewing parameters the projection of each tetrahedron is classified with respect to four different classes and decomposed into triangles correspondingly. For each vertex of the projected profiles color and opacity values are computed taking into account the underlying optical model. Although the bi-linear interpolation of these values across triangles doesn't give accurate results and is restricted to the rendering of pre-shaded samples, it has been established as the most prominent rendering technique for tetrahedral cells due to its efficiency. Different extensions to the cell-projection algorithm have been proposed in order to achieve better accuracy [12, 17] and to enable post-shading using arbitrary transfer functions [7]. The common method here is to pre-compute the volume rendering integral with regard to the current transfer function for a number of different parameter values and to code the results in a texture map. Then, rather than color samples the pre-computed values are interpolated and composited.

The more difficult problem, however, is to determine the correct visibility order of elements. Many algorithms have been proposed during the last couple of years trying to reach interactive frame rates in completely different ways. Probably the most efficient one has been used in [1, 4, 18]. These techniques exploit the fact that for tetrahedral meshes exhibiting a Delauney property the correct order can be found by sorting the tangential distances to circumscribing spheres using any customized algorithm. Although quite impressive frame rates can be obtained using this approach, the most serious drawback is that grids generated in practical applications are usually not Delauney meshes. This might lead to incorrect results and does not allow resolving topological cycles in the data.

A different alternative is the sweep-plane approach [3, 10, 9, 13]. In this approach the coherence within cutting planes in object space is exploited in order to determine the visibility ordering of the available primitives.

In addition, much work has been spent on accelerating the visibility ordering of unstructured elements. Usually, the data is first pre-processed in order to recover topological information, which is stored and used to accelerate the sorting procedure during rendering. Many different variants exist which exploit adjacency information within the grid and construct hierarchical data structures allowing for the efficient traversal of elements in the right order [16, 11, 2]

However, two significant problems are inherent to approaches that rely on pre-computed topological information. First, a considerable increase in memory might be introduced by the data structures necessary to allow for efficient visibility sorting. Second, as soon as changes in the geometry or the topology occur the pre-processing task has to be repeated. Particularly in numerical simulations where time-dependent sequences are generated quite commonly these limitations prohibit interactive frame rates.

In this paper, we present a novel approach for the resampling of scalar data given on unstructured tetrahedral grids. The goal of this approach is twofold: to emphasize the impact of state-of-the-art graphics hardware on current visualization techniques *and* to demonstrate how it can be used for the accurate rendering of large unstructured grids without the need for sorting the elements explicitly. As a direct implication the grid doesn't have to be pre-processed in order to determine topological information thus considerably reducing the memory overhead. Moreover, geometric and topological changes do not implicate expensive re-calculations. Particularly in practical applications where memory issues and topological changes of the grids are of major concern the real strength of our method comes into play.

The reminder of this paper is organized as follows. First, we describe the slicing procedure for tetrahedral cells our technique is based upon and we propose an algorithm in which hardware supported graphics operations are paramount. We then discuss implementation details and we outline two different alternatives for the rendering of unstructured grids taking advantage of our technique. Next, beneficial extensions using consumer graphics cards are demonstrated. We conclude the paper with a detailed discussion, and we show results and timings of our approach applied to real data sets.

2 Slicing tetrahedron

In the previous section we have outlined the general approach for slicing tetrahedra. We mentioned that only if pre-shaded samples are issued per vertex the generated fragments can be directly displayed and blended with pixel values already in the frame buffer. Unfortunately this strategy prohibits the use of arbitrary transfer functions to be applied to the original scalar data. Post-shading, on the other hand, allows the color distribution in the interior of each tetrahedron to be modified non-linearly. This can be achieved by interpreting the scalar material values as one-dimensional coordinates into a linear texture map. During rasterization texture coordinates are bi-linearly interpolated and the color value is finally looked up in a user defined texture lookup table.

The slicing of tetrahedral elements can also be viewed from a different perspective. As stated earlier, along each ray passing through a tetrahedron the material values depend linearly on the values reconstructed at the entry and the exit point. The reconstruction of the scalar field on a slicing plane parallel to the viewing plane thus involves computing the values at both the front and the back faces of the tetrahedron with respect to the current view, and to linearly combine them.

2.1 Hardware support

In order to avoid explicitly computing the sectional polygons and the scalar values used for shading we propose a method that takes advantage of hardware assisted polygon rendering and 2D texture mapping. Therefore let us assume that each tetrahedron to be projected has already been decomposed into triangles based on the ST-classification, named the ST-triangulation hereafter. We note that at the interior vertex of the triangulation two scalar values are stored: one for the point at the front facing edge and one for the point at the back facing edge. We can thus interpolate the data across the front faces and the back faces by scan-converting the ST-triangulation twice: In the second pass the function value at the interior vertex is exchanged (see Figure 2).

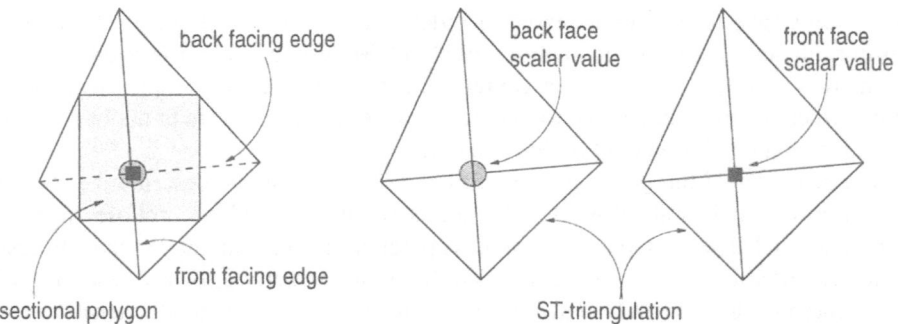

Fig. 2. *The scalar function can be reconstructed across the front and back faces by rendering the ST-triangulation twice with different scalar values at the interior vertex*

Although we already employ graphics hardware to reconstruct the scalar field at each pixel that is covered by the projection of the tetrahedron two problems still need to be addressed. The data should only be drawn to those pixels that are covered by the sectional region between the slicing plane and the tetrahedron *and* values need to be linearly combined in order to obtain the interpolated data samples. Both problems can be solved efficiently by means of a 2D texture map that stores pre-computed weights necessary to perform the linear interpolation. During rasterization the scalar values issued as polygon color are modulated with the texture color that represents the appropriate interpolation weights.

Let Nx, Ny be the size of the texture. Each texture element is described by a luminance and an alpha value. Then, by using $p = 2i/Nx - 1$ and $q = 2j/Ny - 1$ the luminance values in the texture $T1$ are initialized as follows:

$$T1[i,j] = \begin{cases} \frac{p}{p+q} & : \quad i \geq Nx/2 \quad \& \quad j \geq Ny/2 \\ 0 & : \quad otherwise \end{cases}$$

Alpha values are set to one where the luminance values are different from zero, otherwise they are set to zero as well. In addition a second 2D texture map, $T2$, is created. It is similar to $T1$ but non-zero values $T1[i,j]$ are replaced by $1 - T1[i,j]$.

Before we analyze the assignment of texture coordinates to vertices of the ST-triangulation in more detail we should note that texture coordinates are going to be transformed to the range of $(0,1)$ before rendering by means of the OpenGL texture matrix. As a consequence texture entry $(0.5, 0.5)$ is referenced by issuing a texture coordinate $(0,0)$. Thus only for positive coordinates non-zero texture values are mapped.

For every exterior vertex the u coordinate corresponds to the signed distance of that vertex to the current slicing plane. Values are positive if a vertex is located in front of the slicing plane with respect to the point of view. The v coordinate is equal to the u coordinate but its sign is flipped. At the interior vertex, however, the u coordinate equals the signed distance of the front face vertex to the slicing plane, while in the v coordinate the negated signed distance of the back face vertex to the slicing plane is kept. The principal assignment is illustrated in Figure 3.

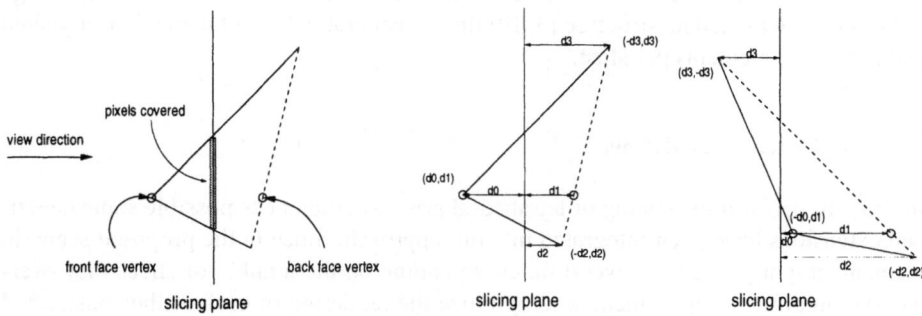

Fig. 3. *Sectional views of different ST-triangulations show the principal assignment of texture coordinates to vertices. Note that texture coordinates are properly transformed in order to address the appropriate texture values.*

Figure 3 tells us that u and v sum up to the thickness of the element along the viewing direction. Since u is just the thickness minus the distance from the slicing plane to the back faces, factors $1 - u/(u+v)$ and $u/(u+v)$ exactly correspond to the weights needed to modulate the scalar values at the front and at the back faces, respectively. These are exactly the factors that are pre-computed in the texture maps. As long as one of the texture coordinates is less than zero, both entry point and exit point lie either behind or in front of the slicing plane. Since negative texture coordinates index alpha values not equal to one, the generated fragments can be discarded before they affect the color buffer by means of the OpenGL alpha test. As a result only those fragments are going to be displayed and blended that are covered by the slicing plane.

In order to obtain for each pixel the accurately interpolated scalar values the ST-triangulation is rendered twice. In the first pass the scalar value at the front facing edge is issued at the interior vertex and fragment color is modulated with the texture $T2$. Blending is disabled in this pass. In the second pass the triangles are rendered using the scalar value at the back facing edge at the interior vertex and texture $T1$ is activated. The blending function is chosen such as to add incoming fragments to pixels, which finally results in the accurately interpolated scalar values.

We should note here the similarity of our method to the one proposed in [7], where a 2D texture was employed to code the interpolation weights needed to render shaded iso-surfaces on a per-pixel basis. In [14] a different alternative using the stencil and the depth test was demonstrated. Using this approach, however, it is necessary to read the color buffer for every single slicing in order to compute the correctly interpolated values. In the current implementation these values are directly obtained by hardware accelerated texture mapping and blending.

3 Tetrahedral grids

Now that we know how to exploit hardware assisted graphics operations for the resampling of scalar values on an arbitrary slice through a single tetrahedron the procedure can easily be extended to tetrahedral grids. First of all, for each slice those elements have to be determined that may intersect that slice. For this purpose we take advantage of an active element data structure [3, 19] that considerably limits the number of visited and processed elements per slice.

3.1 Implementation details

In order to perform the slicing of tetrahedral grids as efficient as possible some beneficial extensions have been integrated into our approach. Since in the proposed scenario elements might get drawn several times depending on the number of slices they overlap, one important requirement is to optimize the rendering of each of the constructed triangle sets. Therefore, in the current implementation we directly create triangle fans for each ST-triangulation [18] thus greatly improving the overall performance.

In addition we exploit coherences between slicing planes thus minimizing the number of numerical calculations to be performed. Although for a certain element the distance of each vertex to the current slice has to be computed, the geometry of the triangulation as well as the scalar values remain unchanged. As a matter of fact the ST-triangulation for every tetrahedron only has to be determined once. As soon as an element is inserted into the active element list the triangle fan and the distance of each vertex to the current slice are computed and stored. For every new slice the distances can now be updated incrementally.

In order to determine the ST-classification and to calculate the resulting triangulations we strictly avoid storing additional information. In his way we minimize the memory overhead, but even more importantly we guarantee that the delay is as short as possible once the grid is temporally modified.

3.2 Resampling vs. rendering

In order to visualize the resampled data we can read the pixel data into main memory and render it by means of any known volume rendering technique. However, in order to avoid the expensive memory transfer we use OpenGL to directly convert the pixel data into a 2D texture map or into a particular slice of a 3D texture map. This allows us to successively construct a stack of 2D textures or one single 3D texture that can be used later on for rendering purposes. Although in the current approach we exclusively exploit 3D texture maps, impressive image quality and performance can be achieved by means of 2D textures [6]. In this case, however, the data has to be resampled three times in order to generate the 2D texture stacks necessary to account for arbitrary changes of the viewing direction.

Additionally our method can be employed to directly render the volume data. Each slice is first rendered into a temporary buffer. Then the results are copied into the color buffer where they are blended appropriately. Pre-shading or post-shading works equally well using this approach. Pre-shaded colors are directly issued as polygon color at the triangle vertices. Post-shading, on the other hand, is achieved by selecting an appropriate lookup table that affects pixel values once they are copied into the color buffer.

Note that perspective projections can not be realized since texture coordinates are computed with respect to an orthographic projection. Even if two different ST-triangulations are computed representing the front faces and the back faces, respectively, and perspectively corrected texture mapping is performed, wrong interpolation weights will be computed due to the perspective distortion.

4 Texture combiners

Our approach can be improved considerably by exploiting the functionality of current PC graphics hardware, i.e. the Nvidia GeForce family GPUs. On these chips programmable fragment arithmetic is available that allows one to compute combinations between the color of incoming fragments and texture samples during rasterization [5]. The hardware is capable of simultaneously performing component-wise products and dot-products between the fragment color, texture samples or user-defined constant RGB values. One of the key features of this chip is that two texture units are available that allow for the simultaneous mapping of different textures to one single triangle. At first glance this functionality doesn't seem to be different to what we can achieve with multitextures as provided in the OpenGL 1.2 function set. In addition, however, texture combiners offer the possibility to multiply texture samples with different polygon colors before they are combined.

In order to take advantage of this functionality textures $T1$ and $T2$ are becoming the multitextures to be combined during rasterization. At each vertex of the ST-triangulation a primary and a secondary color is issued which only differ at the interior vertex. At this vertex the scalar value at the front facing edge and at the back facing edge is coded. In the first register combiner texture samples are simply modulated with the primary and the secondary color, respectively, and the results are added in order to obtain the interpolated scalar values. The correctly interpolated scalar values are thus generated and rendered into the color buffer in one single rendering pass.

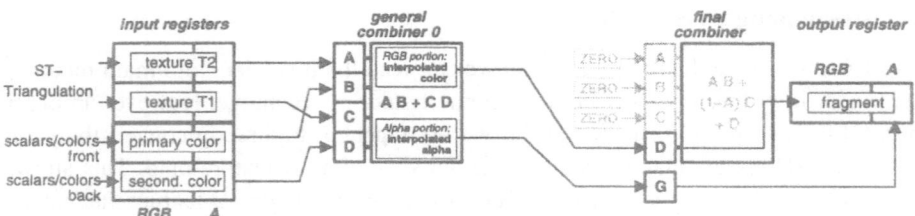

Fig. 4. *The register combiner setup for one pass data resampling.*

Obviously the same procedure can be employed using pre-shaded color samples as primary and secondary color. Since the interpolation is performed in the texture combiner results can be directly blended with pixels already in the color buffer. Neither does the pixel data has to be copied nor do we have to use any additional buffer.

5 Results

In this section we analyze the main modules and features of our system. All tests were run on a SGI Onyx IR2 equipped with one R12000, 300 MHz processor, 64 MB texture memory and 256 MB main memory.

Two tetrahedral data sets were used: A finite-element data set of 180K elements and the bluntfin data set converted to 225K elements. In our first test both data sets were resampled onto differently sized Cartesian grids.

We distinguish between the time needed to initialize and setup the active element list (**Ael**), the elapsed time consumed by the graphics subsystem (**Rnd**), the time it took to read the pixel data into main memory (**ToR**) and the time used to read and build the 3D texture (**ToT**). Since in all our examples we exclusively use 3D texture based rendering the total time is the sum of **Ael**, **Rnd** and **ToT**. Note that by using our approach for the direct rendering of unstructured grids the overall times will be slightly faster since the color buffer only has to be copied and the pixel data doesn't have to be converted into a texture map.

In all experiments the z-coordinate of the Cartesian grids correspond to the number of slices that were processed. No connectivity information was used and space was partitioned into 40 slabs in order to minimize the number of elements to be visited per slice. Table 1 shows explicit timings using hardware assisted color interpolation and 2D texture mapping. All images on the color page below show the resampled data sets rendered via 3D textures. In all cases the frame rate was faster than 8 fps. As can be seen, the time needed to render the tetrahedral elements dominates the overall performance. Since each element has to be rendered twice for every slice it overlaps the overall polygon count increases considerably. In total, the number of triangles rendered for both data sets was of about a factor of 24-39 times higher than the number of elements. On the other hand, as our timings show the additional overhead that is introduced can be absorbed very effectively by the graphics hardware. Since we swap expensive calculations into the graphics subsystem we reach impressive frame rates that are competitive

Resolution	Ael	Rnd	ToR	ToT
64x64x64	0.2	1.4	0.2	0.03
128x128x128	0.24	2.3	0.4	0.1
256x256x256	0.3	3.1	1.1	0.4

Resolution	Ael	Rnd	ToR	ToT
256x128x64	0.15	2.2	0.3	0.1
512x256x128	0.23	3.3	1.0	0.33

Fig. 5. *Timings (seconds) for the heat-sink (left) and the bluntfin (right) data set and differently sized Cartesian grids.*

to those presented in the literature. Although our timings are slower than those proposed in [18], our method keeps the memory overhead low and allows for topologically correct rendering of unstructured grids.

Compared to the most recently published approach [2] we are faster and we need less memory and entirely avoid expensive pre-processing of the data. Particularly for time-dependent data sets we expect our method to be superior to others because constant frame rates are guaranteed even if large parts of the grid are modified.

We have also implemented the proposed method on the nVidia GTS 2 graphics GPU. Using pre-shaded samples it took 2.3 seconds to directly render the heat-sink data set on a 512x512 pixel raster using 256 slices. We didn't realize the resampling into 2D textures due to the immense amount of texture memory necessary to store three copies of the volume.

6 Conclusion

In this paper we have emphasized a novel approach to achieve interactive display of unstructured grids. The major contribution here is that we efficiently exploit standard APIs like OpenGL to perform hardware assisted resampling and direct volume rendering.

In particular we have shown that our resampling approach in combination with texture based volume rendering allows for interactive exploration of large unstructured grids. For direct volume rendering arbitrary transfer functions can be applied to the scalar field.

Our results have shown that the presented method is as fast as any other method that allows for the topologically correct rendering of unstructured grids. Particularly for time-dependent data sets we expect our method to be superior because it does not rely on any pre-computed topological information. Any update of the grid can simply be realized by inserting new elements into the edge list and by removing non-valid elements from this list.

Furthermore, the user can specify arbitrarily sized regions to be resampled in any desired resolution. This allows one to select the resolution that can just be resampled at interactive rates. Once the data has been converted into a 3D Cartesian grid interactive rendering can be achieved using any known algorithm.

We have also shown how to efficiently exploit consumer graphics cards for the rendering of unstructured grids. Taking advantage of per-fragment arithmetic and multi-textures direct rendering of tetrahedral grids using pre-shaded color samples can be achieved. Our timings have shown that we come close to the performance that can be reached using *CellFast* [18]. Although we considerably raise the load in the geometry

74

unit, at the same time we reduce the load in the CPU and minimize memory access. Overall, this leads to a very efficient alternative for the topologically correct rendering of large tetrahedral grids.

References

1. P. Cignoni, C. Montani, D. Sarti, and R. Scopigno. On the optimization of projective volume rendering. In *EG Workshop, Scientific Visualization in Scientific Computing*, pages 59–71, 1995.
2. J. Comba, J. Klosowski, N. Max, J. Mitchell, C. Silva, and P. Williams. Fast polyhedral cell sorting for interactive rendering of unstructured grids. In *Computer Graphics Forum (Proc. EUROGRAPHICS '99)*, pages 369–376, 1999.
3. C. Giertsen. Volume Visualization of Sparse Irregular Meshes. *Computer Graphics and Applications*, 12(2):40–48, 1992.
4. M. Karasick, D. Lieber, L. Nackman, and V. Rajan. Visualization of three-dimensional delaunay meshes. *Algorithmica*, 19(1-2):114–128, 1997.
5. D. Kirk. From Multitexture to Register Combiners to Per-Pixel Shading. http://www.nvidia.com/Developer.
6. C. Rezk-Salama, K. Engel, M. Bauer, G. Greiner, and Ertl. T. Interactive Volume Rendering on Standard PC Graphics Hardware Using Multi-Textures And Multi-Stage Rasterization. In *SIGGRAPH/Eurographics Workshop on Graphics Hardware*, pages 109–119, 2000.
7. S. Roettger, M. Kraus, and T. Ertl. Hardware-accelerated volume and isosurface rendering based on cell-projection. In *Proceedings IEEE Visualization 2000*, pages 109–116, 2000.
8. P. Shirley and A. Tuchman. A Polygonal Approximation to Direct Scalar Volume Rendering. *ACM Computer Graphics, Proc. SIGGRAPH '90*, 24(5):63–70, 1990.
9. C. Silva and J. Mitchell. The Lazy Sweep Ray Casting Algorithm for Rendering Irregular Grids. *Transactions on Visualization and Computer Graphics*, 4(2), June 1997.
10. C. Silva, J. Mitchell, and A. Kaufman. Fast Rendering of Irregular Grids. In *ACM Symposium on Volume Visualization '96*, pages 15–23, 1996.
11. C. Silva, J. Mitchell, and P. Williams. An exact interactive time visibility ordering algorithm for polyhedral cell complexes. In *Proceedings ACM/IEEE Symposium on Volume Visualization 98*, pages 87–94, 1998.
12. C. Stein, B. Becker, and N. Max. Sorting and hardware assisted rendering for volume visualization. In *ACM Symposium on Volume Visualization '94*, pages 83–90, 1994.
13. R. Westermann and T. Ertl. The VSBUFFER: Visibility Ordering unstructured Volume Primitives by Polygon Drawing. In *IEEE Visualization '97*, pages 35–43, 1997.
14. R. Westermann and T. Ertl. Efficiently using graphics hardware in volume rendering applications. In *ACM Computer Graphics (Proc. SIGGRAPH '98)*, pages 291–294, 1998.
15. J. Wilhelms, A. van Gelder, P. Tarantino, and J. Gibbs. Hierarchical and Parallelizable Direct Volume Rendering for Irregular and Multiple Grids. In *IEEE Visualization 1996*, pages 57–65, 1996.
16. P. Williams. Visibility Ordering Meshed Polyhedra. *ACM Transactions on Graphics*, 11(2):102–126, 1992.
17. P. Williams, N. Max, and C. Stein. A high accuracy volume renderer for unstructured data. *IEEE Transactions on Visualization and Computer Graphics*, 4(1):37–54, 1998.
18. C. Wittenbrink. Cellfast: Interactive unstructured volume rendering. In *IEEE Visualization '99 Late Breaking Hot Topics*, pages 21–24, 1999.
19. R. Yagel, D. Reed, A. Law, P. Shih, and N. Shareef. Hardware Assisted Volume Rendering of Unstructured Grids by Incremental Slicing. In *ACM Symposium on Volume Visualization '96*, pages 55–63, 1996.

Editors' Note: see Appendix, p. 338 for colored figures of this paper

Nonlinear Diffusion in Graphics Hardware

M. Rumpf and R. Strzodka

Universität Bonn

Abstract. Multiscale methods have proved to be successful tools in image denoising, edge enhancement and shape recovery. They are based on the numerical solution of a nonlinear diffusion problem where a noisy or damaged image which has to be smoothed or restorated is considered as initial data. Here a novel approach is presented which will soon be capable to ensure real time performance of these methods. It is based on an implementation of a corresponding finite element scheme in texture hardware of modern graphics engines. The method regards vectors as textures and represents linear algebra operations as texture processing operations. Thus, the resulting performance can profit from the superior bandwidth and the build in parallelism of the graphics hardware. Here the concept of this approach is introduced and perspectives are outlined picking up the basic Perona Malik model on 2D images.

1 Introduction

Nonlinear diffusion in multiscale image processing attracts growing interest in the last decade. Methods based on this approach are frequently used tools in image denoising, edge enhancement and shape recovery [1, 10, 12, 9]. Therein the image is considered as initial data of a suitable evolution problem. Time in the evolution represents the scale parameter which leads from noisy, fine scale to smoothed and enhanced, coarse scale representation of the data. The same kind of diffusion models can also be used for the visualization of flow fields through the construction of streamline type patterns [4].

Here our focus is on the efficient implementation of finite element schemes for the solution of the nonlinear diffusion problem. We pick up the regularized Perona and Malik model and rewrite the corresponding linear algebra operations as image processing operations supported by graphics hardware. Thus they act on vectors which are regarded as images. Before we describe the approach in detail let us argue why this unusual approach is expected to ensure superior performance over a standard implementation in software although nowadays CPU performance is superior compared to the computing performance of single operations on a graphics unit.

Memory bandwidth has become a major limiting factor in many scientific computations. Nowadays performance highly depends on the implementation's beneficial use of the hierarchy of caching levels. But automation fails here and the task of optimal use of the memory hierarchy for a given application is very complex. On the other hand PC graphics hardware has undergone a rapid development boosting its performance and functionality and thus releasing the CPU from many computations. Particularly in volume graphics, texture hardware is extensively exploited for a significant performance increase leading to interactive applications [3, 14, 13]. As an example which goes beyond basic graphics operations we cite here Hopf et al. who discussed Gaussian filtering and wavelet transformations in hardware [5, 6].

We proceed along this line and further widen the range of applications even by demonstrating that the functionality of modern graphics cards has reached a state, where the graphics processor unit may be regarded as a programmable parallel fixed-point coprocessor for certain scientific computing purposes. Observing the precision restrictions, it may be used for numerical computations where ultimate precision is not required. Then we benefit from the much higher memory bandwidth and the parallel execution of commands on large data blocks. Partial differential equations in image processing are exactly of this type. They involve large image data and our aim is not to compute exact solutions but to model numerically properties which are known for the continuous model. In case of the nonlinear diffusion these are the decreasing diffusivity in areas of large gradients and the smoothing in image regions which are expected to be apart from edges. Furthermore a maximum principle is regarded as an important property.

We will first review the nonlinear diffusion model and then concentrate on the adaption of the numerical scheme to this graphics oriented setting.

2 Nonlinear Diffusion

We briefly review the model and the discretization of the nonlinear diffusion in image processing, based on a modification of the Perona-Malik [9] model proposed by Catté, Lions, Morel, and Coll [2]. We consider the domain $\Omega := [0, 1]^d$, $d = 2, 3$ and ask for solution of the following nonlinear parabolic, boundary and initial value problem: Find $u : \mathbb{R}^+ \times \Omega \to \mathbb{R}$ such that

$$\tfrac{\partial}{\partial t} u - \operatorname{div}\left(g(\nabla u_\epsilon)\nabla u\right) = 0 \quad, \quad \text{in } \mathbb{R}^+ \times \Omega,$$

$$u(0, \cdot) = u_0\,, \quad \text{on } \Omega,$$

$$\tfrac{\partial}{\partial \nu} u = 0 \quad, \quad \text{on } \mathbb{R}^+ \times \partial\Omega.$$

where in the basic model g is a non negative monotone decreasing function $g : \mathbb{R}_0^+ \to \mathbb{R}+$ satisfying $\lim_{s \to \infty} g(s) = 0$, e. g. $g(s) = (1 + s^2)^{-1}$, and u_ϵ is a mollification of u with some smoothing kernel. The solution $u : \mathbb{R}^+ \times \Omega \to \mathbb{R}$ can be regarded as a multiscale of successively diffused images $u(t), t \in R^+$. With respect to the shape of the diffusion coefficient function g, the diffusion is of regularized "backward" type [7] in regions of high image gradients, and of linear type in homogeneous regions.

We discretize the problem with bilinear, respectively trilinear conforming finite elements on a uniform quadrilateral, respectively hexahedral grid. In time a semi-implicit first order Euler scheme is used, as purely explicit schemes pose very restrictive conditions on the timestep width. In variational formulation with respect to the FE space $\mathcal{V}^h$ we obtain:

$$\left(\frac{U^{k+1} - U^k}{\tau}, \theta\right)_h + \left(g(\nabla U_\epsilon^k)\nabla U^{k+1}, \nabla\theta\right) = 0$$

for all $\theta \in \mathcal{V}^h$. Here $(\cdot, \cdot)$ denotes the L^2 product on the domain Ω, $(\cdot, \cdot)_h$ is the lumped masses product [11], which approximates the L^2 product, and τ the current timestep width. The discrete solution U^k is expected to approximate $u(\tau k)$. Thus in the kth timestep we have to solve the linear system

$$(M_h + \tau L(U^k_\epsilon))\bar{U}^{k+1} = M_h \bar{U}^k \tag{1}$$

where $\bar{U}^k$ is the solution vector consisting of the nodal values, $M_h := ((\Phi_\alpha, \Phi_\beta)_h)_{\alpha\beta}$ the lumped mass matrix, $L(U^k_\epsilon) := ((g(\nabla U^k_\epsilon)\nabla \Phi_\alpha, \nabla \Phi_\beta))_{\alpha\beta}$ the weighted stiffness matrix and Φ_α the "hat shaped" multilinear basis functions. In the concrete algorithm we replace $g(\nabla U^k_\epsilon)$ on elements by the value at the elements' center point.

As the graphics hardware offers only a fixed-point number format, it is important that we separate the small, grid specific element diameter h from the dimensionless diffusion coefficients. Thus both the coefficients and the factor $\frac{\tau}{h^2}$ are close to 1. For an equidistant grid we may rescale the above equation and get

$$\underbrace{\left(I + \frac{\tau}{h^2}\hat{L}(U^k_\epsilon)\right)}_{A^k(\bar{U}^k)} \bar{U}^{k+1} = \underbrace{\bar{U}^k}_{}$$

$$\bar{U}^{k+1} = \bar{R}^k(\bar{U}^k),$$

with the rescaled stiffness matrix $\hat{L}(U^k_\epsilon) := ((g(\nabla U^k_\epsilon)\nabla\hat{\Phi}_\alpha, \nabla\hat{\Phi}_\beta))_{\alpha\beta}$ defined by reference multilinear basis functions $\hat{\Phi}_\alpha$ with support $[-1, 1]^d$.

Any implementation, also that in graphics hardware has to solve the above linear system of equations. In the following section we will therefore first consider the operations involved in solving a general linear system of equations and describe how they can be split into more basic algebraic operations, which are directly supported by graphics hardware.

3 Operations in Linear Iterative Solvers

In fact, many discretizations of partial differential equations lead to a sparse linear system of equations $A(\bar{U}^k)\bar{U}^{k+1} = \bar{R}(\bar{U}^k)$, where the matrix $A \in \mathbb{R}^{n+1,n+1}$ and the right hand side $\bar{R}$ depend on the solution vector $\bar{U}^k$ of the preceding timestep. Frequently an iterative solver is applied to approximate the solution, i.e. we consider an iteration $\bar{X}^{l+1} = F(\bar{X}^l)$ with $\bar{X}^0 = \bar{R}$. Typical smoothers are the Jacobi iteration

$$F(\bar{X}) = D^{-1}(\bar{R} - (A - D)\bar{X}), \qquad D := \text{diag}(A) \tag{2}$$

and the conjugate gradient iteration

$$F(\bar{X}^l) = \bar{X}^l + \frac{\bar{r}^l \cdot \bar{p}^l}{A\bar{p}^l \cdot \bar{p}^l}\bar{p}^l, \qquad \bar{p}^l = \bar{r}^l + \frac{\bar{r}^l \cdot \bar{r}^l}{\bar{r}^{l-1} \cdot \bar{r}^{l-1}}\bar{p}^{l-1}, \qquad \bar{r}^l = \bar{R} - A\bar{X}^l \tag{3}$$

In the above formulas we can easily identify the required operations: matrix vector product, scalar product, componentwise linear combination, componentwise multiplication, application of a componentwise function, vector norm.

The first two of these operations are not directly supported by graphics hardware. Therefore we must split them into more primitive ones. The scalar product may be reformulated using the componentwise multiplication (denoted by '$\bullet$') and a vector norm $\bar{V} \cdot \bar{W} = \|\bar{V} \bullet \bar{W}\|_1$.

The matrix vector product may be expressed in terms of componentwise products with the matrix' subdiagonals $\bar{A}^\gamma := (A_{\alpha-\gamma,\alpha})_\alpha$ which are vectors, and subsequent index shift operations T_γ on vectors, defined by $T_\gamma \circ \bar{V} := (V_{\alpha+\gamma})_\alpha$:

$$(A\bar{X})_\alpha = \sum_\beta A_{\alpha,\beta} X_\beta = \sum_{|\gamma|<n} (\bar{A}^\gamma)_{\alpha+\gamma} X_{\alpha+\gamma}$$

$$A\bar{X} = \sum_{|\gamma|<n} T_\gamma \circ (\bar{A}^\gamma \bullet \bar{X}). \tag{4}$$

Above, $\alpha = 0, \ldots, n$; $\beta = 0, \ldots, n$ range over the matrix' lines or columns respectively, and $\gamma := \beta - \alpha = -n, \ldots, 0, \ldots, n$ indexes the subdiagonals. Elements of the subdiagonal vectors $\bar{A}^\gamma$ indexing matrix elements outside of the matrix A are supposed to be zero, e.g. the first element of the vector $\bar{A}^1$ is $(A^1)_0 = A_{-1,0} = 0$. If most subdiagonals of A are zero, which is always true for FE schemes, then γ ranges only over few nontrivial subdiagonals.

Thus we have successfully split the operations for the linear iterative solvers (2) and (3) into hardware supported functions. Table 1 lists these operations together with their counterparts in graphics hardware.

Table 1. Basic operations in linear iterative solvers.

operation	formula	graphics operation
linear combination	$a\bar{V} + b\bar{W}$	image blending
multiplication	$\bar{V} \bullet \bar{W}$	image blending
general function	$f \circ \bar{V}$	lookup table
index shift	$T_\gamma \circ \bar{V}$	change of coordinates
vector norms	$\|\bar{V}\|_k, k = 1, \ldots, \infty$	image histogram

4 Rewriting the FE Scheme

Now we return to the FE scheme for the nonlinear diffusion introduced in section 2. The general approach to the decomposition of the matrix vector product given in the previous section, is feasible in this case. The matrix A^k consists of only 3^d nontrivial subdiagonals.

Since in this application the vectors $\bar{U}^k = (U_\alpha^k)_\alpha$ represent images, it is appropriate to let α be a 2 or 3 dimensional multi-index, enumerating the nodes of the 2 or 3 dimensional grid respectively. Then the index offset $\gamma := \beta - \alpha$ is the spatial offset from

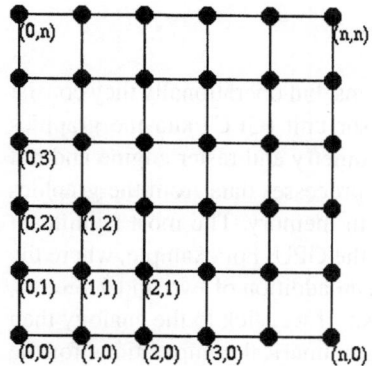
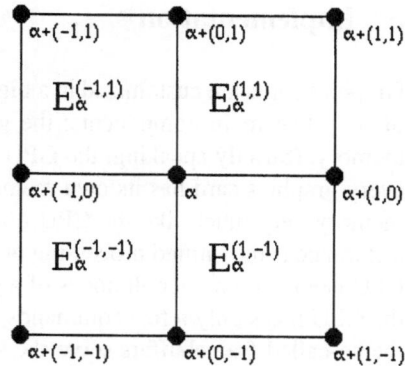

Fig. 1. On the left a 2D grid enumerated by a multi-index, on the right the neighboring elements and the local multi-index offset to neighboring nodes.

node α to node β. Figure 1 shows the enumeration for a 2D grid, and all the 3^2 offsets γ and the neighboring elements E_α^γ for a given node α.

To perform the decomposed matrix vector product (cp. (4)) we need to identify the elements of the subdiagonal vectors $\bar{A}^\gamma$, which now can themselves be regarded as images. For this task it suffices to consider the subdiagonal $\bar{L}^\gamma$ of $\hat{L}(U_\epsilon^k)$, since $\bar{A}^\gamma = \delta_\gamma \bar{1} + \frac{\tau}{h^2}\bar{L}^\gamma$, where δ_γ is the Kronecker symbol. In fact the identity indicated by $\delta_\gamma \bar{1}$ deserves no further attention. By definition we have

$$\bar{L}_\alpha^\gamma = \left(g(\nabla U_\epsilon^k)\nabla\hat{\Phi}_{\alpha-\gamma}, \nabla\hat{\Phi}_\alpha\right).$$

Since we evaluate $g(\nabla U_\epsilon^k)$ by the midpoint rule on elements we may rewrite the matrix element for the node α as a weighted sum over contributions on the neighboring elements

$$\bar{L}_\alpha^\gamma = \sum_{E\in E(\alpha)} G_E^k C_{\gamma,E},$$

where $E(\alpha):= \{E_\alpha^\gamma || \gamma| = d\}$ is defined as the set of elements around the node α (Fig. 1), $G_E^k:= \left.g(\nabla U_\epsilon^k)\right|_E$ is the constant value of the diffusion coefficient at the element's center of mass and $C_{\gamma,E}:= \left(\nabla\hat{\Phi}_{\alpha-\gamma}|_E, \nabla\hat{\Phi}_\alpha|_E\right)$ a local stiffness matrix entry. On an equidistant grid the values $C_{\gamma,E}$ depend only on γ. Since γ takes only 3^d different values, they can be precomputed.

As we have seen, we do not have to store the matrix A^k for the computation of the matrix vector product. Instead we precompute the values G_E^k - again interpreted as an image - only once for every timestep and then split the matrix vector product in the linear solver into few (3^d) products with the subdiagonals $\bar{L}^\gamma$ (cp. (4)):

$$\hat{L}(U_\epsilon^k)\bar{X} = \sum_{|\gamma|\leq d} T_\gamma \circ (\bar{L}^\gamma \bullet \bar{X}).$$

In all these calculations we take care of the natural boundary condition by duplicating the borders of the image $\bar{U}^k$.

5   Implementation

Graphics cards are customized in a big variety of designs, but operationally they consist of only two main components: the graphics processor unit (GPU) and the graphics memory. (Strictly speaking, the GPU splits into a geometry and raster engine and not every graphics card has its own memory.) The GPU processes data from the graphics memory very much like the CPU does from the main memory. The most significant difference is the unified processing of data blocks by the GPU. For example, where the CPU needs to run over all nodes of a grid to perform an addition of two nodal vectors, the GPU takes only a few commands for the same task. If we stick to the analogy then the so-called framebuffers serve the same purpose in numerical computations for the GPU, as the registers do for the CPU. Usually there are at least two such framebuffers: the front buffer which is displayed on the screen and the back buffer where we can perform the calculations invisibly.

An important issue with graphics boards are the number formats supported by the GPU. Resulting from the inherent use, only the range $[0, 1]$ suitable for the representation of intensities is supported (Some GPUs offer extended formats during calculations). In any case we have to encode our numbers to cover a wider range, say $[-\rho_0, \rho_1]$. By nonlinear transformations, also unbounded intervals could be covered, but it is doubtful whether the low precision of the numbers may resolve these intervals sufficiently. Furthermore linear encoding has the advantage that there are many stages in the graphics pipeline where linear transformations can be applied, saving multiple processing of the same operand.

Table 2. Correspondence of operations in numbers and intensities.

Numbers	Intensities
$r : x \rightarrow \frac{1}{2\rho}(x + \rho)$	
$a \in [-\rho, \rho]$	$r(a) \in [0, 1]$
$a + b$	$2\left(\frac{r(a)+r(b)}{2}\right) - \frac{1}{2}$
ab	$\frac{1+\rho}{2} - \rho\left(r(a)(1 - r(b)) + r(b)(1 - r(a))\right)$
$\alpha a + \beta$	$\alpha r(a) + \left(\frac{\beta}{2\rho} + \frac{1-\alpha}{2}\right)$
$f(a)$	$(r \circ f \circ r^{-1})(r(a))$
$\sum_\alpha \alpha a_\alpha$	$\sum_\alpha \alpha r(a_\alpha) + \frac{1}{2}(1 - \sum_\alpha \alpha)$
$\rho(2y - 1) \leftarrow y : r^{-1}$	

In Table 2 we list which operations on the intensities (the encoded values in the images) correspond to the desired operations on numbers (the represented values). The left column shows the operation to be performed, whereas the right column shows which operation must be performed on the encoded operands to obtain the equivalent encoded result. Obviously no other operations than those already discussed are needed to perform these transformed calculations.

By choosing ρ sufficiently large such that any intermediate computations in numbers do not transcend the range $[-\rho, \rho]$, we assure that the corresponding computations in intensities will not transcend the range $[0, 1]$ either. On the other hand a large ρ decreases the resolution of numbers, therefore it should be chosen application dependent as small as possible. The symmetric choice of the interval covers the typical number range of FE schemes and has the advantage of simpler encoded operations on intensities (Table 2).

Below we have outlined the control structures of the algorithm in pseudo code notation.

```
nonlinear diffusion {
    load the original image and the parameters;
    initialize graphics hardware;
    encode the original image in graphics memory Ū⁰;
    for each timestep k {
        store the right hand side image R̄ᵏ = Ūᵏ;
        calculate the image consisting of diffusion coefficients Ḡᵏ = ( g(∇Uₑᵏ)|ₑ )ₑ;
        initialize the solver X̄⁰ = R̄ᵏ;
        for each iteration l
            calculate a step of the iterative solver X̄ˡ⁺¹ = F(X̄ˡ);
        store the solution Ūᵏ⁺¹ = X̄ˡ⁺¹
        decode the solution and display it;
    }
}
```

Now, considering an implementation in OpenGL [8], the basic operations from Table 1 are more or less directly mapped onto OpenGL functionality. The addition and multiplication are achieved by selecting the proper source and destination factors for the blending function (glBlendFunc). Concerning the implementation of a general function of one variable we should keep in mind that the intensities are discretized by m bits, with $m \leq 12$. A general function can thus be represented by 2^m entries in a table. OpenGL can use such a table to automatically output the values indexed by the intensities of an image (glPixelMap), thus applying the designated function to the image. For the index-shift one simply has to change the drawing position for the image. Concerning the vector norms there is a slight difference in implementation, since the OpenGL's histogram extension does not calculate them directly in the GPU, as the other OpenGL methods do, but returns a histogram of pixel intensities (glGetHistogram) from which the CPU has to compute the norm. Let $H : \{0, \ldots, 2^m - 1\} \to \mathbb{N}$ be such a histogram assigning the number of appearances to every intensity of the image $\bar{V}$, then

$$\|\bar{V}\|_k = \left(\sum_{y=0}^{2^m-1} \left(r^{-1}(y) \right)^k \cdot H(y) \right)^{\frac{1}{k}},$$

Fig. 2. Comparison of hardware implemented linear (upper row) and nonlinear diffusion (lower row).

for $k = 1, 2, \ldots$, and for $k = \infty$ we simply pick up the largest $|r^{-1}(y)|$ with $H(y) > 0$, where r^{-1} is the inverse transformation from intensities to numbers. However apart from the overall control structure of the programm, this is the only computation done by the CPU while using the conjugate gradient solver. For the Jacobi solver no CPU calculation at all is required.

6 Results

The computations have been performed on a SGI Onyx2 4x195MHz R10000, with InfiniteReality2 graphics, using 12 bit per color component and the number interval $[-2, 2]$, i.e. $\rho = 2$. Convolution with a Gaussian kernel, which is equivalent to the application of a linear diffusion model is compared to the results from the nonlinear model in Fig. 2. This test strongly underlines the edge conservation of the nonlinear diffusion model.

Figure 3 shows computations with graphics hardware using the Jacobi and the cg-solver and compares them to computations in software. The precision used in the GPU obviously suffices for the task of denoising pictures by nonlinear diffusion. Although the images produced by hardware and software differ, the visual effect is very comparable, and this is the decisive criterion in such applications.

Currently, 100 iterations of the Jacobi, cg-solver for 256^2 images take about 17 sec and 42 sec respectively, which is still slower than the software solution. The reason for this surprisingly weak performance is easily identified in the unbalanced performance of data transfer between the framebuffer and graphics memory. Before, we have already mentioned that the back buffer serves as a register, where auxiliary results are computed before they are stored in a variable in graphics memory. Because nearly all operations effect the back buffer, its access times are highly relevant for the overall performance.

Fig. 3. Comparison of nonlinear diffusion solvers, first row: adaptive software preconditioned cg; second row: jacobi sover in graphics hardware; third row: cg-solver in graphics hardware.

But compared to a computation in software where the reading and writing of a register in the CPU takes the same time, because the operations are supposed to be needed just as often, the graphics of the Onyx2, in contrary, is writing an image from the graphics memory to the framebuffer about 60 times faster than reading it back, because the reading back from the framebuffer to graphics memory is not a very common operation in graphics applications. The histogram extension used for the computation of the scalar products in the cg-solver is even less common, and being even slower than the reading, it further reduces performance. However, the growing use of such extensions in different areas of visualization and image processing will certainly lead to an optimization. There are already graphics cards with less discrepant read/write operations between the framebuffer and the graphics memory and we are working on a respective porting which, however, incorporates some additional difficulties.

7 Conclusions

We have introduced a framework which facilitates the use of modern graphics boards as fixed-point coprocessors for image processing. By showing how common PDE solvers can be split into basic operations, which are directly supported by graphics hardware, we have demonstrated that a wide range of applications could benefit from the large memory bandwidth, which usually is the bottleneck in many scientific calculations. The implementation of nonlinear diffusion has underlined how existing algorithms can quickly be adapted to this graphics oriented setting and that the low precision of numbers does not do any harm to many applications aiming at visual results. The visualization of flow fields based on this approach, for example, is one of our future goals. Finally we have discussed the issue of performance which could not fully unfold itsself yet. But we are very confident that in the near future the application of new graphics cards and drivers will overcome this difficulty, raising the speed of such implementations far beyond pure software solutions.

Acknowledgments

We would like to thank Matthias Hopf from Stuttgart for a lot of valuable information on the graphics hardware programming and Michael Spielberg from Bonn for assistance in the adaption of the finite element code to the graphics hardware.

References

1. L. Alvarez, F. Guichard, P. L. Lions, and J. M. Morel. Axioms and fundamental equations of image processing. *Arch. Ration. Mech. Anal.*, 123(3):199–257, 1993.
2. F. Catté, P.-L. Lions, J.-M. Morel, and T. Coll. Image selective smoothing and edge detection by nonlinear diffusion. *SIAM J. Numer. Anal.*, 29(1):182–193, 1992.
3. T.J. Cullip and U. Neumann. Accelerating volume reconstruction with 3d texture hardware. Technical Report TR93-027, University of North Carolina, Chapel Hill N.C., 1993.
4. U. Diewald, T. Preußer, and M. Rumpf. Anisotropic diffusion in vector field visualization on euclidean domains and surfaces. *Trans. Vis. and Comp. Graphics*, 6(2):139–149, 2000.
5. M. Hopf and T. Ertl. Accelerating 3d convolution using graphics hardware. In *Visualization '99*, pages 471–474, 1999.
6. M. Hopf and T. Ertl. Hardware accelerated wavelet transformations. In *Symposium on Visualization VisSym '00*, 2000.
7. B. Kawohl and N. Kutev. Maximum and comparison principle for one-dimensional anisotropic diffusion. *Math. Ann.*, 311 (1):107–123, 1998.
8. OpenGL Architectural Review Board (ARB), http://www.opengl.org/. *OpenGL: graphics application programming interface (API)*, 1992.
9. P. Perona and J. Malik. Scale space and edge detection using anisotropic diffusion. In *IEEE Computer Society Workshop on Computer Vision*, 1987.
10. J. A. Sethian. *Level Set Methods and Fast Marching Methods*. Cambridge University Press, 1999.
11. V. Thomee. *Galerkin - Finite Element Methods for Parabolic Problems*. Springer, 1984.
12. J. Weickert. *Anisotropic diffusion in image processing*. Teubner, 1998.
13. R. Westermann and T. Ertl. Efficiently using graphics hardware in volume rendering applications. *Computer Graphics (SIGGRAPH '98)*, 32(4):169–179, 1998.
14. O. Wilson, A. van Gelder, and J. Wilhelms. Direct volume rendering via 3d textures. Technical Report UCSC CRL 94-19, University of California, Santa Cruz, 1994.

Voxel Column Culling:
Occlusion Culling for Large Terrain Models

Brian Zaugg   Parris K. Egbert

Computer Science Department, Brigham Young University

Abstract. We present Voxel Column Culling, an occlusion culling algorithm for real-time rendering of large terrain models. This technique can greatly reduce the number of polygons that must be rendered every frame, allowing a larger piece of terrain to be rendered at an interactive frame-rate. This is accomplished by using a form of 3D cell-based occlusion culling. A visibility table for the terrain model is precomputed and stored on disk. This table is then used to quickly bypass portions of the model that are not visible due to self-occlusion of the terrain model. This technique improves performance of real-time terrain simulations by reducing the number of polygons to be rendered.

1 Background

Large terrain models are difficult to render in real-time because so many polygons are needed to accurately represent them. Most such models are based on a digital elevation model (DEM), which is simply a rectangular array of elevation values. The most obvious triangulation of a DEM yields $O(nm)$ triangles, where n and m are the grid dimensions. Although the number of polygons can be greatly reduced using terrain decimation and mesh simplification techniques, the distance to the far clipping plane must be fairly short in order to keep the polygon count low enough to maintain interactive frame rates on typical hardware.

This paper presents Voxel Column Culling, an occlusion culling technique for large terrain models. This technique uses a visibility scheme to cull large portions of the database, thus allowing larger pieces of terrain to be rendered in real-time.

2 Related Work

Algorithms which accelerate real-time terrain rendering fall into two categories, terrain decimation and culling. Terrain decimation algorithms produce a terrain model with fewer polygons than is produced by simple triangulation of a DEM. Culling algorithms quickly identify invisible portions of a scene and skip over them, effectively reducing the scene's polygon count.

2.1 Terrain Decimation Algorithms

Terrain decimation algorithms produce a terrain model that has far fewer polygons than a full-resolution triangulation, but which looks very much like the full-resolution mesh. Several different terrain decimation methods have been used successfully. An overview of these techniques is presented here.

General mesh simplification techniques can be applied to triangulated DEM data. Schroeder *et al.* [15] describe a decimation algorithm that removes a vertex, then re-triangulates the resulting hole. Hierarchical dynamic simplification [12] regenerates the model for every frame, providing a view-dependent simplification. Vertices are clustered together hierarchically, then the hierarchy is queried to supply only the polygons that are important from the current viewpoint. Hoppe [8] introduced progressive meshes, which represent a model as a low-resolution base mesh and a set of vertex split transformations. This method has been extended [9] to provide view-dependent models.

Many terrain decimation algorithms produce Triangulated Irregular Networks, or TINs. These methods minimize the number of triangles created based on some error metric. Examples of TIN methods are given in [5] and [14]. These methods are generally not fast enough to generate a new model each frame, and are not usually view-dependent. They also suffer from popping artifacts if multiple TIN models are used to provide discrete levels of detail.

Semi-regular subdivision methods restrict the triangles in the resulting model to be 45-45-90 triangles. This restriction causes the resulting model to be less optimal, but allows the algorithm to run faster, providing for view-dependent simplification of the model. Examples of terrain decimation algorithms based on semi-regular subdivision include [6] and [11]. Quadtree morphing [4], which also uses semi-regular subdivision, is fast enough to create a view-dependent model in real-time, and morphs between models based solely on the viewpoint.

2.2 Culling Algorithms

Culling algorithms accelerate rendering by quickly removing geometry that will not contribute to the final image. Back-face culling and view frustum culling are well known and widely used techniques [3]. Clustered culling algorithms decide with a single test whether larger sets of polygons should be rendered [10].

Cell-based culling, introduced by Airey *et al.* [1], divides a scene into 2D cells. For each cell, a preprocessing step determines a potentially visible set (PVS) containing all geometry which may be visible from some location within that cell. Luebke and Georges [13] define portals as portions of cell boundaries which don't obstruct a line-of-sight (such as doors and windows), and use these to determine the PVS.

Greene *et al.* [7] introduced an occlusion culling technique known as hierarchical z-buffers. This algorithm uses an object-space octree together with an image-space z-pyramid to determine visible portions of the scene. Zhang *et al.* [16] proposed hierarchical occlusion maps, an image-space culling algorithm. Both hierarchical z-buffers and hierarchical occlusion maps are slow enough that they must be able to cull a large percentage of the model to achieve speedup, and so work best for models with very high depth complexity.

3 Voxel Column Culling

This section gives an overview of Voxel Column Culling, the culling technique described in this paper. This technique determines terrain visibility in a viewpoint dependent fashion. Significant reduction in rendering time is achieved through the use of this algorithm.

3.1 Partitioning the Space

Voxel Column Culling is a cell-based object-space algorithm. The object-space surrounding the terrain model is partitioned into tiles, columns, and cells. This is illustrated in figure 1.

The terrain model itself is divided into a uniform grid of square *tiles*. The length of a tile side must be a power of two. Varying the tile size provides a mechanism for adjusting a performance trade-off. Selecting a larger tile size means that the offline visibility computations take less time, but it also means that less of the terrain can be culled away at runtime.

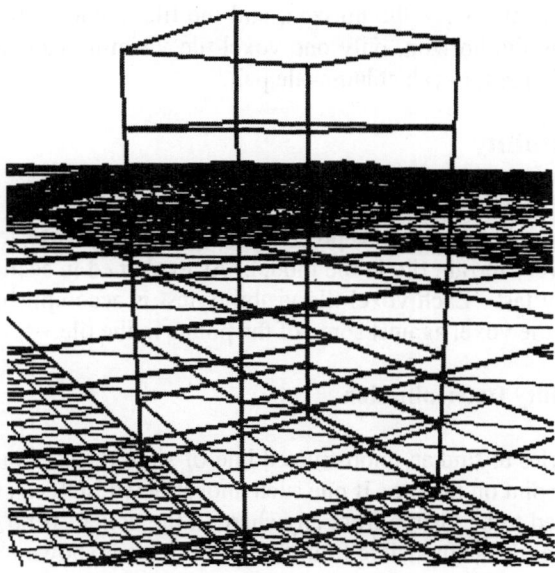

Fig. 1. Partitioned object-space. Terrain tiles are displayed on the grid, and one column of voxels is shown. One such column exists above every tile.

Extending up vertically from each tile is a viewpoint *column*. Visibility from each viewpoint column is computed and stored separately. Each column is sliced into *voxels* at a user-selected number of altitudes. The same trade-off is involved here as was described for the selection of tile size. Visibility is computed from each viewpoint voxel in object-space to every terrain tile in the model. All viewpoints within a given voxel are considered the same viewpoint by the culling algorithm.

3.2 Reducing the Problem Size

General 3D visibility computation for large models is not feasible. Each potential viewpoint must be tested for visibility with every point of geometry in the scene, and there is no good way of knowing what objects in the scene might cause an occlusion. Visibility for terrain models can be computed much faster than visibility can be computed for arbitrary models for two reasons. First, heightgrid-based terrain models exist at only one altitude z for each (x, y) point. This makes it much easier to determine whether the model is self-occluding from a given viewpoint. Second, a significant amount of computation is only required for one viewpoint voxel in each column. For each viewpoint column and terrain tile, there is a special voxel which we will call the *horizon* voxel from which the tile is invisible. The horizon voxel is the voxel in that column below which the tile is always invisible and above which that tile is always visible. Computation of visibility for voxels above the horizon voxel is very fast, usually requiring only one point-point test (See section 4). Visibility determination for each column is done in top-to-bottom order. Since this calculation is very fast for the voxels above the horizon, and the tile is known to be occluded from all voxels below the horizon, only one voxel-tile visibility calculation takes a significant amount of time for each column-tile pair.

4 Computing Visibility

Determining visibility from any viewpoint around the terrain model to any piece of terrain is slow enough that it must be precomputed. Visibility is tested from each viewpoint voxel to every terrain tile in the model. As explained in section 3.2, many of these tests are very fast. Each voxel-tile visibility test is accomplished by testing some of the points in the voxel against some of the points in the tile.

4.1 Voxel-Tile Visibility Determination

Voxel-tile visibility tries to find an unobstructed line-of-sight (LOS) from some point in the voxel to some point on the tile. If one such unobstructed LOS can be found, the entire tile is considered to be visible from anywhere in the voxel. If no unobstructed LOS can be found, the tile is not visible from the voxel. This is shown in figure 2.

If a viewpoint from which a tile is visible is moved down towards the terrain, at some point that tile will become occluded by another piece of terrain. However, if a viewpoint from which a tile is not visible is moved down towards the terrain, the tile will never become visible. Thus, we need test only the four edges of the top face of

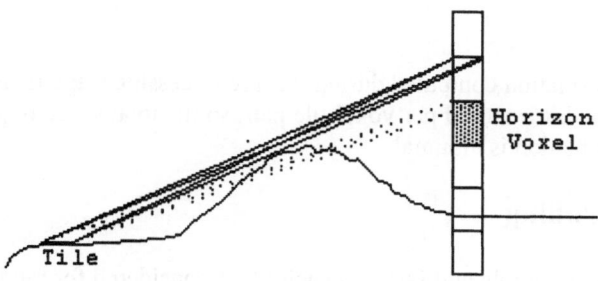

Fig. 2. The horizon voxel is the highest voxel in the current column from which the current tile is occluded.

the voxel. If the tile is not visible from those edges, it will not be visible from anywhere in the voxel.

In order to find an unobstructed LOS as quickly as possible, point-point visibility tests are done repeatedly on successively finer sub-samplings of the edges of the top of the voxel and the terrain. On the first iteration, every 32nd point along the top voxel edges will be tested against every 32nd point in the tile. If no unobstructed LOS is found, this process is repeated for every 16th point in the edges and tile. This is repeated until an unobstructed LOS is found or every edge point has been tested against every point in the tile.

Voxel-tile visibility tests are done in top-to-bottom order for each column and tile. For most voxels above the horizon, this test will be very fast because the first point-point test will find an unobstructed LOS. After the horizon voxel is found, the tile is marked as being occluded for viewpoints in what remains of the column below the horizon. This means that only the horizon voxel and sometimes the voxel above it take much time to compute.

4.2 Point-Point Visibility Determination

The point-point visibility test is fairly simple. The heightgrid data that was used to create the terrain model is treated as a 2D image, across which a line is drawn from one point to the other. The altitude of the start and end points is known, and altitudes are interpolated along the line at each point. We use a modified Bresenham line algorithm [2] to do the interpolation using integer arithmetic. This provides for a very fast 3D line. At each point along the line, the current elevation of the line is compared with the elevation of the terrain at that point. If the line ever drops below the terrain elevation, the test fails; otherwise the points are mutually visible.

4.3 Storage Format

The visibility information computed during this preprocessing stage is stored for each column. Only one bit is stored per voxel-tile pair, so the total space required to store the visibility information is minimal.

5 Real-Time Culling

Culling at runtime is simple and fast. As each tile is considered for rendering, a table lookup is done to determine whether that tile is visible from the current viewpoint. If so, it is added to a set of visible tiles which are then rendered. For subsequent frames, a quick check is done to determine whether the new viewpoint is still within the same voxel. If it is, the same set of visible tiles is used to render the new frame. Otherwise, a new set must be created by performing the table lookup for each tile.

Because of the speed of this method, the simulation is not noticeably slowed even if all of the terrain is visible and there is no reduction in polygon count.

6 Results

We tested the Voxel Column Culling algorithm using terrain data covering the Wasatch Front in northern Utah, including most of the sites that will be used for the 2002 Winter Olympics. For view-dependent terrain decimation, we used the Quadtree Morphing algorithm developed by Cline and Egbert [4]. The Wasatch Front DEM used contains 2070x2637 samples and covers about 5500 square kilometers. A tile size of 32x32 samples (about one square kilometer) was used, and the columns were divided vertically into 7 voxels at 500, 1000, 1500, 2000, 2500, and 3000 meters above the lowest point in the DEM. All data was collected with the far clipping plane beyond the far end of the terrain model. Figure 5 shows the large portion of the model that is culled. Figure 6 shows a frame rendered in three modes, textured, wireframe without Voxel Column Culling, and wireframe with culling enabled. Note the distant terrain that is rendered when culling is disabled, but disappears when culling is enabled.

6.1 Polygon Count

The results discussed in this section are from a flyby of the Wasatch Front dataset including both high- and low-altitude flight over both flat and mountainous terrain. As can be seen in figure 3, the number of polygons sent to the rendering pipeline was significantly decreased when the Voxel Column Culling algorithm was used. As was expected, the biggest improvement was observed for low-altitude flight over mountainous terrain. These portions of the flyby can be identified in the graph as the areas where there is a large difference between the two polygon counts. However, as can be seen in the graph, high-altitude and flat-terrain portions of the flyby have very low polygon counts to start with.

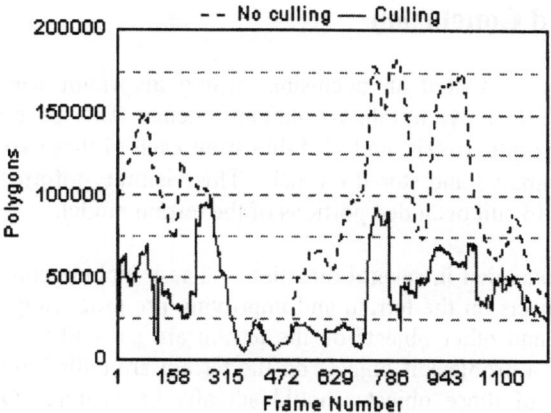

Fig. 3. Polygon count for flyby of Wasatch Front dataset at 1600x1200 resolution with and without occlusion culling enabled.

6.2 Framerate

Although the framerate is mostly a function of polygon count, we wanted to verify that the culling algorithm wasn't slowing the framerate significantly in portions of the flyby where little or no culling was possible. Figure 4 shows the framerate attained on an SGI 320 Visual Workstation with a Pentium II 350 MHz processor. Also note that the framerate isn't noticeably decreased by the culling algorithm even in high-altitude, flat-terrain portions of the flyby.

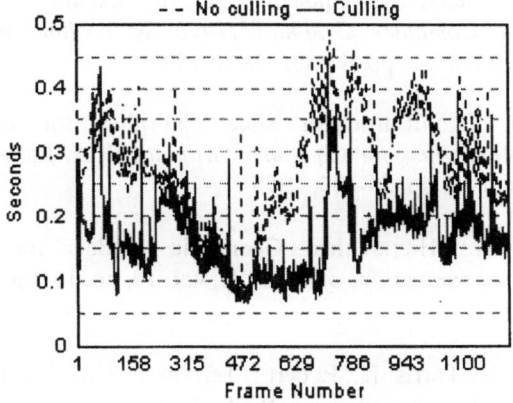

Fig. 4. Rendering time required per frame for Wasatch Front dataset flyby.

7 Future Work and Conclusion

In this paper we have presented an occlusion culling algorithm which achieves significant speedup of interactive terrain simulation systems. The space surrounding the terrain is partitioned into voxels, and visibility from each of these voxels to each tile of terrain is precomputed and stored on disk. This visibility information is then used during simulation to cull occluded portions of the terrain model.

This research can be extended in several directions. These include consideration of buildings and other objects on the terrain and improving pre-processing speed using coherence. Buildings and other objects on the terrain are currently ignored by the algorithm. This poses a problem if objects on the terrain are culled along with the terrain when the tops of those objects should actually be visible. One possible solution to this problem is to add the heights of the objects to the values of the heightgrid at the appropriate locations.

Computation of visibility from voxels to tiles is currently performed without taking previous visibility results into account. The preprocessing step which computes visibility could probably be sped up by exploiting coherence between neighboring voxels and tiles.

Acknowledgments

Thanks to David Cline and Kirk Duffin for building the terrain-rendering infrastructure which was used to test this algorithm. This work was funded in part by a grant from the Utah State Center of Excellence.

References

1.   Airey, John M., John H. Rohlf, and Frederick P. Brooks Jr.: Towards Image Realism with Interactive Update Rates in Complex Virtual Building Environments. *Computer Graphics (1990 Symposium on Interactive 3D Graphics)*, vol. 24, no. 2, pp. 41-50, March 1990.

2.   Bresenham, J. E.: Run-length slice algorithm for incremental lines. *Fundamental Algorithms for Computer Graphics*, R.A. Earnshaw, ed., NATO ASI Series, Vol. F17, pp. 59-105, 1985.

3.   Clark, James H.: Hierarchical Geometric Models for Visible Surface Algorithms. *Communications of the ACM*, vol. 19, no. 10, pp. 547-554, October 1976.

4.   Cline, David and Parris K. Egbert.: Terrain Decimation through Quadtree Morphing. To be published in *IEEE Transcations on Visualization and Computer Graphics*.

5. DeFloriani, L. and E. Puppo.: Hierarchical Triangulation for Multiresolution Surface Descriptions. *ACM Transactions on Graphics*, vol. 14, October, 1995.

6. Falby, John S. *et al.* NPSNET: Hierarchical Data Structures for Real-Time Three-Dimensional Visual Simulation. *Computers and Graphics*. vol. 17, pp. 65-69, 1993.

7. Greene, N., M. Kass, and G. Miller.: Hierarchical Z-Buffer Visibility. *Computer Graphics (SIGGRAPH '93 Proceedings)*, vol. 27, pp. 231-238, 1993.

8. Hoppe, Hugues.: Progressive Meshes. *Computer Graphics (SIGGRAPH '96 Proceedings)*, vol. 30, pp. 99-108, 1996.

9. Hoppe, Hugues.: View-Dependent Refinement of Progressive Meshes. *Computer Graphics (SIGGRAPH '97 Proceedings)*, vol. 31, pp. 189-198, 1997.

10. Kumar, S., D. Manocha, B. Garrett, and M. Lin.: Hierarchical Back-Face Computation. *Proceedings of Eurographics Rendering Workshop 1996*, pp. 235-244, June 1996.

11. Lindstrom, Peter *et al*.. Real-Time Continuous Level of Detail Rendering of Height Fields. *Computer Graphics (SIGGRAPH '96 Conference Proceedings)*, vol. 30, pp. 109-118, 1996.

12. Luebke, David, and Carl Erikson.: View-Dependent Simplification of Arbitrary Polygonal Environments. *Computer Graphics (SIGGRAPH '97 Proceedings)*, vol. 31, pp. 199-208, 1997.

13. Luebke, David P. and Chris Georges.: Portals and Mirrors: Simple, Fast Evaluation of Potentially Visible Sets. *Proceedings 1995 Symposium on Interactive 3D Graphics*, pp. 105-106, April 1995.

14. Schroder, F. and P. RossBach.: Managing the Complexity of Digital Terrain Models. *Computers and Graphics*, vol. 18, pp. 65-70, 1994.

15. Schroeder, William J., Jonathan A. Zarge, and William E. Lorensen.: Decimation of Triangle Meshes. *Computer Graphics (SIGGRAPH '92 Proceedings)*, vol. 26, pp. 65-70, July 1992.

16. Zhang, Hansong, Dinesh Manocha, Tom Hudson, and Kenneth E. Hoff III.: Visibility Culling using Hierarchical Occlusion Maps. *Computer Graphics (SIGGRAPH '97 Proceedings)*, vol. 31, pp. 77-88, 1997.

Editors' Note: see Appendix, p. 339 for colored figures of this paper

Stream Surface Generation for Fluid Flow Solutions on Curvilinear Grids

Allen Van Gelder

Computer Science Department
University of California, Santa Cruz, USA avg@cs.ucsc.edu

Abstract. A *stream surface* in a steady-state three-dimensional fluid flow vector field is a surface across which there is no flow. Stream surfaces can be useful for visualization because the amount of data presented in one visualization can be confined to a manageable quantity in a physically meaningful way.

This paper describes a method for generation of stream surfaces, given a three-dimensional vector field defined on a curvilinear grid. The method can be characterized as *semi-global*; that is, it tries to find a surface that satisfies constraints over a region, expressed as integrals (actually sums, due to discreteness), rather than locally propagating the solution of a differential equation.

The solution is formulated as a series of quadratic minimization problems in n variables, where n is the cross-wind resolution of the grid. An efficient solution method is developed that exploits the fact that the matrix of each quadratic form is tridiagonal and symmetric. Significant numerical issues are addressed, including degeneracies in the tridiagonal matrix and degeneracies in the grid, both of which are typical for the applications envisioned.

1 Introduction

Fluid flows include both gases and liquids, and are important in the design of many machines. Because of their complexity in the neighborhood of solid objects (even greatly simplified objects), almost all fluid flows are computed numerically on a grid or grids of some kind. The solutions involve both scalar and vector quantities, and can run to gigabytes of data. Significant simulations usually require super-computers.

Visualization of fluid flows is one of the most challenging visualization problems. Among the reasons for this are the fact that the grids on which the solutions have been computed are shaped to fit the boundaries of important solid objects. Some examples are aircraft fuselages, propellers, turbine blades, ship hulls, internal combustion cylinders, and ocean floors.

A second aspect that makes the visualization task challenging is that three-dimensional vector fields contain an overwhelming amount of information. One of our primary motivations for producing stream surfaces is to reduce the amount of information to a manageable, yet informative, quantity.

A third visualization challenge arises in many engineering applications because the flow needs to be studied in the close neighborhood of so-called *no-slip* boundaries: that is, boundaries of solid objects where the flow velocity is constrained to be zero

[KHL99]. So another motivation for stream surfaces is to produce a surface *near* the no-slip boundary such that the flow along this surface provides the needed insights.

In this paper we concentrate on *curvilinear grids*, whose regularity makes them attractive for computation. A curvilinear grid can be thought of as a continuous deformation of a rectangular parallelepiped, with *cells* that are warped cubes. Each vertex is identified by a triple index (i, j, k) and edges connect vertices that differ by one in exactly one index.

Methods to generate stream surfaces can be broadly grouped into two categories, *local* and *global*. The *local* methods generate stream lines from various points (usually on an upwind boundary face of the grid) by solving the differential equation implied by the vector field of the flow. Then somehow these stream lines need to be connected to make a surface. Substantial difficulties have been reported, including numerical inaccuracies that grow as the stream lines are propagated downwind.

Global methods attempt to overcome some of the drawbacks of local methods. The general idea of global methods is to choose points on the surface that are connected up as surface patches in such a way that the normal vectors of the patches are orthogonal to the flow field; error is measured as the degree of nonorthogonality. By applying the error same criterion at every patch, it is hoped to avoid large errors downwind due to accumulated inaccuracies from upwind. However, global methods have their own computation difficulties, and the hoped-for accuracy might not be achieved. In this paper we describe a limited approach, point out some of the problems that arose, and explain how we addressed those problems.

Fluid flows satisfy several conservation properties, one of the most important being conservation of mass: In a steady-state flow the net mass flowing across a closed surface is zero. There is no clear-cut way to exploit this property by locally propagating stream lines. However, it can be exploited in several ways with global methods, providing another motivation for a global approach.

The flow field can be transformed into parameter space (also called computational space) while preserving the mass-conservation property, provided that the vector field represents *momentum* (which is velocity weighted by density). Therefore a stream surface in the geometrically regular computational space can be mapped directly into physical space and it will remain a stream surface.

We shall demonstrate that, although this transformation into computational space may have a geometric singularity that causes the velocity to become undefined at certain points in computational space, the *computational space momentum* remains well-defined and finite. Such geometric singularities arise in practice when the longitudinal lines of a hemisphere converge to a single point at the pole. Such grid shapes are needed to fit around the nose of an aircraft fuselage and similarly shaped objects.

The paper is organized as follows. The problem we address is specified in Section 2. The methodology we developed is reported in Section 3. Experimental results are reported in Section 4. Related work is discussed briefly in Section 5, and conclusions are discussed in Section 6. Many details that we are forced to omit here due to space limitations, some additional images, and animations can be accessed on the Internet at `ftp://ftp.cse.ucsc.edu/pub/avg/Scivi`.

2 The Problem

The data given consists of a three-dimensional array of grid points, $p(i, j, k)$, also called grid vertices, and momentum vectors $\mathbf{m}(i, j, k)$, where $0 \leq i < m$ and $0 \leq j < n$ and $0 \leq k < K$. Each grid point has components $p = (x, y, z)$. Each momentum vector has components $\mathbf{m} = (u, v, w)$, and $\mathbf{m}(i, j, k)$ denotes the flow at $p(i, j, k)$. Also part of the data are scalar fields for density, $\rho(i, j, k)$, and stagnation energy, $E(i, j, k)$, but E will not play a direct role in this paper. Notice that (u, v, w) denote momentum, not velocity, for this paper; the relationship is $(u, v, w) = \rho(\dot{x}, \dot{y}, \dot{z})$, where $(\dot{x}, \dot{y}, \dot{z})$ is velocity.

We assume that the flow is generally in the direction of $\partial p / \partial x$, in accordance with the convention of fluid mechanics. We also assume that the surface $p(i, j, 0)$ is (at least in part) a no-slip boundary, where $\mathbf{m}$ is zero. The latter assumption is not required, but no-slip boundaries present important difficulties, so we include it in the problem.

The upwind face of the grid is comprised of the vertices $p(0, j, k)$. The problem we study is to find a stream surface in the flow whose intersection with this upwind face is specified. That is, suppose a one-dimensional family of "heights" is given as $\psi(0, j)$. Associate a point $p(0, j, \psi(0, j))$ with each given "height" by interpolation among the given points $p(0, j, k)$. Then we look for the stream surface that passes through the points $p(0, j, \psi(0, j))$. For this paper, trilinear interpolation is used generally, so that if k is the integer such that $k \leq \psi(0, j) < k + 1$, then the intersection point computed by linear interpolation along the edge between $p(0, j, k)$ and $p(0, j, k + 1)$.

Grids arise frequently in which the points near $i = 0$ approximate a spherical or ellipsoidal cap. In this case, for a fixed k, the points $p(0, j, k)$ are all in the same location for varying j, as shown in Figure 1. Grid cells adjacent to $i = 0$ become wedges. A stream surface can be continuous in this region only if the "heights" $\psi(0, j)$ are all equal. This is one motivation for specifying the intersection of the desired stream surface with the upwind face.

Actually, we expect to want a series of stream surfaces that are near the no-slip surface and nonintersecting. They should be varying "distances" from the no-slip surface. By adjusting the "heights" at the upwind face, surfaces at various distances can be generated. If the surfaces do not intersect at the upwind face, they will not intersect anywhere, in theory (or mass would be destroyed).

3 Methodology

The general approach we followed was to define the conditions that the dot product of the surface normal with the flow vector field should be zero, or close to zero. When the problem is viewed this way, it can be transformed into computational space. This section first describes the transformation into computational space, and how certain degeneracies are handled. Then it gives the constraints to be satisfied by the stream surface. Finally, it describes the methods we implemented to satisfy those constraints.

3.1 Transformation into Computational Space

Suppose a rectilinear region in parameter space (r, s, q), is considered, where $0 \leq r \leq m - 1, 0 \leq s \leq n - 1, 0 \leq q \leq K - 1$ are continuous parameters. Trilinear transformations can be defined in each unit cube with integer boundaries for each of the quantities, x, y, z, u, v, w, as well as ρ and E. Consider a fixed triple of integers (i, j, k), such that the cell is the region $i \leq r \leq i+1, j \leq s \leq j+1, k \leq q \leq k+1$.

To make the presentation independent of (i, j, k) we define a local coordinate system (α, β, γ) for each cell, where the local coordinates vary from 0 to 1.

Notation: Partial derivatives with respect to α, β, and γ are denoted by subscripting: $p_\alpha = \partial p / \partial \alpha$, etc. Vectors are considered to be column vectors; the superscript T denotes *transpose*, converting a column vector into a row vector. A matrix is sometimes denoted by a row of column vectors or a column of row vectors. The usual 3D cross product (vector product) is denoted by "$\times$" and the usual dot product (inner product) is denoted by "$\cdot$".

Let T be the 3-vector of trilinear transformations that maps $(0, 0, 0)$ into $p(i, j, k)$, $(1, 0, 0)$ into $p(i+1, j, k)$, $(0, 1, 0)$ into $p(i, j+1, k)$, $(0, 0, 1)$ into $p(i, j, k+1)$, etc. We write $p(\alpha, \beta, \gamma) = T(\alpha, \beta, \gamma)$ within this cell. The Jacobian matrix of T is

$$J = [p_\alpha \ p_\beta \ p_\gamma] \tag{1}$$

where p is regarded as a column vector. Let $\|J\|$ denote the determinant of J, which we assume is nonnegative. Both J and $\|J\|$ vary with (α, β, γ), but this dependence is suppressed in the notation.

For trilinear transformations, recall that the partial derivatives that appear in Eq. 1 are simple vector differences at the cell corners. For example,

$$p_\alpha(0, 0, 0) = p_\alpha(1, 0, 0) = p(i+1, j, k) - p(i, j, k) \tag{2}$$

and so on. Elsewhere in the cell p_α is a bilinear function of β and γ.

We want to define computational-space analogs of the basic quantities given in the data: momentum, density and energy. With some abuse of notation we indicate computational-space quantities with the same symbols and their physical-space counterparts, and use the parameters (α, β, γ) or (x, y, z) to distinguish which is intended. Details of the derivations are omitted due to lack of space.

It is known, at least in the folklore of CFD, that *computational-space density* is given by

$$\rho(\alpha, \beta, \gamma) = \|J\| \, \rho(x, y, z). \tag{3}$$

However, the following expression for *computational-space momentum* has not appeared elsewhere to the best of our knowledge. It relies on the relationship between cross-products and the inverse of a 3x3 matrix, which is not "well advertised."

$$\mathbf{m}(\alpha, \beta, \gamma) = \begin{bmatrix} (p_\beta \times p_\gamma)^T \\ (p_\gamma \times p_\alpha)^T \\ (p_\alpha \times p_\beta)^T \end{bmatrix} \mathbf{m}(x, y, z) \tag{4}$$

A very important property of this definition is that the mass-conservation law holds in computational space, as well as in physical space. If we define a stream surface in computational space to be a surface whose normal vector is everywhere orthogonal to $\mathbf{m}(\alpha, \beta, \gamma)$, then the mapping of this surface into physical space is also a stream surface, since no mass crosses either surface.

We also have an expression for the determinant in terms of the triple scalar product of the columns of the matrix:

$$\|J\| = (p_\alpha \times p_\beta) \cdot p_\gamma \tag{5}$$

As mentioned in connection with Eq. 1, for trilinear transformations, Eqs. 4 and 5 can be evaluated at cell corners with simple vector differences, cross products, and dot products.

3.2   Degeneracies in the Jacobian Matrix

It is normal for the Jacobian matrix to have singularities in common modeling situations. A typical case is when the grid face for $i = 0$ collapses into a line (or polygonal line) as edges shrink to 0 length in the j direction. The effect is like the end of a cylinder being pinched down to a single point. Therefore, it is quite useful to have a representation in computational space that does not depend on the transformation being 1–1, that is, does not depend on the Jacobian matrix having an inverse. Equations 3 and 4 described such a representation.

3.3   The General Scheme

We construct stream surfaces in computational space on the assumption that they should not intersect the $q = 0$ face of the grid. The stream surface will be defined as a set of quadrilateral patches in (r, s, q) space (called (i, j, k) space at integers), where the vertices of the quadrilaterals have integer values for r and s, and the values of q can be thought of as a height field, $\psi(r, s)$, over the $q = 0$ plane. The edges of these patches run from $(i, j, \psi(i, j))$ to $(i+1, j, \psi(i+1, j))$ or from $(i, j, \psi(i, j))$ to $(i, j+1, \psi(i, j+1))$. See Figure 2.

Clearly this scheme is not sufficiently general to generate any well-defined stream surface. However, we are primarily interested in the case that $q = 0$, or $k = 0$, is the location of a no-slip surface. The region near such a surface is called the *boundary layer* and flows within the boundary layer are often the most important for design purposes, because they determine the impact of the environment on the machine being designed. This representation should be adequate when the flow is "somewhat parallel" to the no-slip surface.

However, there are important exceptions where the representation as a height field is not adequate. These occur, for example, where the flow becomes generally toward the no-slip surface (an *attachment* flow) or generally away from the no-slip surface (a *separation* flow). Normally, the u component of computational-space momentum is negative somewhere near such regions. Our basic method of generating a stream surface breaks down in such regions, and some alternative is needed, as discussed in Section 3.6.

In general terms, our goal is to choose a height field $\psi(i, j)$ for $0 \le i < m$ and $0 \le j < n$ such that the mass under this height field is a specified amount, and the flux

100

(flow normal to the surface defined by the height field) is as close to zero as possible, in some sense. In addition, the upwind edge of the surface, $\psi(0, j)$ must satisfy some further shape constraint to make the solution unique; the shape constraint we have used is that $\psi(0, j)$ must be constant.

The method that we have implemented extends the surface in strips. Each new strip runs across j for the next higher value of i. Thus the first strip defines values for $\psi(1, j)$ and once they are defined, the second strip defines values for $\psi(2, j)$, and so on (see Figure 2).

Without going into detail on the mathematics, the i-th strip is generated as follows (values of $\psi(i - 1, j)$ have already been decided):

1. Choose provisional values for $\psi(i, j)$; for example, $\psi(i, j) = \psi(i - 1, j)$ to start with.
2. In each quadrilateral patch running from $i - 1$ to i and from $j - 1$ to j, compute the coefficients of a bilinear model of $\mathbf{m}(\alpha, \beta)$.
3. In each quadrilateral patch as above, express the normal vector throughout the patch, treating $\psi(i, j - 1)$ and $\psi(i, j)$ as variables.
4. In each quadrilateral patch as above, compute the integral of the squared flux, still treating $\psi(i, j-1)$ and $\psi(i, j)$ as variables, but using the bilinear model of $\mathbf{m}(\alpha, \beta)$. The flux is the dot product of the momentum and the normal vector. The integral can be found in closed form and is a quadratic expression in the variables $\psi(i, j-1)$ and $\psi(i, j)$.
5. The sum of all such quadratic expressions over the whole strip is the squared flux to be minimized. Define x_j to be the n-vector of values of $(\psi(i, j) - \psi(i - 1, j))$, for $0 \leq j < n$, that minimizes the quadratic form. The x_j's can be found by solving the linear system that represents the gradient of the quadratic form. Because x_j is only involved in constraints with x_{j-1} and x_{j+1}, the matrix of the linear system is tridiagonal. Because the linear system is the gradient of a quadratic form the matrix is symmetric.

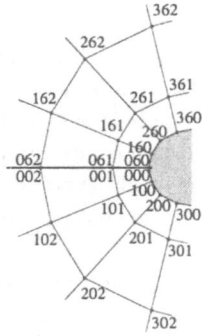

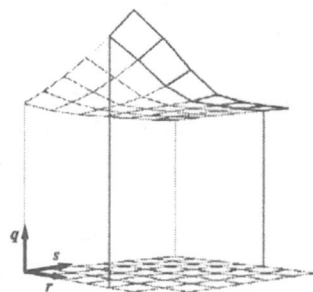

Fig. 1. To fit a curvilinear grid over a spherical cap, the $i = 0$ grid face collapses to the heavy line. Indexes are shown as ijk. For this example j varies from 0 to 12. The no-slip boundary is at $k = 0$.

Fig. 2. Patches that make up the stream surface in computational space are shown above the $q = 0$ plane, with the newest strip on the right.

6. Use the solution x_j to compute new provisional values $\psi(i, j) = \psi(i - 1, j) + x_j$. Compute an updated bilinear model of $\mathbf{m}(\alpha, \beta)$. Repeat the process until the squared flux is no longer decreasing or the solution stabilizes within a specified tolerance.

After all all the strips have been generated according to the above outline, compute the total mass under the surface and compare it with the amount that was required. Adjust the height of the upwind edge up or down until the required total mass is under the surface, within a specified tolerance.

Thus the procedure involves nested iterations. The inner iteration reduces the squared flux, and the outer iteration adjusts the overall surface height. The only saving grace is that the tridiagonal system can be solved very efficiently, in time that is linear in n. A routine matrix inversion would be order n^3.

3.4 Details of the Quadratic Form

This section sketches the development of the quadratic form that is minimized. The basis is the analysis of one patch, i.e., one quadrilateral, whose vertices are at $(i - 1, j + 1, \psi(i - 1, j + 1))$, $(i - 1, j, \psi(i - 1, j))$, $(i, j + 1, \psi(i, j + 1))$, and $(i, j, \psi(i, j))$. Switching to the local coordinate system, the surface is modeled as the bilinear function $\psi(\alpha, \beta)$. The squared flux for this patch is:

$$F_j^2 = \int_0^1 \int_0^1 \left(-u(\alpha, \beta)\, \psi_\alpha - v(\alpha, \beta)\, \psi_\beta + w(\alpha, \beta) \right)^2 \, d\alpha \, d\beta \qquad (6)$$

The integral has a closed form, in which the heights $\psi(i, j + 1)$ and $\psi(i, j)$ appear quadratically. Let $x_j = (\psi(i, j) - \psi(i - 1, j))$. The quadratic form to be minimized is

$$F^2 = \sum_{j=0}^{n-2} F_j^2 \qquad (7)$$

which is nonnegative definite, so a minimum must exist.

The program contains an option to scale the squared flux inversely by the squared momentum; intuitively this makes the sine of the angle between the momentum and the surface patch the error, rather than the flux itself.

We define a series of terms leading to a tridiagonal linear system whose solution minimizes F^2. Momentum terms always refer to computational-space momentum and are values at the center of the patch (which are also averages over the patch). Subscripts of α and β denote partial derivatives (of the appropriate bilinear function). We introduce local symbols $g, G, h, q, Q, r, R, s, S$ in the equations below; their purpose is to obtain expressions for a_j, b_j, and c_j.

$$g_j = (-u + v)/2 + (u_\beta + v_\alpha)/12 \qquad G_j = (-u - v)/2 - (u_\beta + v_\alpha)/12$$
$$q_j = -u + (u_\beta - v_\beta)/2 \qquad Q_j = -v + (u_\alpha - v_\alpha)/2$$
$$r_j = u + (u_\beta + v_\beta)/2 \qquad R_j = v + (u_\alpha + v_\alpha)/2$$
$$a_j = g_j^2 + (q_j^2 + Q_j^2)/12 + G_{j-1}^2 + (r_{j-1}^2 + R_{j-1}^2)/12 \qquad (8)$$
$$b_j = G_j g_j + (r_j q_j + R_j Q_j)/12 \qquad (9)$$

$$h_j = v(\psi(i-1,j+1) - \psi(i-1,j)) - w$$
$$s_j = -v_\beta(\psi(i-1,j+1) - \psi(i-1,j)) + w_\beta$$
$$S_j = -v_\alpha(\psi(i-1,j+1) - \psi(i-1,j)) + w_\alpha$$
$$c_j = h_j g_j + h_{j-1} G_{j-1} + (s_j q_j + S_j Q_j) + s_{j-1} r_{j-1} + S_{j-1} R_{j-1})/12 \quad (10)$$

Any terms containing subscripts outside the range 0 through $n - 1$ are considered 0.

A linear system is defined using Eqs. 8–10. The vector c has c_j as its components. Finally, we define the matrix A whose entries are 0, except for three diagonals:

$$A_{jj} = a_j \qquad A_{j,j+1} = A_{j+1,j} = b_j. \qquad (11)$$

The linear system to be solved is $Ax = c$.

3.5 Solving the Tridiagonal System

The results of this section are obtained with standard methods of linear algebra and vector calculus, which may be found in many college texts on the subject. Let $A = LL^T$ be the Cholsky factorization of A. That is, L is lower triangular. Then L has the form $L_{jj} = d_j$, $L_{j+1,j} = e_j$, and all other entries are 0. With the convention that $e_{-1} = 0$, the d_j and e_j are defined by the recurrence relations:

$$d_j^2 = a_j - e_{j-1}^2 \qquad (12)$$
$$e_j = b_j/d_j \qquad (13)$$

Some care is required to avoid numerical problems. Equations 8 and 10 allow us to infer special properties of linear system. The key is to interpret them in terms of inner products of certain combinations of the vectors: (g_j, q_j, Q_j), $(G_j r_j, R_j)$, and (h_j, s_j, S_j), $0 \le j < n$. It can be shown that the right-hand side of Eq. 12 must be nonnegative, so if a negative value is obtained, it is due to numerical inaccuracies and may be replaced by 0. Also, it can be shown that if $d_j = 0$ in Eq. 13, then b_j must also be 0, and the equations of rows 0 through j separate from (have no variables in common with) those in rows $j + 1$ through $n - 1$. This independence can be implemented by setting e_j to 0, rather than using Eq. 13. Similarly, if any $a_j = 0$, then Eqs. 8 and 9 imply that b_{j-1} and b_j are 0, and a similar separation occurs.

To solve $Ax = c$, we compute the intermediate result $f = L^{-1}c$, followed by $x = (L^T)^{-1}f$. The procedure uses a forward recurrence equation, followed by a backward recurrence equation. As usual, f_{-1} and x_n are considered to be 0.

$$f_j = (c_j - e_{j-1}f_{j-1})/d_j \qquad (14)$$
$$x_j = (f_j - e_j x_{j+1})/d_j \qquad (15)$$

Again, due to separations mentioned above, whenever $d_j = 0$, then f_j and x_j can be set to 0, rather than using the recurrence equations. This rule delivers a valid solution of $Ax = c$, although the solution is not unique because A is singular.

3.6 Additional Numerical Issues

If the w component of the computational-space momentum dominates the u component, then the inner iteration described in Section 3.3 does not converge, for essentially the same reasons that a nonadaptive Runge-Kutta procedure does not converge when the derivative is too large. Because it is common for cells to be very thin in the k direction near a no-slip boundary, even a vector that is nearly parallel to the no-slip surface in physical space can become nearly orthogonal to the no-slip surface in computational space. We address this problem in two ways.

1. If the solution vector x does not lead to a smaller squared flux than the previous provisional value, then some fraction of x between 0 and 1 is searched for that *does* reduce the squared flux. If no satisfactory fraction is found, the iteration terminates with the previous provisional value becoming the final value.
2. If the values of ψ in question are less than 1, where 0 is the location of a no-slip surface, then we interpolate $w(i, j, \psi) = \psi^2 w(i, j, 1)$, rather than linearly. There is physical justification for this: Boundary layer theory going back to the classical work of Blasius shows that w varies quadratically in the neighborhood of the no-slip boundary (while u remains linear). This method of interpolation allows the interpolated momentum vector to approach tangency with the no-slip surface, as is required by conservation of mass.

The moral is that linearizing an inherently nonlinear system does not guarantee good results.

If the u component of the computational-space momentum changes sign between $i-1$ and i then there might be an asymptotic surface that passes between $(i-1, j, \psi(i-1, j))$ and $(i, j, \psi(i, j))$. That is, the surface we have propagated to $i-1$ approaches but does not cross this asymptotic surface. A rigorous treatment of this situation would involve an eigenvalue and eigenvector analysis of the linearized 3D vector field in the neighborhood, which is beyond the scope of this paper. The main idea is to identify the two eigenvalues whose real parts have the same sign (all three cannot agree on sign or mass is not conserved). Then the plane that "corresponds to" these two eigenvalues is the asymptotic plane.

For now we have implemented a stop-gap solution whenever u is negative. We assume that u will be positive if we get far enough away from the no-slip boundary. So if $u(i, j, \psi(i, j)) < 0$ for some value of ψ that is being considered, we increase $\psi(i, j)$ until it is positive, or leaves the grid. If it leaves the grid, then we assume that u is a positive "free stream" value there, and compute accordingly.

4 Experimental Results

We have implemented programs to convert CFD solutions into computational space and to generate stream surfaces. We used the NASA/NAS FAST program to produce images. This section reports some experimental results. Timing results are all based on an SGI Onyx 2 using one 195 MHz R10000 processor. Solution quality is measured by the root-mean-square (RMS) sine of the angle between the momentum vector and the tangent plane of the surface. Figures cited in this section appear in the color plates.

We omit detailed results on a 9x9x9 test dataset with a quadratic vector field, mass-conservative flow, and no-slip surface. It confirmed that the program converges quickly on "nice" data.

The first dataset we report on is a simulation of the space shuttle launch vehicle during ascent. The simulation used 9 zones in its grid, but we analyzed only the third zone, and further limited that to the part having a no-slip boundary. This comprised an 80x77x48 grid, 295,680 vertices in all. We generated 15 stream surfaces in 799 CPU seconds. The RMS error varied in a narrow range from 0.286 to 0.304.

Several images are shown in Figure 3. Part (a) shows the no-slip fuselage for reference. Part (b) shows the third stream surface in a family of 15 in purple. Notice that the surface needs to begin well in front of the nose in order to be tangent to the flow. The grid degeneracies in front of the nose did not cause any computational difficulties. Part (c) shows the stream surface separating from the underneath of the fuselage.

Part (d) is another view. Notice the "ear" that appears where we might expect a tail fin (but there is no tail fin). This is a region of turbulent flow due to the bulge of the engine housing. The lowest part of the stream surface (furthest from the fuselage) is in a region where the computational-space value of u is negative, and the stop-gap method mentioned in Section 3.6 came into play. Although the stream surface might be inaccurate in these regions, it indicates visually that "something is happening," so other tools can be used to investigate more closely.

The final dataset we report on is a steady-state delta wing simulation with 15-degree angle of attack. The entire grid is 67x209x49 but again we limited it to the portion over a no-slip boundary, the size of which was 47x209x49, or 481,327 vertices. We generated 3 stream surfaces in 2725 CPU seconds. The RMS error varied from 0.320 to 0.336.

Figure 4 shows the third stream surface, counting from the no-slip "fuselage." Part (e) shows an overview from the rear and above. Notice where the stream surface separates from the "wing" surface near the outer edges forming a few channels. Part (f) shows a closer view of one side. These generally parallel lines of separation and attachment confirm and supplement the observations of "open" separation lines by Kenwright, Henze, and Levit [KHL99].

5 Related Work

Lack of space prevents an extensive review of the literature; please see cited works for additional bibliography [Ken93,vW93,KM96]. Hultquist described the construction of stream surfaces by patching together *stream ribbons*, which in turn are produced by generating stream lines [Hul92].

Three-dimensional stream functions, also known as *Clebsch potentials*, are scalar functions whose isosurfaces are stream surfaces *Clebsch potentials* [Lam32]. They have been studied computationally by Kenwright and Mallinson [KM92], by Kenwright [Ken93], by van Wijk [vW93], by Knight and Mallinson [KM96], by Feng *et al.* [FWJ98], and others. The method of Knight and Mallinson is global, but it applies to tetrahedral grids and solenoidal (curl-free) flows. The method of van Wijk is also global, but (as reported) it is limited to incompressible flows, is apparently implemented only for regular grids, and only handles certain flow topologies. This method propagates stream lines backward through the velocity field, so it is unclear how it would handle no-slip surfaces, where the velocity is 0.

In general, the methods reported in the literature do not address the special problems in CFD flow data that our program tries to address, such as no-slip surfaces, grids with degeneracies, and somewhat noisy data that results from numerical PDE solutions. They often have restrictions that prevent them from operating on the data at all, so a detailed comparison is not practical.

6 Conclusion

We have presented a new method for generating stream surfaces. It is a "semi-global" method in the sense that it simultaneously solves constraints over a large region of space, rather than working in one local region at a time, yet there is an element of downstream propagation. Its efficiency depends on a fast procedure for solving tridiagonal linear systems. The implementation so far has limited flexibility. Work in progress addresses situations in which u is negative, and will be the subject of a future report. Future work should also deal with multiple no-slip surfaces.

Stream surfaces are best used in combination with other cues for visualization. In this paper they were usually supplemented with arrows at various points in the surface. The surface gives shape information and the arrows give direction within the surface and magnitude information.

Acknowledgments This work was supported in part by NASA-Ames grant NAG2-1239 and NSF grant CCR-9503829. The datasets are from the NASA/NAS group at Ames, except the test. We thank the Zentrum für Angewandte Informatik Köln within the Regionales Rechenzentrum (Regional Computing Center) of the University of Cologne for making their extensive computer resources available during the author's visit.

References

[FWJ98] D. Feng, C. Wenli, and S. Jiaoying. Stream surface construction using mass conservative interpolation. *J. Computer Science and Technology*, 13 (suppl.issue):45–53, 1998.

[Hul92] J. P. M. Hultquist. Constructing stream surfaces in steady 3D vector fields. In *Proceedings of Visualization '92*, pages 171–178, Boston, MA, October 1992. IEEE.

[Ken93] D. N. Kenwright. *Dual Stream Function Methods for Generating Three-Dimensional Streamlines*. PhD thesis, Department of Mechanical Engineering, University of Aukland, New Zealand, August 1993.

[KHL99] D. N. Kenwright, C. Henze, and C. Levit. Feature extraction of separation and attachment lines. *IEEE Trans. Visualization and Computer Graphics*, 5(2):135–144, 1999.

[KM92] D. N. Kenwright and G. D. Mallinson. A 3-D streamline tracking algorithm using dual stream functions. In *Visualization '92*, pages 62–68. IEEE, October 1992.

[KM96] D. Knight and G. D. Mallinson. Visualizing unstructured flow data using dual stream functions. *IEEE Trans. Visualization and Computer Graphics*, 2(4):355–363, 1996.

[Lam32] Horace Lamb. *Hydrodynamics*. Dover, 6th edition, 1932.

[vW93] J. J. van Wijk. Implicit stream surfaces. In *Visualization '93*, pages 245–252, San Jose, CA, 1993. IEEE Comput. Soc. Press.

Editors' Note: see Appendix, p. 340 for colored figures of this paper

Vector and Tensor Field Topology Simplification on Irregular Grids

Xavier Tricoche, Gerik Scheuermann, Hans Hagen, and Stefan Clauss

University of Kaiserslautern
P.O. Box 3049, D-67653 Kaiserslautern, Germany
E-mail: {tricoche|scheuer|hagen|clauss}@informatik.uni-kl.de

Abstract. Topology-based visualization of planar turbulent flows results in visual clutter due to the presence of numerous features of very small scale. In this paper, we attack this problem with a topology simplification method for vector and tensor fields defined on irregular grids. This is the generalization of previous work dealing with structured grids. The method works for all interpolation schemes.

1 Introduction

Turbulent flows are characterized by the presence of many close vortices of different scales. Such datasets typically result in very complicated pictures when visualized with topology-based methods. This is due to the existence of numerous singularities and corresponding separatrices in the graph depiction: The resulting image is cluttered and the most meaningful features cannot be efficiently extracted. Earlier, we attacked this problem in the case of 2D vector fields defined on a bilinear interpolated curvilinear grid [3]. This method has been extended to second-order symmetric 2D tensor fields over curvilinear grids [6]. In both cases, we achieved a reduction of complicated structures by merging close singularities. The motivation for this operation is the equivalence in the large of several close simple singularities and a single, higher order, singular point. In the present paper, we use the same basic principle to achieve a topology simplification of planar tensor or vector fields defined on arbitrary grids, with arbitrary interpolation schemes. The major improvement resides in the way of determining the groups of close singularities that are merged: The new method works independently from the underlying cell structure, focusing only on the singularities.

The paper is structured as follows. First, a brief review of existing techniques for simplifying vector fields is given. An overview of the general concepts of vector and tensor field topology is proposed in section 3. Then, we describe the clustering strategy used to partition the field into subdomains where only close singularities are present: This is the purpose of section 4. The fusion of these close singularities is explained in section 5. As last step, the topological structure of these new higher order singularities must be identified: This is done in section 6. Finally, results are shown for an analytic field and a numerical simulation (section 7).

2 Related Work

In [3], we proposed a method for simplifying the topology of planar vector fields defined on curvilinear grids. The grid structure is used to determine cell clusters of close singularities. This enables local deformations of the grid based upon the piecewise linear nature of the interpolant on the edges.

W. de Leeuw et al. have addressed the issue of vector field topology simplification [4]. Their method simplifies the topological graph by successive removal of connected pairs of critical points. This leads to a significant reduction of the number of critical points and clarifies the topological structure. Yet, no analytic description is provided for the simplified vector field and the method cannot handle arbitrary topological configurations. A cluster based simplification of vector fields has also been presented by B. Heckel et al. [5]. No cell connectivity information is needed and the simplification process reduces substantially the size of large 3D vector data sets. However, possible changes in the vector field topology after simplification can result in structural inconsistency.

3   Vector and Tensor Field Topology

In this section, we briefly review the notion of vector and tensor field topology in the piecewise linear case. Then, we turn to the general case of critical points encountered in continuous piecewise analytic fields.

3.1   Vector Field Topology

The linear topology of planar vector fields is well-known to the visualization community [1]. It consists of critical points (zero locations where streamlines can meet as opposed to any other point) of first order and particular streamlines, called separatrices, that delimit subdomains of the flow where every streamlines are, to some extent, similar. The possible critical points, classified according to the eigenvalues of the Jacobian matrix at their position, are given in Fig. 1. Separatrices are, in this approximation, streamlines that reach or emanate from a saddle point.

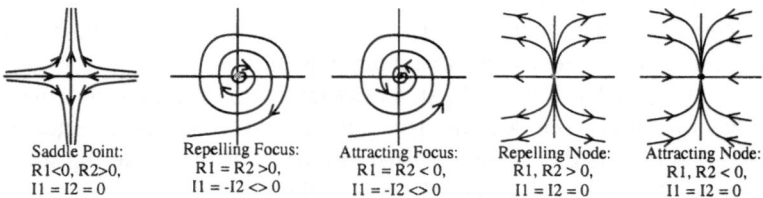

Saddle Point:
R1<0, R2>0,
I1 = I2 = 0

Repelling Focus:
R1 = R2 >0,
I1 = -I2 <> 0

Attracting Focus:
R1 = R2 < 0,
I1 = -I2 <> 0

Repelling Node:
R1, R2 > 0,
I1 = I2 = 0

Attracting Node:
R1, R2 < 0,
I1 = I2 = 0

Fig. 1. First order critical points

3.2   Tensor Field Topology

Similar to vector field topology, symmetric second-order 2D tensor field topology has been introduced [2]. Basically, it is defined as the topology of one of the two (bidirectional) eigenvector fields: One defines major (resp. minor) tensor lines as the curves everywhere tangent to the eigenvector associated with the major (resp. minor) eigen-value. The singularities are locations where both eigenvalues are equal. These degen-erate points appear in two possible types (see Fig. 2). (Note that there exists a wedge point configuration where $S_1 = S_2$.)

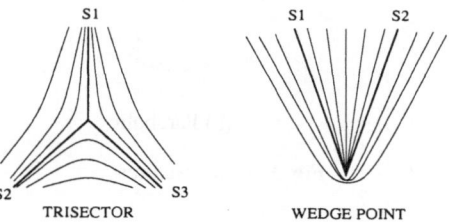

Fig. 2. First order degenerate points

Here, the classification is based upon the so-called δ-invariant: If the symmetric second-order tensor field is denoted by the matrix

$$\begin{pmatrix} T_{11} & T_{12} \\ T_{12} & T_{22} \end{pmatrix},$$

we use the linear approximation in the vicinity of a position (x_0, y_0)

$$\begin{cases} \frac{T_{11}-T_{22}}{2} \approx a(x - x_0) + b(y - y_0) + ... \\ T_{12} \approx c(x - x_0) + d(y - y_0) + ... \end{cases},$$

and define $\delta = ad - bc$.

A trisector point is characterized by a negative value of δ and a wedge point by a positive value. Separatrices are defined as the tensor lines that reach a trisector point along S_1, S_2 or S_3 and a wedge point along S_1 or S_2.

3.3 General Case

In the general case, a singularity (critical or degenerate point) can have one of two types: *center* type (every streamline in the vicinity of the singularity is closed and ro-tates around without reaching it) or *non-center* type (at least, one streamline reaches the singularity, forming one or more curvilinear sectors). In the non-center case, the char-acterization is based upon position and type of the curvilinear sectors. These sectors have three possible natures: Hyperbolic, parabolic or elliptic (see Fig. 3). Separatrices are defined as streamlines that bound hyperbolic sectors.

4 Singularity Clustering

Suppose that the singularity positions have been found on an irregular grid. We take them as input for a clustering process that provides us with groups of close singularities. Earlier [3], we proposed a method that preserves the structure of a curvilinear grid. In the present method, we just take the given singularity locations into account.

Let $P_1, ..., P_m$ be the positions of the m singularities lying inside a particular cluster. We

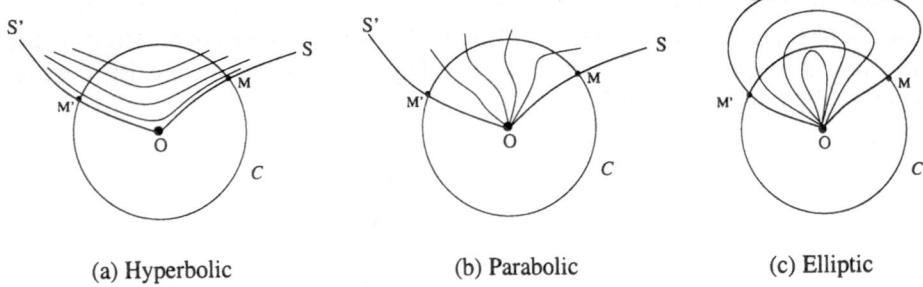

(a) Hyperbolic   (b) Parabolic   (c) Elliptic

Fig. 3. Sector Natures

want to minimize the approximation error (for a given norm) of these m singularities by a single point, where this point (or cluster *center*) Q is the singularities' mean point (see

Fig. 4). The corresponding error is $S = \dfrac{\sum_{j=1}^{m} \omega_j \|P_j - Q\|}{\sum_{j=1}^{m} \omega_j}$, where ω_i is the weight

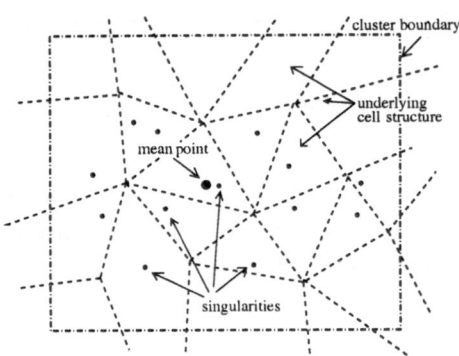

Fig. 4. Cluster singularities and cluster center

associated with the *ith* singularity. (We use here a uniform weighting but one could take into account any measure of singularity strength to determine the weights ω_i.)

The aim of the clustering process is to get a set of clusters that all have an error value smaller than a specified threshold and enclose all singularities.

If a cluster does not satisfy the given error criterion, we split it into sub-clusters. To do this, we introduce the projected variances associated with a given cluster,

$$V_i = \sum_{j=1}^{m} \omega_j (P_j^i - Q^i)$$

where $i \in 0, 1$ is the considered coordinate axis ($P_j = (P_j^0, P_j^1)$).

Now, considering the grid bounding box as initial cluster and putting all cells inside it, the method is as follows.

Step 1.   Take as cluster center Q the singularities' mean point.

Step 2.   Compute the approximation error S.
If (S > THRESHOLD) go to step 3.
Otherwise stop.

Step 3.   Compute the coordinate axis with largest
projected variance (i.e. max(V0, V1)).

Step 4.   Split the cluster by a line through Q perpendicular
to the selected coordinate axis, creating 2 sub-clusters.
For each cell contained in the present cluster, check if
it partially lies in each of both sub-clusters (bounding
box intersection test):
Add the indices in the corresponding sub-clusters.
For each sub-cluster, go to step 1.

Remark that the splitting strategy used in the present method is not the only possible one. A cluster can be subdivided in an arbitrary way if the convexity of the sub-clusters is preserved. Nevertheless, our choice leads to clusters with edges parallel to the coordinate axes which enables fast processing.

At the end of this process, we are left with a set of clusters that contain close singularities (according to the metric used) and know what cells are (at least partially) contained in them. The next step is to compute, for each final cluster and for each contained cell, the possible intersections of an edge with the cluster boundary. (This can be done very efficiently because the cluster edges are parallel to the coordinate axes). Adding the 4 cluster corner points to the intersection positions we found, we get a list of positions that we sort next, in a counterclockwise order. Now we isolate the interior domain of the cluster from the rest of the grid. This can be done by removing all cells entirely contained in the cluster (without intersection with cluster boundary) and cutting away the part of every cell intersecting the cluster interior domain: This corresponds to superimposing locally a new small grid on the initial one (see Fig. 5(a)). The cut cells get a modified geometry but keep their interpolation scheme to ensure continuity and consistency with the original field. In particular, the field value along the cluster boundary is unchanged.

5 Singularity Fusion

The technique used is similar to our earlier ideas [3]. We want to get a continuous piecewise analytic field description after modification that ensures the presence of a unique singularity located at the singularities' mean point. Furthermore, we want this description to preserve the field value on the cluster boundary. Consequently, we cover the cluster interior domain as follows: Inserting an additional vertex at the mean point position, we build a triangle star strip connecting this point with every position on the cluster boundary. Furthermore, we associate the new vertex with a singular value. In the vector case, it is a simple zero vector. In the tensor case, every isotropic matrix (of the form λI_2, $\lambda \in \mathbb{R}$) is a valid choice. Actually, one can show that the isotropic component of a 2D second-order tensor field does not influence the topology of its

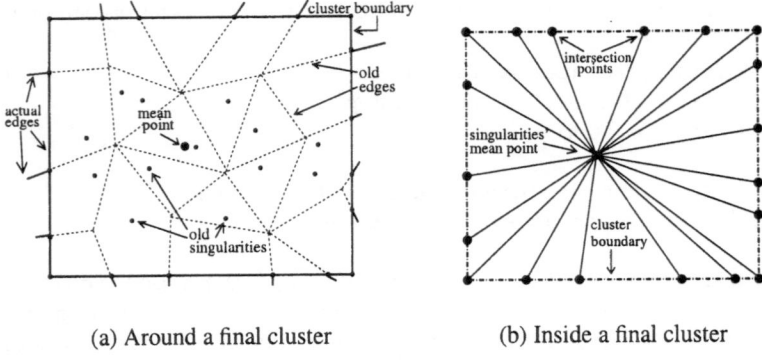

(a) Around a final cluster   (b) Inside a final cluster

Fig. 5. New local grid structure

eigenvector fields. For this reason, we take a zero matrix as new artificially created degenerate point (see Fig. 5(b)).

In each of these triangles, we have to define an interpolation scheme that preserves the field value on the cluster boundary. This can be done by using a simple side-vertex interpolation scheme: The position of every point inside such a triangle is determined as shown in Fig. 6, so we get $Q(t) = (1-t)A + tB$ and $P(t,u) = (1-u)\Omega + uQ(t)$. The interpolated value is (with f denoting the considered field)

$$f(P)(t,u) = u\,f(Q(t)), \quad \text{since } f(\Omega) = 0$$

where $f(Q(t))$ is the original value on the cluster boundary. This ensures that the field on the boundary is preserved which guarantees continuity for the new piecewise analytic description. We can also claim that the new artificial singularity is the only one contained in the cluster after modification (otherwise, we would have, for some t, $Q(t) = 0$, and we would have a singular value on the boundary which cannot occur because such a case is rejected during the clustering process).

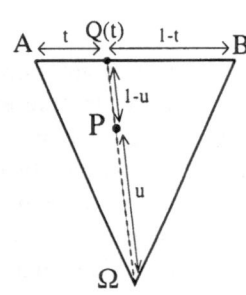

Fig. 6. Side vertex interpolation

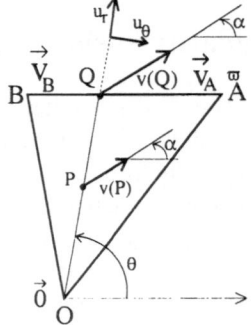

Fig. 7. Vector field in polar coordinates

6 Local Structure Identification

Once such an artificial singularity has been created, its structure must be identified to enable the drawing of the topology. In particular, according to section 3.3, the boundary curves of the hyperbolic sectors must be detected to serve as starting positions for separatrix integration. But first, we need the following property of the side/vertex interpolation scheme: Taking the artificial singularity as polar coordinate origin, the direction of the vector (resp. eigenvector) field does not depend on the radius (see Fig. 7). Practically, it means that separatrices are joining singularity and cluster boundary along straight lines and furthermore, that the search for separatrices positions can be restricted to the cluster boundary. (The proof is straightforward.) In the following, we consider successively vector and tensor cases for the purpose of separatrix position detection.

6.1 Vector Case

We are looking for separatrices that emanate from the singularity along straight lines up to the cluster boundary. In other words, we seek, for each edge on the cluster boundary, positions where the vector field is parallel to the polar coordinate vector u_r. At this stage, the positions do not all correspond to an actual separatrix location: They can lie in the middle of a parabolic sector. For this reason, we also detect the positions where the vector field is orthogonal to u_r (that is parallel to u_θ) and check, in the parallel case, the sign of $< v.u_r >$ and, in the orthogonal case, the sign of $< v.u_\theta >$, where v is the vector field value at the considered position. Note that the search for such positions consists of finding the roots of a polynomial equation. The order of this equation depends on the type(s) of interpolation scheme(s) used on the irregular grid. In particular, this order may be different for different edges on the cluster boundary. Typically, if the interpolant is a polynom of degree n, the system to solve will be a polynomial problem of degree $n + 1$. When no algebraic solution can be found, a specific root finder could be applied, based upon the knowledge of the interpolant. Once the interesting positions have been found (marked PARALLEL+, PARALLEL-, ORTHOGONAL+ or ORTHOGONAL-) and sorted in counterclockwise order along the cluster boundary, we make use of the graphs shown in Fig. 8 to determine the sector types and thus the actual separatrices positions.

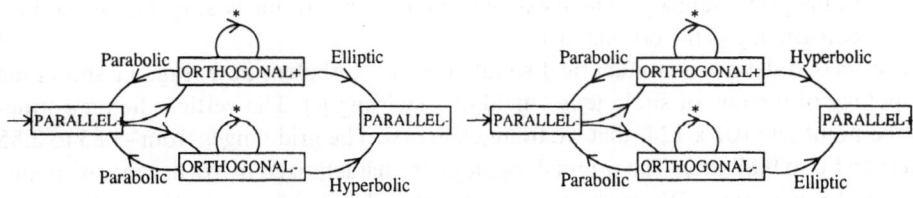

Fig. 8. Sectors discrimination graphs

6.2 Tensor Case

In the tensor case, the principle is basically the same. Nevertheless, because of the sign indeterminacy of the eigenvectors (the eigenvector fields are bidirectional), the sectors discrimination graph cannot be applied (the ORTHOGONAL positions are of no help to distinguish hyperbolic from elliptic sectors). Furthermore, the computation of the eigenvectors induces a higher order of the polynomial equation to solve. Typically, if the tensor field interpolant has order n, then the determination of the PARALLEL positions leads to a polynomial equation of order $n + 2$. After that, we sort the PARALLEL positions in counterclockwise order and look at the angle variation of the eigenvector field between two consecutive PARALLEL positions as follows. Depending on the sector type, one gets

- $\Delta\alpha = \theta$ in the *parabolic* case
- $\Delta\alpha = \theta - \pi$ in the *hyperbolic* case
- $\Delta\alpha = \theta + \pi$ in the *elliptic* case

which enables a sector type identification (see Fig. 9).

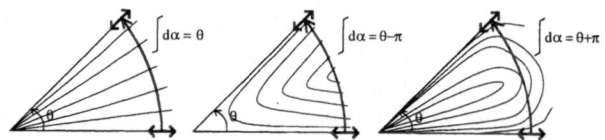

Fig. 9. Angle variation in the parabolic, hyperbolic and elliptic case

7 Results

We present here the results of our method applied to an artificial turbulent vector field and to a tensor field provided by a numerical simulation.

The first dataset is a 2D vector field defined on the Delaunay triangulation of an unstructured point set. The interpolation scheme is piecewise linear. The grid has 400 vertices, ranging from -5 to +5 in x and y. The original topology contains 189 critical points and 380 separatrices (see Fig. 10). We first simplify this complicated topology with a clustering threshold of 0.5. The graph has now 114 critical points and 286 separatrices (see Fig. 12(a)). To ease the interpretation, higher order singularities are depicted as big circles. Using a threshold of 1.5, there are 81 critical points and 188 separatrices remaining (Fig. 12(b)). Note that the simplification process does not affect the topology close to the grid boundary (which explains the presence of many singularities) to preserve consistency to the original data.

The second dataset is a numerical simulation of a turbulent flow: Fig. 11 shows the topology of the rate of strain tensor field of a swirling jet. The vertices lie on a structured point set (101 x 124) that we triangulate first. The grid ranges from -3.85 to 3.85 in x and 0 to 9.87 in y. The original topology is characterized by 67 degenerate points and 144 separatrices. We start simplifying with a threshold of 0.5: We get 34 degenerate points and 78 separatrices. Once again, the clusters are shown (Fig. 13(a)). If we move to a larger value for the clustering threshold (1.0), the topology simplifies dramatically as shown in Fig. 13(b): 18 degenerate points are present and only 42 separatrices remain.

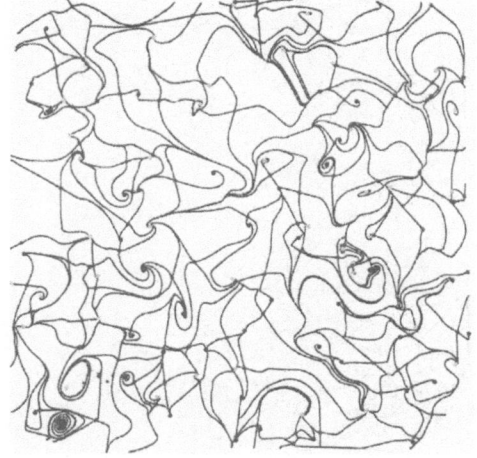

Fig. 10. 1st example: Vector case

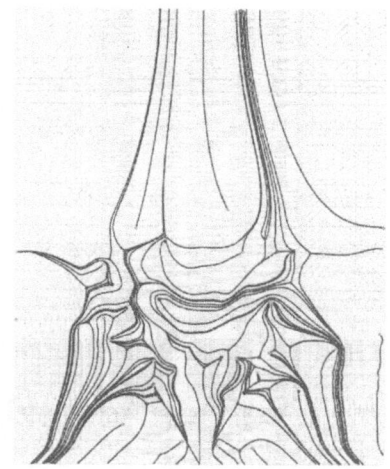

Fig. 11. 2nd example: Tensor case

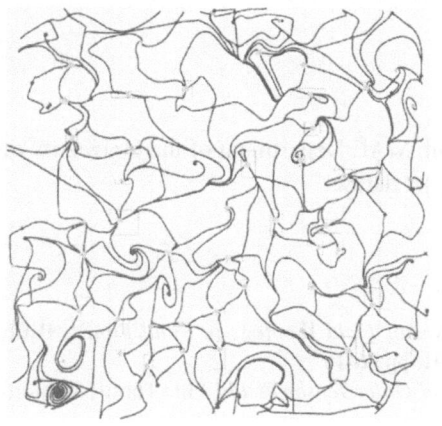

(a) threshold = 0.5

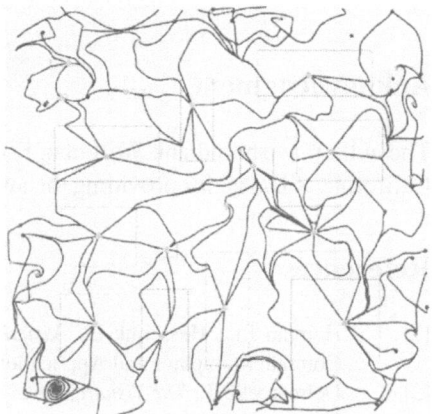

(b) threshold = 1.5

Fig. 12. 1st example: Simplified topologies

8 Conclusion

We have presented a method for simplifying the topology of vector and tensor fields defined over irregular grids with arbitrary interpolation schemes. Our technique is based upon a clustering strategy that handles the singularities (critical or degenerate points), omitting the underlying cell structure. It permits a flexible local topology simplification by merging the close singularities lying in the same final cluster while providing piecewise analytic description for the field after simplification. The method produces clarified depictions and preserves topological consistency with the original data.

116

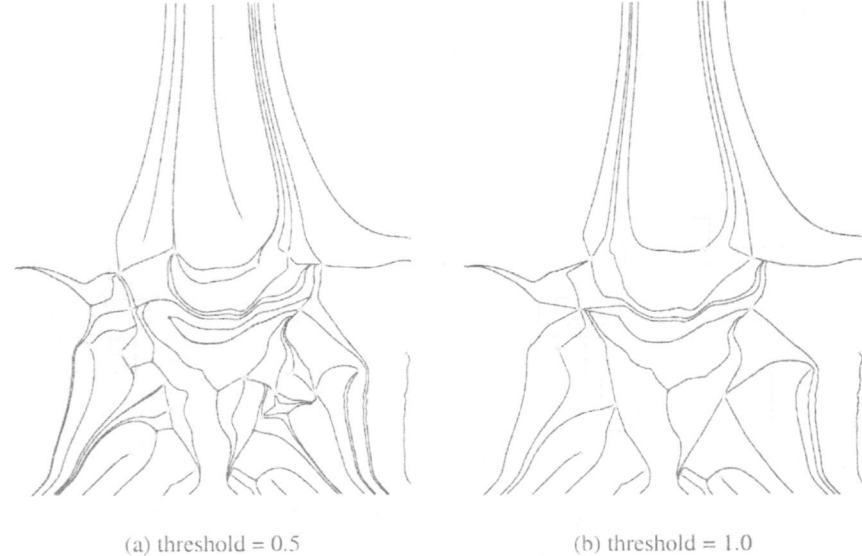

(a) threshold = 0.5   (b) threshold = 1.0

Fig. 13. 2nd example: Simplified topologies

Acknowledgment

The authors wish to thank Wolfgang Kollmann, MAE Department of the University of California at Davis, for providing the swirling jet dataset.

References

[1]   Helman J.L., Hesselink L., *Visualizing Vector Field Topology in Fluid Flows*. IEEE Computer Graphics and Applications, 1991, pp.36-46.

[2]   Delmarcelle T., *The Visualization of Second-Order Tensor Fields*. PhD Thesis, Stanford University, 1994.

[3]   X. Tricoche, G. Scheuermann, H. Hagen, *A Topology Simplification Method for 2D Vector Fields*. In Proceedings IEEE Visualization'00, IEEE Computer Society, Los Alamitos CA, 2000, pp.359-366.

[4]   de Leeuw W., van Liere R., *Collapsing Flow Topology Using Area Metrics*. In Proceedings of IEEE Visualization '99, IEEE Computer Society, Los Alamitos CA, 1999, pp.349-354

[5]   Heckel B., Weber G., Hamann B., Joy K.I., *Construction of Vector Field Hierarchies*. In Proceedings of IEEE Visualization '99, IEEE Computer Society, Los Alamitos CA, 1999, pp.19-25.

[6]   Tricoche X., Scheuermann G., Hagen H., Clauss S., *Scaling the Topology of Symmetric Second-Order Tensor Fields* In NSF/DoE Lake Tahoe Workshop on Hierarchical Methods for Scientific Visualization, October 2000.

Editors' Note: see Appendix, p. 341 for colored figures of this paper

Topology-Based Visualization of Time-Dependent 2D Vector Fields

Xavier Tricoche, Gerik Scheuermann, and Hans Hagen

University of Kaiserslautern
P.O. Box 3049, D-67653 Kaiserslautern
Germany
E-mail: {tricoche|scheuer|hagen}@informatik.uni-kl.de

Abstract. Topology-based methods have been successfully applied to the visualization of instantaneous planar vector fields. In this paper, we present the topology-based visualization of time-dependent 2D flows. Our method tracks critical points over time precisely. The detection and classification of bifurcations delivers the topological structure of time dependent vector fields. This offers a general framework for the qualitative analysis and visualization of parameter-dependent 2D vector fields.

1 Introduction

Topology-based visualization of steady 2D vector fields was first introduced by Helman and Hesselink [1]. The location and type identification of the critical points associated with particular streamlines enables a domain partition into subregions where the flow is structurally uniform. This permits a simple, synthetic depiction of vector datasets while preserving the qualitative properties of the flow. This method has been widely applied in the last decade. Yet, in the case of unsteady vector fields, the original method must be extended to offer insight into the qualitative and structural evolution of the 2D flow over time. Helman and Hesselink [2] proposed a solution that consists of computing the topological graph for each given discrete time step. After this, the resulting curves are connected graphically from one time step to the next, based upon qualitative criteria. Here, time is visualized as third dimension. The major drawback of this method is that time is handled in a discrete manner, which prevents the resulting display from identifying and properly depicting the qualitative changes that may occur between two consecutive time steps. As a matter of fact, the introduction of time as an additional parameter has a fundamental consequence: Bifurcations take place that correspond to a transition from a stable structure to another through a punctual unstable state called bifurcation point. Now, the visualization of these features is of major interest because they show how and when a vector field evolves to the structures that are observed on discrete time samples. Their depiction requires the analysis of the vector field as a continuous map defined over a space/time domain.

The paper is structured as follows: First, we briefly review works dealing with the visualization of time-dependent flows. Then we recall fundamental definitions of vector field topology and give an overview of the bifurcation types we have to deal with (section 3). Next, we describe in section 4 how we integrate time in our data structure to get the required continuous space/time domain. This enables the precise tracking of critical points over time, as explained in section 5. As a last step, the surfaces spanned by the curves of the topological graph must be constructed (see section 6). Results are shown in section 7.

2 Related Work

From the mathematical point of view, the *Visual Math Project* [3] has presented the basic mathematical ideas in an understandable way by the use of discerning sketches. In particular, for depicting bifurcations in 2D, an additional space dimension is used to represent the parameter domain and to visualize the structure evolutions related to it. Furthermore, the practical significance of unsteady flow fields has led to several techniques for the visualization of time-dependent vector fields. In [4], Dickinson has described a method for the interactive analysis of the topology of time-dependent 2D or 3D vector fields. The focus is the correlation of the separatrices between consecutive time steps. This is achieved by solving symbolically the eigensystem of the Jacobian. A method for computing streaklines in 3D unsteady flow fields has been proposed in [5]. The basic principle is to integrate streaklines thanks to an interpolation over space and time. The technique works also with moving grids. Using this scheme, a method for displaying unsteady flow volumes has been presented in [6]. Based upon an adaptive subdivision strategy, the authors arrive at integrating streaklines starting on a generating polygon. For the purpose of feature visualization, in [7], one tracks and correlates the extracted structures by detecting the following fundamental events: Continuation, bifurcation, amalgation, creation and dissipation.

3 Time-Dependent Topology

3.1 Topology

The topology of steady planar vector fields is formed by critical points, some particular streamlines that connect them called separatrices and closed orbits (or cycles) that act as sinks or sources. If one focuses on singularities of first order, there are 5 common types of singularities (see Fig. 1). Note that attracting nodes and foci constitute sinks while

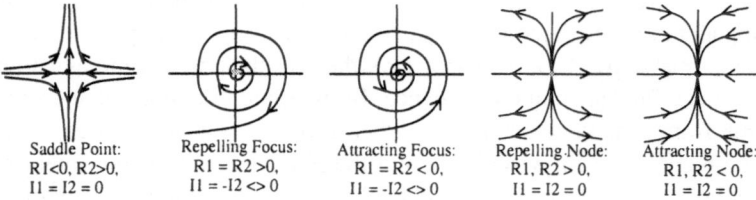

Saddle Point:
R1<0, R2>0,
I1 = I2 = 0

Repelling Focus:
R1 = R2 >0,
I1 = -I2 <> 0

Attracting Focus:
R1 = R2 < 0,
I1 = -I2 <> 0

Repelling Node:
R1, R2 > 0,
I1 = I2 = 0

Attracting Node:
R1, R2 < 0,
I1 = I2 = 0

Fig. 1. Common first order singularity types

repelling nodes and foci are sources. A fundamental invariant is the so-called index of a critical point, defined as the number of field rotations while traveling around the critical point along a closed curve, counterclockwise. Note that all sources and sinks mentioned above have index +1 while saddle points have index -1. In a first order approximation, separatrices are defined as those streamlines that start or end at a saddle point.

3.2 Bifurcations

If we turn now to time dependent vector fields, we have to face topological features that are fundamentally new: When the time parameter varies, the qualitative structure of the topology may change from one stable state to another. Such changes are called *bifurcations*. One distinguishes between two types of bifurcations: On one hand, some bifurcations only affect the nature of a singular point or a closed orbit and the corresponding new stable state is to be found in a neighborhood: These bifurcations are called *local bifurcations*. On the other hand, bifurcations that change the global structure of the flow and cannot be deduced from local information are called *global bifurcations*. In the following, we focus on typical 2D local bifurcations and say a few words about global bifurcations.

Local Bifurcations. There are actually two main types of local bifurcations affecting the nature of a singular point in 2D vector fields. The first one is the so-called *Hopf bifurcation* and the second one is the pairwise annihilation or creation of a saddle and a source or sink, called fold bifurcation. Other local bifurcations (like those where an attracting and a repelling closed orbit appear simultaneously for instance) may also occur. Yet, for our purpose, we only pay attention to those that entail a nature transition for a singularity.

Hopf Bifurcation. We start with an attracting focus. If the attracting effect of this sink weakens, the number of streamline rotations around this critical point increases (the convergence "slows down"). This corresponds mathematically to an increasing negative real part of the complex conjugate eigenvalues of the Jacobian matrix at the critical point. When the real part is zero, we get a center point, which is an unstable structure: A bifurcation has occurred. If the real part increases further, a new stable structure appears which consists of an attracting closed orbit typically moving away from the critical point. The critical point itself has transformed into a repelling focus (i.e. a source). So, a sink has changed into a source with the emission of a cycle: This is a Hopf bifurcation. An illustration of this evolution is given in Fig. 2. Inverting the direction of time, we get the transition from a cycle containing a repelling focus into an attracting focus over an instantaneous center. Similar transitions are obtained by inverting the direction of the flow (that is by replacing sources by sinks and vice versa).

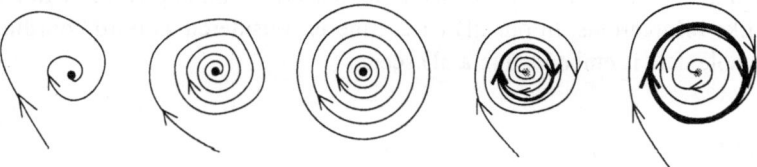

Fig. 2. Hopf bifurcation

Fold Bifurcations. At the beginning, there are a saddle point and a singularity of index +1, say a sink, that are linked by a separatrix. If the attraction/repulsion relation between both singularities along the separatrix weakens (i.e. the real positive eigenvalue of the saddle point, corresponding to the considered separatrix, decreases and the

120

negative real part of one of both sink's eigenvalues increases), both critical points become closer and closer until they meet. At this point, an unstable critical point appears with a zero eigenvalue: A fold bifurcation has occurred. As time goes on, this unstable structure vanishes: The new stable structure contains no singularity: This is a pairwise annihilation or fold catastrophe (see Fig. 3). Inverting the direction of time, we get a pairwise creation. Similar transitions are obtained by inverting the direction of the flow (the sink becomes a source).

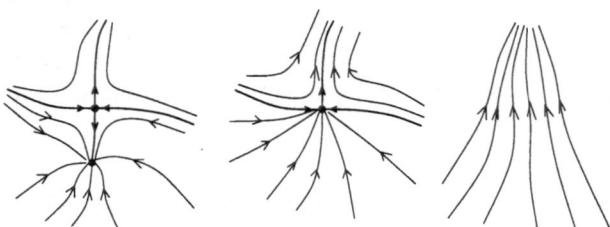

Fig. 3. Pairwise annihilation

Global Bifurcations. As opposed to the cases mentioned above, global bifurcations do not take place in a small neighborhood of a singularity but entail significant changes in the flow structure and involve large domains. We shall not go into details but present an example known as *basin bifurcation*. The essential aspect of such a bifurcation is the saddle/saddle connection as unstable intermediate configuration. If we focus on the basin bifurcation, the situation is as follows. We start with two saddle points that are not connected. This is a stable structure. Progressively, two separatrices become closer and closer until they meet: Both saddle points are connected through this heteroclinic separatrix. This is the bifurcation point. Right after this moment, the separatrix splits into two distinct separatrices which constitutes a swap, compared to the initial configuration.

4 Data Representation

The visualization of time dependent vector data has to deal with a higher dimensional mathematical space where time constitutes an additional dimension. This space must be handled as a continuous one to enable the detection and depiction of fundamental features like bifurcations. In our 2D case, time is considered as third coordinate axis and the whole data is embedded in a 3D scene.

4.1 Grid Construction

Practically, we process a 2D vector field lying on a curvilinear grid with constant positions through time: We dispose the several instantaneous states of the field parallel to another (each of them is called *time plane* in the following), corresponding to their position along the time line. A 3D grid evolves by connecting these planar grids together with 3D cells as shown in Fig. 4(a). There are several possible cell structures for connecting the successive time planes. Since the data points are (in 2D) lying on a

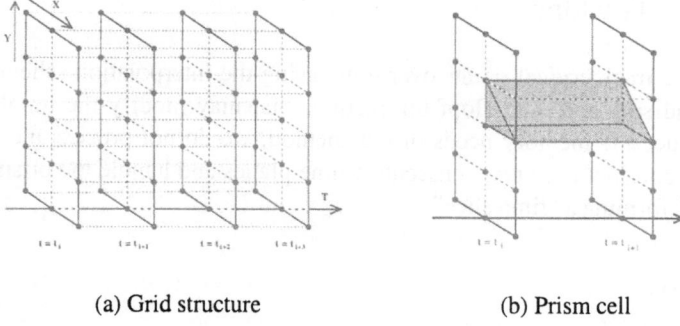

(a) Grid structure   (b) Prism cell

Fig. 4. Space/Time grid

structured grid, a natural choice would be to adopt a 3D structured grid made of hexahedron cells. Nevertheless, for the simplicity and efficiency of singularity tracking (see section 5.2), we use prism cells: We first triangulate each time plane and connect the corresponding triangles by translation through time, as in Fig. 4(b). Note that we prefer this cell structure to a simple tetrahedrization of the structured point set because it preserves the structure continuity of each triangle over time.

4.2 Interpolation Scheme

The input data are 2D vectors. Our task consists in interpolating a 2D vector field on a 3D grid. Mathematically, we consider the following map

$$f : I\!R^3 \supset G \longrightarrow TI\!R^2 \approx I\!R^2 \subset I\!R^3$$
$$(x, y, t) \mapsto v(x, y, t) \equiv (v_1(x, y, t), v_2(x, y, t))$$

where G is the 3D grid. Since we need consistency with the piecewise linear interpolation applied on the 2D triangulation, we must ensure that the restriction of the 3D interpolant to each time plane is piecewise linear. This means that if we fix the time coordinate and take it as a parameter, the interpolant must be linear. This is the reason why we choose the following interpolant inside each prism cell.

For a given prism cell lying between $t = t_i$ and $t = t_{i+1}$, let $f_j(x, y) = a_j + b_j x + c_j y$, $j \in \{i, i + 1\}$ be the affin linear interpolants corresponding to the prism triangle faces lying in the planes $\{t = t_i\}$ and $\{t = t_{i+1}\}$ respectively. Then we define the interpolant over the whole prism cell by linear interpolation over time:

$$f(x, y, t) = a(t) + b(t)x + c(t)y$$

where

$$a(t) = \frac{t_{i+1} - t}{t_{i+1} - t_i} a_i + \frac{t - t_i}{t_{i+1} - t_i} a_{i+1}$$

(idem for b and c). This formula obviously ensures, for each fixed value t, that f_t is affin linear in x and y.

5 Singularity Tracking

In this section, we track critical points over time, using the interpolation scheme presented above, and seek fold and Hopf bifurcations that may modify the topological structure. To reduce the memory needs of our method, we do not process the whole grid at once: We consider only two consecutive time planes and handle the prism cells connecting them, forming a "time slice".

5.1 Cell Analysis

Singularity Path The interpolant has been chosen so that it contains at most one critical point for a fixed value of t in a cell. This is due to the affin linear nature of its restriction to any time plane. Straightforward calculus leads to the following singularity coordinates $(x(t), y(t))$.

$$\begin{cases} x(t) = |-a(t)\ \ c(t)|\ \ |b(t)\ \ c(t)|^{-1} \\ y(t) = |b(t)\ \ -a(t)|\ \ |b(t)\ \ c(t)|^{-1} \end{cases}$$

If t moves from t_i to t_{i+1}, the singularity position describes a 3D curve. Yet, we are only interested in the curve sections that intersect the interior domain of the considered prism cell. A simple way to determine them is to consider the singularities lying on the side faces of the prism: Two triangular faces lying in $t = t_i$ and $t = t_{i+1}$ and 3 quadrilaterals connecting them. Finding the position of a singularity in each prism face requires the solution of a simple linear (resp.quadratic) system. If this is done, we sort the found (3D) positions in ascending time order and associate them pairwise. Indeed, since only one critical point can be present in a prism cell at a given time t, we know that a critical point must first leave the cell before it reenters it later. So, for each pair, we identify an entry and an exit position.

Path Type We now need to pay attention to the singularity types to complete the topological information. The generic types of linear singularities are *source*, *sink* or *saddle* (omitting *center* points that correspond to a transition between source and sink). The transition from one type to another constitutes a bifurcation. In section 3, we have shown that one needs to focus on two bifurcation types and their inverses: fold and Hopf bifurcation. Since a prism cell contains at most one critical point for a given t, we can assert that no fold bifurcation (involving two critical points) can occur inside it: *A fold bifurcation must occur at the cell boundary*. Consequently, we only seek Hopf bifurcations occurring between two consecutive entry and exit points. Practically, the determination of a singularity type is based upon the eigenvalues of the Jacobian at its position. In our case, decomposing the vectors b and c in the canonical basis ($b = b^0 e_0 + b^1 e_1$, idem for c), this matrix is

$$J(t) = \begin{pmatrix} b^0(t) & b^1(t) \\ c^0(t) & c^1(t) \end{pmatrix}.$$

Thus we compute the Jacobian matrix and its associated eigenvalues at each entry and exit point and check if they are the same. If this is the case, we can assert that no Hopf bifurcation has occurred since, otherwise, at least two bifurcations have taken place, which is impossible since our interpolant varies linearly over time. If the type has changed, a Hopf bifurcation is on the way. Because a Hopf bifurcation corresponds to a transition from a repelling to an attracting nature, the instantaneous nature of the intermediate singularity is a center and the trace of the Jacobian is zero at this point, which we use as criterion to determine the exact time location of the bifurcation.

5.2 Tracking

At this stage, every cell knows the successive entry and exit positions of the singular paths through its interior domain, as well as the possible presence of a Hopf bifurcation on the way. Yet, this information is scattered and must be reconnected to offer a global view of the topology evolution over time. A fundamental aspect of this task is to detect and identify the bifurcations that may take place on the faces. They are detected when connecting singular path sections lying in neighbor cells as we show next.

We start in the time plane $\{t = t_i\}$. The restriction of the grid to this plane is a triangulation. In a boolean array, we indicate for each triangle if its corresponding prism must be further inspected or not. If it has to be inspected, we check the information collected during the cell analysis for a path section starting at an entry point located on the considered triangle face. This provides us with an exit point at which the tracking must be proceeded. This exit point can either lie on a side (quadrilateral) face of the prism or on the triangle face lying in $\{t = t_{i+1}\}$. In the latter case, we signify in the boolean array of the next time plane $\{t = t_{i+1}\}$ that this triangle must be further inspected. If the path left the cell at $t^* < t_{i+1}$ (side face), then it entered a neighboring cell at the same time. Back to a 2D representation, we get in time plane $\{t = t^*\}$ a singular point lying on the common edge of both neighboring triangle cells. Due to the discontinuity of the Jacobian matrix through this edge, the singularity may have another type when considered from the other side. This simple argument explains the possible existence of a bifurcation in such cases. Two situations may actually occur.

- There is a simple crossing of a critical point from one cell to another cell where no singularity was present so far. In this case, the type may change but, because of the index invariance, a saddle remains a saddle (index -1). Practically, a source can become a sink and vice versa (both index +1). As mentioned previously, the transition source/sink or sink/source is a Hopf bifurcation. We easily detect it by checking if the type of a source (resp. a sink) path changes after passing the face.
- The second situation corresponds to a merging of two coexisting critical points in neighboring cells on their common face (common edge in 2D). Now, an important property of the piecewise affin linear interpolant over a triangulation is that two neighboring triangles cannot contain two singularities with same index [8] which implies that only 2 singularities with opposite index can merge in that way: A saddle and a source or a saddle and a sink. This type of bifurcation has been considered in section 3: It is a fold bifurcation or its inverse.

When we leave the current cell through a side face, and a possible bifurcation has been detected, we identify this position. We proceed with tracking the path by asking the neighbor cell for the next exit position. We get three possible answers.

1. The exit position lies in the time plane $\{t = t_{i+1}\}$: We have reached the next time plane. The corresponding triangle face is marked `true` in the boolean array corresponding to the next time plane.
2. The "exit" position lies in the time plane $\{t = t_i\}$: If we were tracking the current path forward so far, we have found a pairwise annihilation. In this case, the corresponding triangle is set to `false` in the boolean array of $\{t = t_i\}$, indicating that this cell has been processed now.
3. The exit position lies on a side face for $t = t^{\#}$. If $t^{\#} < t^*$ and we were tracking the current path forward so far, we have found a pairwise annihilation and proceed with backward tracking. If $t^{\#} > t^*$ and we were tracking the current path backward so far, we have found a pairwise creation and proceed with forward tracking: We are back to the current situation.

6 Separating Surfaces

In the 2D steady case, separatrices are curves that emanate from saddle points. Now, as the saddle points move through the 2D/time grid, so do their corresponding separatrices that describe *separating surfaces*. To depict them, we use the information gained about the singular points in the tracking step.

Practically, given a saddle path, we save, at its first position in time and for each of its four separatrices, the following information: Starting vector, associated direction and index of the singular path reached if any or last position obtained by numerical integration and corresponding case: `boundary` (one left the grid) or `cycle` (one reached a closed orbit). At every position along the saddle path, we associate each separatrix with one at the previous position by taking its best approximation in term of starting vector and direction. After that, if both corresponding separatrices reach the same singular path, the same cycle or close positions on the grid boundary, we add the new separatrix to the surface spanned by the old one (we add a new "ribbon" to this surface). Otherwise, we check if the path reached previously has ended at a bifurcation point. In this case, we end the previous surface at the bifurcation by integrating a separatrix at the exact time position of the bifurcation and start a new one at the current separatrix. If no bifurcation has occurred but the connectivity has changed, then we face a global bifurcation that could not be detected while tracking the singularities: We simply end the surface at the previous separatrix and start a new surface at the current separatrix. Doing this for each discrete position along the saddle path, and that for each saddle path, we depict eventually all separating surfaces in the domain.

7 Results

To test our method, we have created an analytic vector field containing four critical points, of which position is a function of time, describing closed curves in the plane. We have sampled this vector field on a rectilinear point set for several values of the time parameter. The rotation of the critical points (each with a specific frequency) entails many structural changes for the topology, which is very interesting for our purpose since bifurcations are present and may be observed with our method. We show first the results of the singularity tracking step: The path of each critical point through time has

125

been tracked as well as all the local bifurcations taking place (indicated by small balls). The grid is displayed to give an impression (see Fig. 5). If one focuses on a particular bifurcation, one can observe how the separatrices evolve through the bifurcation point. In the case of a Hopf bifurcation for instance, we consider the picture (without surfaces drawn to ease interpretation) in Fig. 6: The creation of a closed orbit can be easily seen. At last, as an illustration of the topological pictures our method can produce, Fig. 7 proposes an overview of the whole topology evolution in the same perspective as in Fig. 5. The breaks that can be observed on the surfaces correspond to structural transitions associated with bifurcations. The two colors used for the surface depiction refer to the stable and unstable directions of the saddle points, respectively.

8 Conclusion

We have presented a topology-based method for the visualization of time-dependent planar vector fields. The topological graph is tracked over time. The interpolation of the discrete planar data over time enables the precise detection of local bifurcations that result in dramatic structural changes. Thus, singularity paths are displayed as well as the bifurcations that affect them. These paths are connected in 3D by separatrix surfaces that provide the structure of the visualized topology.
We think that this technique offers a general framework for the visualization of parameter-dependent planar vector fields.

References

[1]   Helman J.L., Hesselink L., *Representation and Display of Vector Field Topology in Fluid Flow Data Sets*. Visualization in Scientific Computing, G.M. Nielson & B. Shriver, eds., 1989.

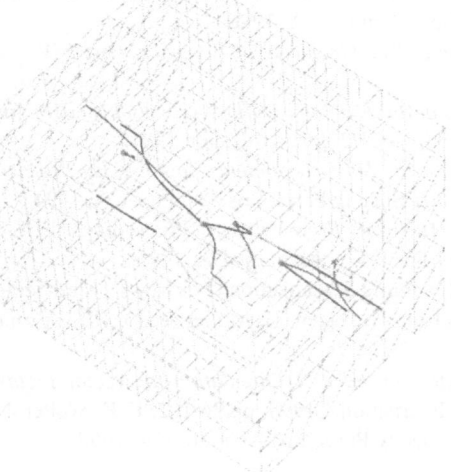

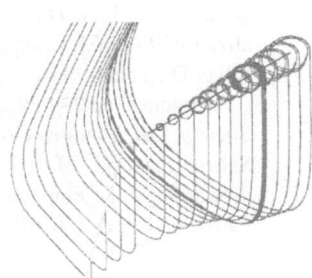

Fig. 5. Singular paths and bifurcations   **Fig. 6.** Hopf bifurcation (lines)

126

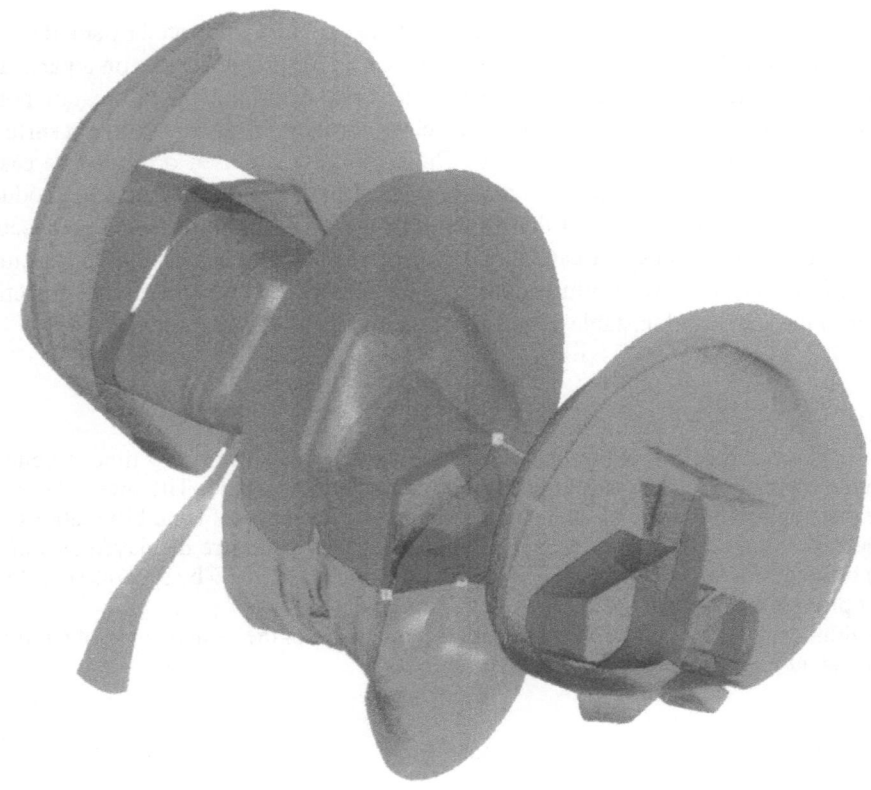

Fig. 7. Overview of the topology evolution

[2]   Helman J.L., Hesselink L., *Surface Representation of Two- and Three-Dimensional Fluid Flow Topology.* Proceedings of the First IEEE Conference on Visualization, pp.6-13, IEEE Computer Society Press, Los Alamitos CA, 1990.

[3]   Abraham R.H., Shaw C.D., *Dynamics The Geometry of Behavior, Part I-IV.* Aerial Press, Santa Cruz, CA, 1982-1988.

[4]   R.R. Dickinson *Interactive Analysis of the Topology of 4D Vector Fields.* IBM Journal of Research and Development, Vol. 35, No. 1/2, January/March 1991.

[5]   Lane D.A., *UFAT - A Particle Tracer for Time-Dependent Flow Fields.* Proceedings IEEE Visualization'94, IEEE Computer Society Press, Los Alamitos CA, 1994.

[6]   Becker B.G., Lane D.A., Max N.L., *Unsteady Flow Volumes.* Proceedings IEEE Visualization'95, IEEE Computer Society Press, Los Alamitos CA, 1995.

[7]   Silver D., *Feature Visualization* In Scientific Visualization Overviews - Methodologies - Techniques, pp.279-293, G.M. Nielson, H. Hagen, H. Müller (eds.), IEEE Computer Society, Los Alamitos CA, 1997.

[8]   Scheuermann G., Hagen H., *A Data Dependent Triangulation for Vector Fields.* In proceedings of Computer Graphics International 1998, pp.96-102, F.-E. Wolter, N. M. Patrikalakis (eds.), IEEE Computer Society Press, Los Alamitos CA, 1998.

Editors' Note: see Appendix, p. 342 for colored figures of this paper

Virtual Colon Flattening

A. Vilanova Bartrolí[1], R. Wegenkittl[2], A. König[1], E. Gröller[1], and E. Sorantin[3]

[1] Institute of Computer Graphics and Algorithms
Vienna University of Technology
[2] Tiani Medgraph
[3] Section of Digital Information and Image Processing,
Department of Radiology, University Hospital Graz
Austria

Abstract. We present a new method to visualize virtual endoscopic views. We propose to flatten the organ by the direct projection of the surface onto a set of cylinders. Two sampling strategies are presented and the introduced distortions are studied. A non-photorealistic technique is presented to enhance the perception of the images. Finally, an approximate but real-time endoscopic fly-through is possible by using the data obtained by the projection technique.

1 Introduction

Virtual endoscopy deals with the inspection of hollow organs and anatomical cavities using medical imaging (e.g. CT and MRI) and computer visualization techniques. Virtual endoscopy has the potential of becoming a substitute of real endoscopy for some diagnostic procedures. A real endoscopy is invasive and, furthermore, involves a certain degree of risk for the patient.

Most of the virtual endoscopy techniques presented in the last years [1–3] concentrate on simulating the view of a real endoscope. This is the view that the endoscopists are used to. It can be useful for certain applications, like in an intraoperative scenario, but it is not necessarily the best way to inspect the inner surface of an organ. Actually, a real endoscope and organ are subject to physical limitations that a virtual endoscope and organ do not have. This paper considers virtual colonoscopy, which focuses on the examination of the colon. Physicians are mainly interested in visualizing the inner surface of the colon which is where polyps can be detected with endoscopy. It is important that the physician can estimate the size of polyps, since large polyps are more likely to develop into malignities. The usual endoscopic view visualizes just a small part of the surface. Furthermore, it is difficult to detect polyps that are situated behind the folds of the colon. An efficient way to inspect the inner surface would be to open and flatten the colon and then examine its internal surface. Unfortunately, this cannot be done in reality if we want that the patient survives. On the other hand, there is no patient damage if this dissection of the organ can be achieved virtually with the medical data obtained by CT or MRI (i.e. the virtual organ).

Some authors proposed a technique to straighten and unravel an organ virtually [4][5]. Their approach starts with defining a path which is placed as close to the center of the object as possible. Then a sequence of frames is calculated. For each frame, a cross-section orthogonal to the path tangent is calculated. Then the central path is straightened and the cross-sections are piled to form a stack. As a last step the straightened colon is flattened obtaining a volume model of the flattened colon. The model is displayed afterwards using standard volume rendering techniques. This method allows to visualize the complete surface at once. One of the main problems of this technique appears in high curvature areas of the central path, i.e. at path locations where the radius of curvature is bigger than the organ diameter. In such cases orthogonal cross-sections intersect each other or are far apart in some other regions (see figure 1). As a consequence, a polyp can appear twice in the flattened model or it can be missed completely. Wang et al. in a later work [6] try to overcome this problem. They use electrical field lines generated by a locally charged path to govern curved cross-sections instead of the planar sections. The cross-sections tend to diverge avoiding conflicts, but the technique cannot ensure that they will not intersect. Haker et

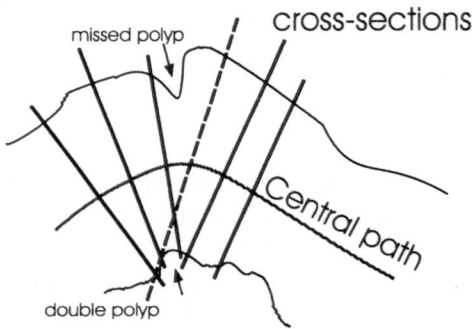

Fig. 1. Illustration of the possible undersampling and double counting of polyps due to intersections of the cross-sections in high curvature areas. The dash cross-section line produces a double counting of the polyp.

al. [7] use conformal mapping which is angle preserving to project the colon surface colored with the gaussian curvature to a plane. One of the main problems of this method is that a highly accurate segmentation is necessary to ensure good results in case they are used for diagnosis.

Paik et al. [8] propose other kinds of camera projections for virtual endoscopy. With a normal endoscopic view just 8% of the solid angle of the camera is seen in each frame. Paik et al. project the whole solid angle of the camera by map projection techniques used to map the world globe in charts. They suggest the use of the Mercator projection to map the solid angle to the final image. This technique samples the solid angle of the camera, then the solid angle is mapped onto a cylinder which is mapped finally to the image.

All of these methods introduce some kind of deformation since it is mathematically impossible to perform a mapping between two surfaces preserving angles and area at the same time if the two surfaces do not have the same gaussian curvature.

In section 2, we propose a method to flatten the colon using a new camera projection technique. Section 3 presents different sampling options that cause different deformation problems. A minimization of the rotation for the camera movement is described in section 4. Section 5 describes a non-photorealistic technique that enhances the perception of the image. Then it is presented how an approximate but fast endoscopic view can be generated with the data calculated for the projection method. Finally, some results and studies with colon data are presented.

2 Method Overview

The methods proposed by Wang et al. [6] generate a flat model of the colon that later on will be carefully inspected by the physician. Our method will not generate a flat model of the whole colon, but allows to inspect locally flattened regions such that double counting of polyps does not occur.

The presented method involves moving a camera along the central path of the colon. The central path can be calculated using one of the common techniques used to skeletonize an object. We used a thinning algorithm which ensures topological preservation of the object (see Vilanova et al. [3]). The path is smoothed and finally approximated by a B-spline curve.

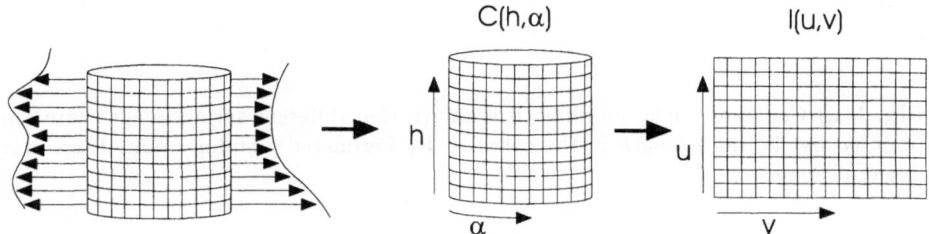

Fig. 2. Illustration of the projection procedure. A region of the surface is projected to the cylinder $C(h, \alpha)$. Then the cylinder is mapped to the image $I(u, v)$

For each camera position a small cylinder tangent to the path is defined. The middle point of the cylinder axis corresponds to the camera position. The length of the cylinder axis has a constant value for all camera positions. The length is defined by inspecting the camera path, and calculating the distance following the tangent between the path position and the colon surface. The length is defined by the distance that in any camera position ensures that the axis will not get out of the colon.

Rays starting at the cylinder axis and being orthogonal to the cylinder surface are traced (see figure 2). For each ray, direct volume rendering compositing is used to calculate the color which corresponds to the cylinder point where the ray was projected. Finally, the colored cylinder with the sampled rays is mapped to a 2D image. This is done by simply unfolding the cylinder.

The result is a video where each frame shows the projection of a small part of the inner surface of the organ onto the cylinder. If the camera is moved slowly enough the coherence between frames will be high and the observer will be able to follow the movement of the surface.

In high curvature areas also the problem which corresponds to the intersection of cross-sections (see figure 1) appears. In the presented method, possible double sampling of polyps emerges just between frames. However, it does not cause a double counting of polyps since the human brain is able to track the polyp movement due to the coherence between frames. Moving along the central path in such a high curvature area, a polyp might move up and down (due to double sampling) but is clearly identified as a single object.

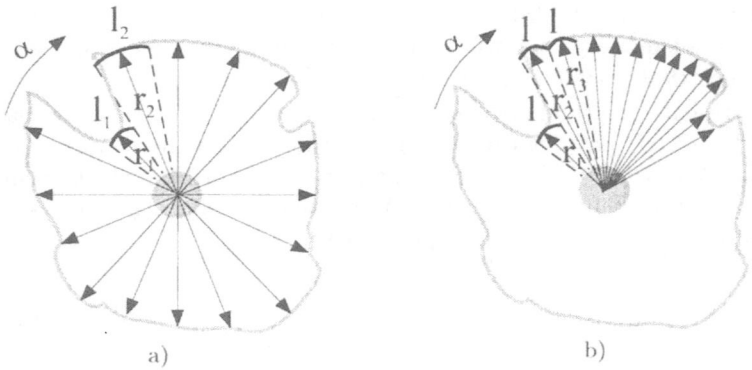

Fig. 3. a) Constant angle sampling: it is shown that different surface lengths are represented by the same length in the cylinder. **b)** Perimeter sampling: same length but different angle.

3 Projection onto a Cylinder

The proposed projection is illustrated in figure 2. A cylinder $C(h, \alpha)$ is defined for each camera position. This cylinder is colored by tracing rays orthogonal to the cylinder surface (i.e. projecting a region of the surface onto the cylinder). Then the cylinder is mapped to the final image $I(u, v)$ by a simple mapping function $f : (h, \alpha) \rightarrow (u, v)$.

The sampling distance (i.e. the distance between two consecutive rays) in the h-direction is constant, and it must be at most half of the size of a voxel (see figure 2). In this way the Nyquist frequency in the h-direction is preserved.

For each h-value the rays are traced in radial directions with respect to the cylinder axis. The rays are diverging from each other, so the volume data is not sampled uniformly if the incremental angle is constant (see figure 3a). In the next section, two methods are described which project the organ surface onto the cylinder depending on the sampling of angle α : constant angle sampling and perimeter sampling.

3.1 Constant Angle Sampling

Constant angle sampling means that the angle between consecutive rays in α direction is constant for rays with the same h-value. Figure 3a illustrates how this sampling is done. Using this method, the cylinder is sampled uniformly but the surface itself is not uniformly sampled.

The advantage of this method is that the relation between both directions is locally preserved. Therefore the angles are locally preserved. An image generated by this method can be seen in figure 4a.

On the other hand, the area of the projected region is not preserved (see figure 3a). Therefore, the dimension of the projected polyps depends on the proximity of the cylinder axis and the diameter of the cavity. Consequently, the physicians cannot trust the sizes of the polyps.

Polyps can be missed with constant sampling (see figure 3a), if the angle increment is too large. If the sampling distance is too small the rays are traced where it would not be necessary. This makes the method inefficient.

3.2 Perimeter Sampling

In this section we propose a sampling strategy in which the rays are calculated so that the surface length that they represent is constant. A constant sample

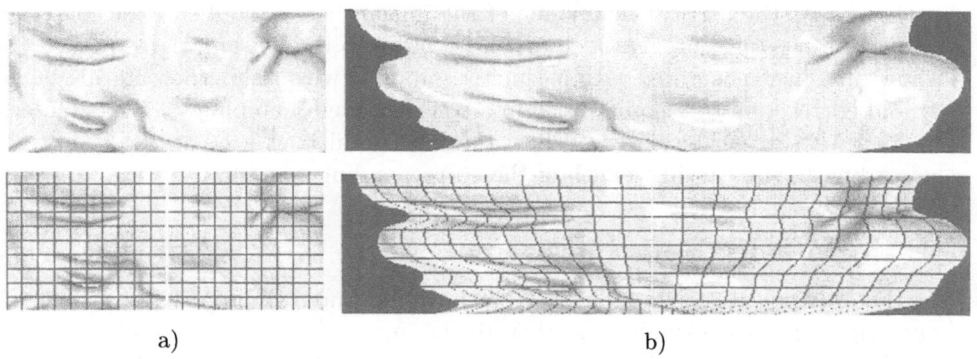

a)   b)

Fig. 4. a) Resulting image of the projection technique using constant angle sampling. **b)** Same camera position as a) but with a perimeter sampling. The bottom images show a grid which was generated by fixing a constant angle value.

length l is defined. l must be at most half the size of a voxel to keep the Nyquist frequency and therefore not to miss any important feature. l should have the same value as the sampling distance in the h-direction to preserve the ratio, or proportion, in the final mapping.

The algorithm incrementally calculates the ray directions which are in the plane defined by a certain value of h. The angle between the current ray and the next one is computed such that the length of the surface sample that the current ray represents is l in the α-direction (see figure 3b). The first ray is traced along an arbitrary angle α_0. α_0 must be the same for each value of h. r is defined as the distance from the cylinder axis to the surface point hit by the ray. The surface sample length in the α-direction that a ray represents is approximated by the arc with radius r. Therefore, the value of the angle increment for the next ray is estimated as $\frac{l}{r}$ radians.

This projection method projects the organ surface to a generalized cylinder whose radii are not constant within the cylinder. In this case the mapping function f maps uniformly the contours and also the surface of the generalized cylinder. Moving along the central path, contours of varying length are represented by varying numbers of rays. In the mapping to the image plane this results in the fact that in the v-direction (horizontal scanlines) typically only part of the pixels are covered by an unfolded contour. Therefore, the generalized cylinder is not mapped to the complete domain of the image (see figure 4b). The function f maps each sampled ray to a pixel in the image (i.e. each pixel corresponds to an area of $l \times l$ of the surface). The projected points that correspond to the rays at angle α_0 are positioned on a vertical line in the center of the image. Then from left to right the ray values are mapped to the image until the perimeter length is reached.

This projection has the area preservation property. So the relative sizes of surface elements are preserved in the image plane and do not depend on the proximity of the cylinder axis to the surface. On the other hand, a distortion is introduced with respect to the h and α-direction, so the angles are not preserved anymore. At the vertical center line of the image there is no distortion, but the distortion increases progressively when we move to the left and right. Figure 4b shows an image generated with perimeter sampling with a superposed grid which would correspond to a regular grid in a constant angle sampling of the cylinder. In this way, it can be observed how the horizontal lines are varying in extent according to the varying length of the corresponding contour.

4   Minimally Rotating Frame

In the previous section, a technique has been presented to project the surface of the organ onto the cylinder and then to the image.

At each position of the camera in the central path an orthogonal coordinate system is taken which specifies the location and orientation of the cylinder. One coordinate axis is given by the tangent vector of the central path. The other axes are in the plane orthogonal to the central path at the camera position. Taking the Frenet frame is not a good choice for this coordinate system. Firstly, the Frenet frame is not defined in linear portions of the central path. Secondly, by moving along the path the two vectors orthogonal to the tangent vector are rotating more than necessary, thus reducing coherence between adjacent frames. Therefore we investigate a rotation-minimizing coordinate frame.

We implemented the rotation-minimizing coordinate frame presented by Klok [9]. The coordinate frame is obtained by solving the following differential equation:

$$z(s) = \frac{c'(s)}{\|c'(s)\|}$$

$$x'(s) = -(c''(s) \cdot x(s))\frac{c'(s)}{\|c'(s)\|} \tag{1}$$

$$y'(s) = -(c''(s) \cdot y(s))\frac{c'(s)}{\|c'(s)\|}$$

where $c(s)$ represents the parametric central path and $(x(s), y(s), z(s))$ is the coordinate frame we are looking for. An initial orthogonal frame (x_0, y_0, z_0) is defined. Then the differential equations are solved using fourth order Runge-Kutta method. Theoretically, equations 1 produce orthogonal coordinate frames. To avoid accumulation of numerical errors (i.e. orthogonality is not ensured anymore), we take the following approach: $z(s)$ and $x'(s)$ are calculated according to the above formulas. Then $y(s)$ is taken as the cross-product of $z(s)$ and $x(s)$ $(y(s) = z(s) \times x(s))$. Finally we also correct $x(s)$ by taking it as the cross-product of $y(s)$ and $z(s)$ $(x(s) = y(s) \times z(s))$

5 Level Lines Enhancement

Using the distance of the hit surface point to the cylinder axis r, we can generate a depth image (see figure 6a). The depth image together with the shaded image represent a high field, similar to a landscape in topography. A good way to visualize landscapes in topographical maps is showing level lines, where each line correspond to a level of depth. The level lines improve the perception of depth and surface changes of the map.

The level lines are generated from the depth image. Firstly, the gradient of the depth image is calculated using a first derivative of the Gauss filter. The level lines are drawn based on the technique described by Saito et al. [10].

In order to improve the perception, a hue shift is applied to the level lines color. The colors of the lines are coded depending on the level of depth (see figure 6c). Hue shift has the advantage that it does not interfere with the highlights and dark areas of a shaded image. Technical illustration artists commonly use the temperature of colors in their drawing. The temperature of a color is defined as warm (red, orange and yellow) and cool (blue, violet and green). The temperature also gives depth cue information since the perception of the cool colors recedes whereas the perception of the warm colors advances. The hue shift has been chosen yellow-blue since these colors have a large shift range, and red-green is undesirable due to color blindness. Yellow corresponds to closer level lines and blue to level lines far away.

Once the level lines have been obtained they can be combined with the original shaded image (see figure 6d). The color of the shaded image should not be in the range between yellow and blue to achieve a good contrast. The level lines provide information about the surface change and the distance of the surface to the cylinder axis.

6 Endoscopic View Generation

Once a polyp has been detected in the video of the flattened colon, the physician would be interested in seeing its location with an endoscopic view. Using the already calculated shaded images and depth images for each frame, a fast fly-through can be generated efficiently.

To generate the interactive navigation, the horizontal center lines of the depth images of the movie of the flattened colon are used. Knowing the camera location for each frame, the center lines can be backprojected to the 3D space (see figure 5a). A polygonal surface can be easily generated using triangle stripes. Each stripe corresponds to the triangles generated between one center line of the depth image and the center line of the next frame. We obtain a fast rendering since we use stripes and we render just the surface in the neighborhood of the camera (see figure 5). This can be achieved easily since the stripes are sorted by path position. With this method we achieve frame rates around 30f.p.s. with a Pentium II at 400 MHz with common OpenGL graphics hardware acceleration.

Each triangle can be colored by OpenGL with a correct lighting. Another option is to assign to each triangle vertex its color value calculated in the corresponding shaded image of the video of the flattened colon. The last option produces incorrect lighting but it has a better matching with the flattened colon images.

Obviously the resulting images are approximations and some artifacts appear due to the cross-section problem (see figure 1). However, it gives a good impression of the structure and it can be used by the physician to position the camera to the area that they want to visualize with a better quality but slower rendering.

7 Results

The images presented in this paper correspond to a CT data set of a cadaveric colon with a resolution of 381x120x632. The colon is 50 cm long and 13 artificial polyps were physically created in the cadaveric colon. These polyps had a size between 3.5x2.5 mm and 11.8x9.0 mm. Figure 6e shows an outside view of the segmented cadaveric colon and its central path together with the camera. The camera position corresponds to the image in figure 6g. Figure 6f is an endoscopic view moving the camera a bit backwards to show the same region as in figure 6g. It can be observed that the shape of the polyps is much more clear in the flattened images than in the endoscopic view figure 6f (please refer to http//www.cg.tuwien.ac.at/research/vis/vismed/ColonFlattening/) for the videos). The physicians who collaborate in this project could easily identify the polyps in the colon flattened images. Figures 4, 6b, 6d, 6g, and 6h show some of the polyps.

We also tested this method with a 256x256x311 CT data set of a colon. Figure 6i and 6j show the images generated from this data set. These figures were

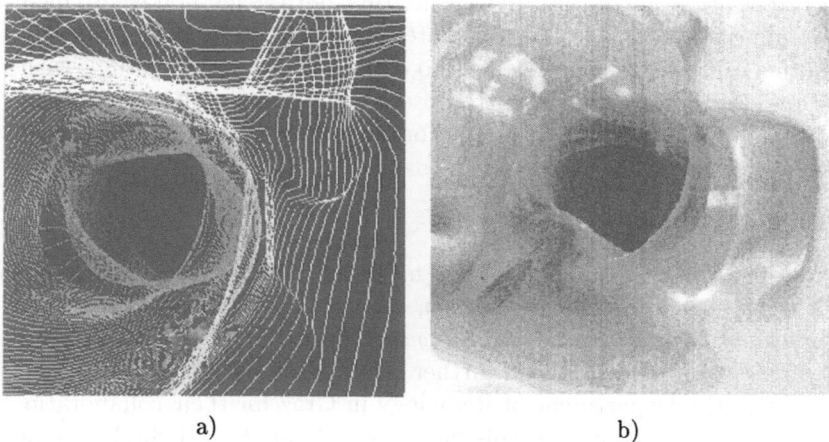

a)   b)

Fig. 5. a) Endoscopic view backprojecting the lines using the depth information and the camera frame. **b)** Endoscopic view using the backprojection of the generated shaded stripes.

generated from the same camera position but the projection was done with constant sampling and perimeter sampling respectively. Both sampling techniques can be useful to the physician. Perimeter sampling preserves the area and allows the physician to evaluate the size of the polyps. While constant angle sampling preserves the angles and allows a better evaluation of the shape.

8   Conclusions and Future Work

We presented a new technique for virtual colonoscopy which does not simulate the usual endoscopic view. The images are generated with a projection technique that allows the physician to visualize most of the surface, and to easily recognize polyps that in an endoscopic view would be hidden by folds or would be hard to localize. The presented method avoids double counting of polyps. We presented two sampling strategies that respectively preserve the angle or area of the projected surface elements. We maximized the coherence between frames by minimizing the camera rotation. The images are also enhanced by calculating level lines which represent the depth. Finally we presented a technique to generate a real-time endoscopic view navigation by using the data of the video of the flattened colon.

As future work it is planned that the doctor is able to go back to the original data once the polyp has been detected. This can easily be done using the camera position of each frame of the video of the flattened colon.

It is also important that the cylinder axes do not get outside the organ. The cylinder height could be optimized for each data set and even adaptively defined depending on the camera position. The camera movement has to be specified in a way that we do not miss any surface region. In the current implementation, the camera steps are so small that they ensure this, but the method is also inefficient because unnecessary images are calculated.

Another subject of future work is to extensively test the method with data of real patients with pathologies, to observe how the algorithm performs. The method might also be applied to other organs.

Acknowledgements

The work presented in this publication has been funded by the $V^{is}M^{ed}$ project. $V^{is}M^{ed}$ is supported by *Tiani Medgraph*, Vienna (http://www.tiani.com), and the *Forschungsförderungsfonds für die gewerbliche Wirtschaft*, Austria. See http://www.vismed.at/ for further information on this project.

We thank the Department of Radiology in Graz for their collaboration and for providing the data used in this paper. We thank Jiří Hladůvka for his collaboration concerning image processing techniques.

References

1. L. Hong, S. Muraki, A. Kaufman, D. Bartz, and T. He. Virtual voyage: Interactive navigation in the human colon. In *SIGGRAPH 97 Conference Proceedings*, Annual Conference series, pages 27–34. ACM SIGGRAPH, Addison Wesley, August 1997.
2. M. Wan, Q. Tang, A. Kaufman, Z. Liang, and M. Wax. Volume rendering based interactive navigation within the human colon. In *IEEE Visualization '99*, pages 397–400. IEEE, nov 1999.
3. A. Vilanova, A. König, and E. Gröller. VirEn: A virtual endoscopy system. *Machine GRAPHICS & VISION*, 8(3):469–487, 1999.
4. G. Wang and M.W. Vannier. GI tract unraveling by spiral CT. In *Proceedings SPIE.*, volume 2434, pages 307–315, 1995.
5. E. Sorantin, E. Balogh, K. Palagy, G. Werkgartner, E. Spuller, and S. Loncaric. *MEDICAL RADIOLOGY - Diagnostic Imaging*, chapter "Technique of Virtual Dissection of the Colon based on Spiral CT data". Springer Verlag Press, 2001.
6. G. Wang, S.B. Dave, B.P. Brown, Z. Zhang, E.G. McFarland, J.W. Haller, and M.W. Vannier. Colon unraveling based on electrical field: Recent progress and further work. In *Proceedings SPIE*, volume 3660, pages 125–132, May 1999.
7. S. Haker, S. Angenent, Allen Tannenbaum, and R. Kikinis. Nondistorting flattening maps and the 3d visualization of colon CT images. *IEEE Transactions on Biomedical Engineering*, 19(7):665–671, July 2000.
8. D.S. Paik, C.F. Beaulieu, R. B. Jeffrey, Jr. C.A. Karadi, and S. Napel. Visualization modes for CT colonography using cylindrical and planar map projections. *Journal of Computer Tomography*, 24(2):179–188, 2000.
9. F. Klok. Two moving coordinate frames for sweeping along a 3D trajectory. *Computer Aided Geometry Design*, 3:217–229, 1986.
10. Takafumi Saito and Tokiichiro Takahashi. Comprehensible rendering of 3-D shapes. In Forest Baskett, editor, *SIGGRAPH'90 Conference Proceedings*, Annual Conference Series, pages 197–206, August 1990.

Editors' Note: see Appendix, p. 343 for colored figure of this paper

Improved visualization in virtual colonoscopy using image-based rendering

Iwo Serlie[1,2], Frans Vos[1,3], Rogier van Gelder[4], Jaap Stoker[4],
Roel Truyen[5], Frans Gerritsen[5], Yung Nio[4], Frits Post[2]

[1]Department of Applied Physics, Delft University of Technology,
Lorenzweg 1, 2628 CJ Delft, The Netherlands

[2]Faculty of Information Technology and Systems, Delft University of Technology,
Zuidplantsoen 4, 2628BZ Delft, The Netherlands

[3]Department of Radiology, Erasmus University Medical Center,
P.O. Box 1738, 3000 DR, Rotterdam, The Netherlands

[4]Department of Radiology, Academic Medical Center,
P.O. Box 22700, 1100 DE Amsterdam, The Netherlands

[5]EasyVision Advanced Development, Philips Medical Systems Nederland B.V.,
P.O. Box 10.000, 5680 DA Best, The Netherlands

Abstract

Virtual colonoscopy (VC) is a patient-friendly alternative for colorectal endoscopic examination. We explore visualization aspects of VC such as surface in view, navigation and communication of a diagnosis. A series of unfolded cubes presents an animated full 360-degree omnidirectional field-of-view to the physician, to facilitate thorough and rapid inspection. For communication between physicians a tool has been designed that uses image-based rendering. Clinical evaluation has shown a reduction in inspection time from 19 minutes to 7 minutes without loss of sensitivity. With current virtual colonoscopy using a 2-sided view only 94% of the surface is available for exploration. In our approach the surface in view is increased to potentially 100%. Thus, the entire colon can be explored with better confidence that no regions are missed.

1 Introduction

Colon cancer is the second cause of cancer deaths in the Western world [1]. Early detection of colonic polyps has proven to lead to a decrease in incidence [2]. However, the traditional technique for detection and removal of polyps, optical colonoscopy, causes serious discomfort. In the past few years, virtual colonoscopy (VC) has been developed as a more patient-friendly screening alternative [4][5][6][7].

In general, the VC procedure consists of the following stages. First, the patient's colon is cleansed and inflated with air. Next, a 3D image volume of the abdomen is acquired using CT imaging. The bowel surface is then extracted from the volume and visualized in a way similar to camera colonoscopy. Finally, the physician navigates through the virtual colon and examines the surface for abnormalities from intra-luminal perspective. The last step may take as long as 30 minutes.

For practical application, virtual colonoscopy must meet general requirements regarding:

- efficiency (time spent by the radiologist)
- effectiveness (part of the surface area in view)
- sensitivity (number of polyps detected)

Because of the trade-offs between accuracy, speed and hardware costs, volume rendering is not always feasible at the required quality level. Current systems often use fly-through sequences to reduce the time needed for diagnostic inspection. Such fly-throughs are often based on an automatically generated path in the center of the lumen, with evenly-spaced views in both forward and backward viewing directions. This approach still yields selective imaging of the inner surface. It may well be however that important parts of the surface are missed, while insignificant parts are reviewed twice (Figure 1). In addition, the inspection/navigation is restricted by the central path through the colon.

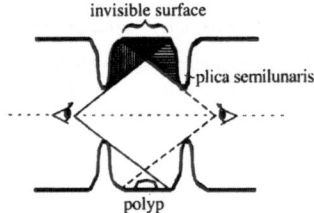

Figure 1. Surface parts missed in two sided views.

Identified polyps can be removed by the endoscopic surgeon. For proper communication of the diagnosis to the surgeon it is required that properties of detected polyps such as their location, size and appearance are reported.

So-called panoramic views have been considered to solve the described missed surface parts. Such an approach implies folding out the colon, and spreading the inner wall on a flat plane, so that the full inner surface is exposed for examination. To do this, several techniques are described in the literature, ranging from simple cylindrical coordinate transforms, to complex conformal mappings [8][9]. The cylindrical transformation produces excessive distortions, while local geometry is much better preserved in a conformal transformation. We are not aware of any work on tools for communication of diagnosis.

In this paper, a number of visualization aspects from virtual colonoscopy will be analyzed. We will introduce a novel visual representation. A series of unfolded cubes presents an animated full 360-degree omnidirectional field-of-view to the physician, to facilitate a more thorough and rapid inspection. Consequently, almost the entire colon surface is visible for examination. For annotation and communication of a diagnosis, a tool has been designed using a selected series of views that are generated by image-based volume rendering. To test our technique, a clinical evaluation study has been conducted.

The paper is organized as follows. In Section 2.1, we will review image-based rendering, and its use in virtual colonoscopy. Subsequently, we will consider the visible part of the surface and navigation aspects (Section 2.2). In Section 2.3 we will discuss the communication of a diagnosis. Results including the clinical evaluation will be presented in Section 3. Finally, the paper will draw conclusions and describe suggestions for future work.

2 Methods

2.1 Image-based rendering

The common approach to image generation in virtual endoscopy is by rendering of polygonal iso-surfaces extracted by the marching cubes algorithm [10]. Alternatively, direct perspective volume rendering of the acquired CT volume can be used [11][12] as in this study. To acquire fast rendering speed needed for screening purposes, concessions are often made to quality and scene complexity.

An alternative approach to the classical type of scene modeling is provided by image-based rendering [13]. Image-based rendering is an approach in which pre-rendered (or captured) images are used to compose a virtual scene. The scene is modeled by a collection of intermediate images, which are used to generate different views. The images are warped in real time to simulate camera panning and zooming. Thus, rendering time does not depend on the complexity of the scene, but is only determined by the number of intermediate images and their resolution.

As a more formal description, consider the intensity and chromaticity of light observed from every position and direction in space. As a result, the environment viewed from a fixed position is represented by a two-dimensional map, which can be projected on the surface of a virtual viewing space. The camera is placed at the center, and the images are projected on the walls of the viewing space.

Typical examples of viewing spaces are a cylinder, a sphere, or a cube. We have used a cubic mapping, with the viewpoint at the center, and on each face an image is projected with a 90-degree viewing angle, to cover the full omnidirectional projection of the scene. The six images are warped in such a way that the pixel grid appears regular for each view, regardless of the viewing direction, and no irregularities appear at the edges or corners of the cube.

The VRML language [14] allows the modeling of an image-based scene description using a cube. Rendering this representation consists of generating six images, volume rendered in perspective (Figure 2), and texture mapping these images on the faces of the viewing cube.

In general, a limited amount of disk space and main memory is needed for reconstruction and archiving such a cubic representation. Consequently, image-based rendering gives the opportunity to communicate information via a low-bandwidth network and displayed using standard software on a low-end PC. Real-time interactive exploration is supported, but the full environment is not visible in a single view, and the user has to look around to inspect the whole scene. From visual perception research it is well known that active exploration leads to better information transfer. Therefor, it

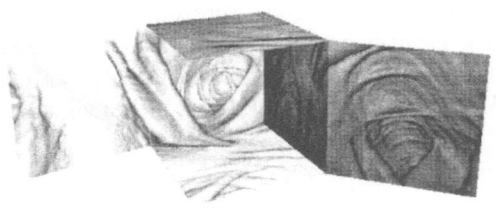

Figure 2. Image based rendering using a cubic environment map.

is desirable that visualization of the environment is invariant to the orientation of the camera.

2.2 Visible surface, orientation/navigation

Increasing the viewing angle enlarges the amount of surface inspected. However, this approach goes at the expense of increased distortions towards the edges of the rendered image. Clearly, at an angle of 180 degrees, half the scene is projected onto the viewing plane, but the distortion at the edges goes to infinity (see Figure 3).

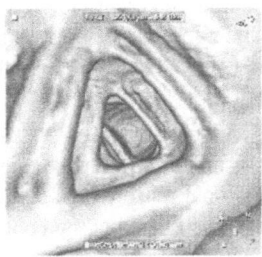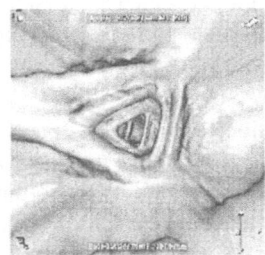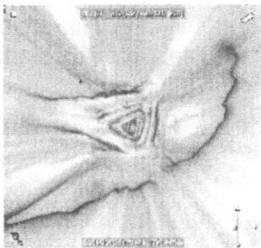

Figure 3. The distortion near the edges increases with a larger viewing angle (the images are rendered at 90, 120 and 140 degrees).

An approach to increase the visible surface area is by unfolding the cubic viewing space introduced in 2.1, and project all six images onto a single plane for simultaneous inspection. Small images are added to resolve discontinuities. We present such a visualization to the physician as an animated image sequence (Figure 4). In Section 3 we will study the increase in surface visibility using this method.

The sequence is generated from a limited set of points sampled on the central path through the colon. For obtaining this central path we refer to [16]. However, it may be that surface areas are still invisible. We have developed a technique to identify the invisible parts, and to allow these parts to be inspected from a different viewpoint, so that potentially 100% of the surface can be examined. 100% is achievable if new views are created for partly invisible surface features as small diverticula.

Invisible regions of the surface are determined given a central path and camera geometry. Our algorithm proceeds in two stages. Initially, the voxels of the colonic sur-

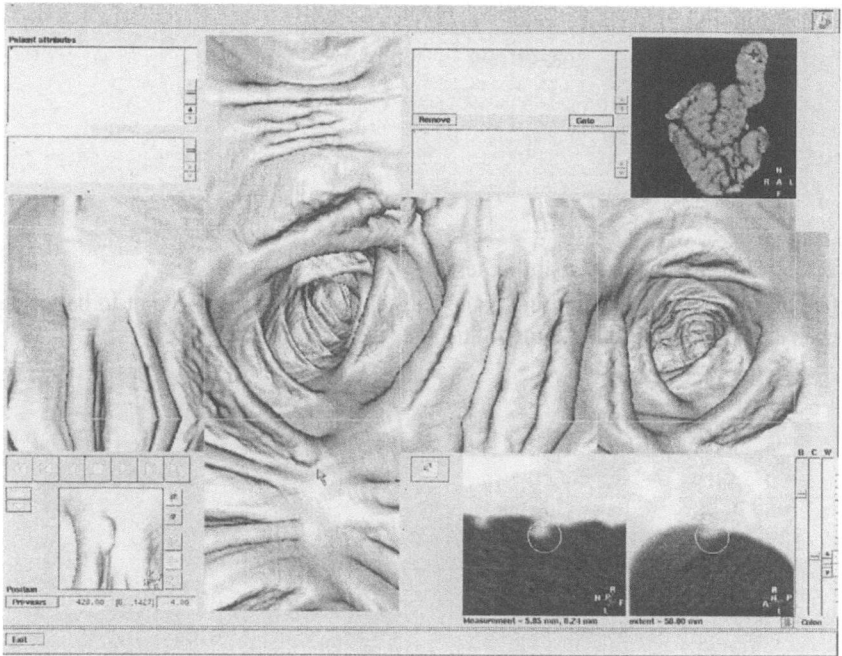

Figure 4. A sequence of outfolded cubes is presented to improve efficiency

face are extracted; these must satisfy the following conditions:

- Inner faces of surface-voxels must closely follow the tissue-air interface.

- At least one face of surface voxels must be potentially visible from any interior viewing position.

- The surface must not contain holes.

The colon's interior volume is obtained by a thresholding of the image volume, and region growing from the centerline. Then, the surface is identified by the voxels with faces adjacent to the interior. Finally, a dilation of the grown object(s) is performed, with the restriction of 6-connectivity. The contour added as a result of the latter operation contains the surface with an edge connectivity property.

Next, the visibility of a voxel is determined by ray-casting from all viewpoints to the center of all potentially visible voxel faces. If another surface voxel intersects this ray, the voxel face is invisible (Figure 5). The visible voxels are expressed as a percentage of the total number of surface voxels.

Because only a small percentage of the colonic surface is not visible using the cubic viewing space, which is in most cases restricted to a few areas, we can calculate a number of extra viewpoints to resolve the invisible patches. As such surface elements often appear in clusters, a hierarchical clustering is performed. For every cluster the center is projected on the surface. Using the surface normal, a new viewing position is determined. When it is not possible to generate a viewing position at a large enough

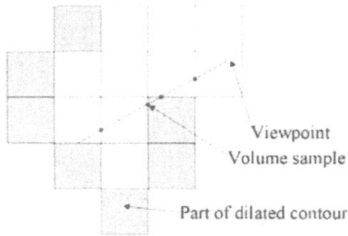

Viewpoint

Volume sample

Part of dilated contour

Figure 5. The surface voxel visibility test.

distance from the cluster, a cutting plane is used, to allow the viewpoint to be located outside the colon's interior. Figure 6 illustrates these alternative views.

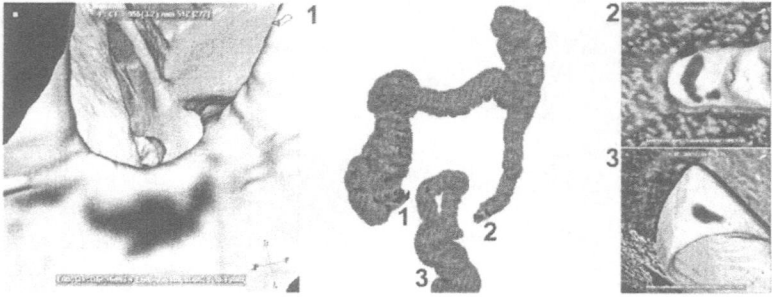

Figure 6. Generated alternative views of invisible surface parts.

Besides using the centerline as a camera path, we facilitate navigation by some additional resources. For easy orientation, the normal to the central image in the unfolded cube (Figure 4) is always parallel to the path tangent at the current location. In addition, an overview image shows the current global position in the colon (top right in Figure 4). Finally, we designed a mode in which two reformatted slice views allow easy exploration of the original CT data (bottom right). Both images are generated automatically when the user clicks in the unfolded cube. The center of the two images is determined by projecting the selected point on the colon wall (Figure 7). The first image is perpendicular to the path where it is closest to the selected position. The second image, an orthogonal cross-section is parallel to the centerline tangent. The physician can scroll interactively through these images and rotate them for closer inspection of suspicious areas.

2.3 Annotation and communication of a diagnosis

Previously, we argued that image-based rendering uses only a limited amount of resources (memory/disk space). Moreover, generally available tools such as VRML can be used for viewing the 3D image-based scenes, also remotely via the Internet. Thus, a practical way to communicate a diagnosis is by way of image-based rendering. A similar approach is described in [17] for mere exploration of bronchial data.

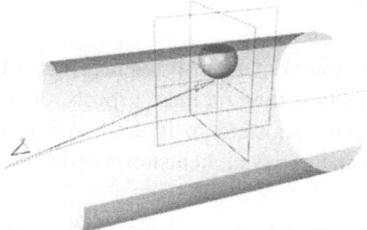

Figure 7. Two planar images are centered at a selected position: perpendicular and parallel to the tangent in the nearest pathpoint.

We have implemented the communication as follows. Once a polyp has been identified, the radiologist can create a 'diagnosis record'. Such a record contains a small series of unfolded cubes that are saved as JPEG image files. Any annotations (e.g. measurement of polyp size) are just pasted in the images (see figure below). Additionally, the overview image as well as the multiplanar reformats are stored. The latter images are created in the manner described above for each path point and also for the clicked positions. A typical record requires approximately 10 Mb of disk space.

For review by the endoscopic surgeon we created a tool in VRML. Running the code in a web browser such as MS Internet Explorer requires a simple plugin module such as the Cosmo Player. Figure 8 shows a typical view. The physician can move forward and backward along the central path and interactively look around. For orientation one may refer to the overview image or to the reformatted views.

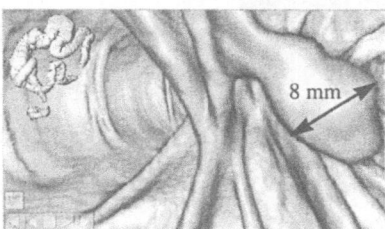

Figure 8. Typical view of our tool for communication of a diagnosis.

The exploration is implemented in VRML by transformations applied to the cube. Movement along the path only requires loading new images to be pasted on the walls. When looking around, the cube is rotated. The user is hardly aware of such transformations because the camera is centered at the position from where the cube textures were rendered. Only extremely small artefacts may become visible near the edges where pixels referring to objects in scene space are sqrt(2) larger than the central pixels. The resulting deformation is invisible as long as the texture sampling is above the Nyquist limit. The tool was tested on a 550 MHz Pentium III processor using 320 MB of main memory and equipped with an HP Visualize fx graphics board. For movement along the path the frame rate was 5 Hz, while rotations for looking around could be done in real time (24 Hz).

3 Clinical evaluation

Any virtual endoscopy system must meet requirements of efficiency, effectiveness and sensitivity (Section 1). For testing on these aspects, data from 20 patients visiting the Academic Medical Center Amsterdam is used. An informed-consent procedure was followed, in which the main patient inclusion criterium was the referal for conventional colonoscopy.

For these patients 3D image data were acquired by computed tomography (CT) in supine and prone position. Subsequently, each patient underwent optical colonoscopy. Identified polyps were resected and their size, location and morphology were annotated to define the ground truth.

In 10 patients 15 colonic polyps were identified with optical colonoscopy. 10 patients did not contain polyps. The resulting 40 sets of 3D-image data were retrospectively examined by one radiologist using a 'conventional' virtual endoscopic method as well as the unfolded cube method. Both methods were implemented on an experimentally enhanced version of the EasyVision Endo-3D software package (Philips Medical Systems). The more conventional tool displayed forward and backward movies that had been created offline. The viewing angle was set to 120 degrees. During exploration the radiologist could stop at any position and change the camera orientation and position for local inspection of the surface. In addition, axial slices of the CT data-volume could be examined. The unfolded cube system was used as described in Section 2.2. (Figure 9 shows typical consecutive views).

Figure 9. Three consecutive views (http://www.ph.tn.tudelft.nl/~iwo).

Table 1 collates the mean time spent on reviewing the conventional and unfolded cube sequences (both patient positions are considered individually). The difference between paired sequences is significant ($P < 0.05$) according to Student's t-test. We attribute the increased reviewing time for the conventional technique to the distortion at the edges. Due to the large viewing angle, features near the borders of the viewing plane appear deformed. For closer observation of such surface parts the physician may want to interactively adjust the viewing angle, which inherently takes rendering time.

Technique	Two-sided view	Unfolded cube	Paired difference
Mean time (+ std. dev.):	18'58" (6'59")	7'04" (2'34")	11'53" (4'53")

Table 1. Efficiency of image sequence reviewal expressed time.

To monitor the effectiveness (defined as area in view), surface voxels in each volume were identified in the same manner as described in Section 2.2. Subsequently, we determined which fraction of them was visible from the given set of viewing positions. For a distended colon with a central path with a length 1500 mm, our visibility calculations take approximately five minutes of CPU-time on a SUN with 440 MHz ULTRA SPARC 2 processor and 512 MB of internal memory. The results are summarized in Table 2. The single plane entry is defined by considering the backward and forward views of the conventional technique individually (n = 80). The surface percentage of 0.5% not seen using the unfolded cube corresponds to approximately 1500 voxels. This is significantly more than the surface of a polyp, which is about 500 voxels (modeling it by a sphere of 10 mm in radius at a voxel size of 0.3^2 x 2 mm.). The latter discrepancy clearly justifies our approach of visualizing uninspected surface at the end of the unfolded cube movie.

Technique:	Single-sided view	Two-sided view	Unfolded cube	Unfolded cube + extra views
Surface in view (+ std. dev.):	73.2% (2.8%)	93.7% (0.5%)	99.5% (0.1%)	99.9% (0.01%)

Table 2. Effectiveness in terms of surface in view.

Finally, the sensitivity was higher for the unfolded cube then for the two-sided view. One polyp was missed using two-sided views. Thus, more polyps are detected with the unfolded cube technique, which we attribute to a larger part of the colon surface coming into view.

4 Conclusion

This paper reports on important progress that was made in virtual colonoscopy. Our "unfolded cube" method complies with requirements regarding efficiency, effectiveness and sensitivity, whereas the more conventional bi-directional fly-through does not comply sufficiently. By using our apporach, interactive inspection of almost the complete inner surface of the colon can be done in 10-15 minutes. Currently, we focus on such topics as (a) segmentation that is robust in the presence of remaining stool, (b) automatic detection of polyps to speed-up navigation and to enhance visualization, and (c) improved facilities for taking measurements. Such further improvements are aimed at enabling virtual colonoscopy to become the method of choice to screen for colorectal polyps.

5 References

[1] J.D. Potter, M.L. Slattery, R.M. Bostick, 'Colon cancer: a review of the epidemiology', *Epidemiol. Rev.*, vol. 15, pp. 499-545, 1993.

[2] N.W. Toribrara, M.H. Sleisenger, 'Screening for colorectal cancer', *N Engl J Med*, vol. 332, pp. 332, pp. 861-867, 1995.

[3] S.J. Winawer, R.H. Fletcher, L. Miller, 'Colorectal cancer screening: clinical guidelines and rationale', *Gastroenterology*, vol. 112, pp. 594-642, 1997.

[4] D.J. Vining, 'Virtual endoscopy: is it reality?', *Radiology*, vol. 200, pp. 49-54, 1996.

[5] L. Hong, S.Muraki, A. Kaufman, D. Bartz, T. He, 'Virtual voyage: interactive navigation in the human colon', *Proceedings A.C.M. Siggraph Conf*, pp. 27-34, ACM Press, 1997.

[6] W.E. Lorensen, F.A. Jolesz, R. Kikinis, 'The exploration of cross-sectional data with a virtual endoscope', *Interactive echnology and the new health paradigms*, vol.459, pp. 221-230, 1995.

[7] P. Rogalla, J. Terwisscha van Scheltinga, B. Hamm, 'Virtual Endoscopy and Related 3D Techniques', *Springer-Verlag, Berlin Heidelberg New York*, 2000, ISBN 3-540-65157-8.

[8] C.F. Beaulieu, R.B. Jeffrey, S.Napel, 'Display model for CT colonography: Part II', *Radiology*, vol. 212, pp. 203-212, 1999.

[9] S.Haker, S. Angenent, A. Tannenbaum, R. Kikinis, 'Non distorting flattening for virtual colonoscopy', *Proc. MICCAI conf.*, pp. 358-366, Springer, 2000.

[10]W.E. Lorensen, H.E. Cline, 'Marching cubes: a high resolution 3D surface construction algorithm', *Proceedings A.C.M. Siggraph Conf*, vol. 21, pp. 163-169, 1987.

[11]M.K. Bosma, 'Iso-Surface Volume Rendering, Speed and Accuracy for Medical Applications', *Ph.D. thesis Twente University*, PrintPartners Ipskamp, Enschede, 2000, ISBN 90-365-1397-5.

[12]B. Lichtenbelt, R. Crane, S. Naqvi, 'Introduction to Volume Rendering', *Hewlett-Packard Professional Books*, Prentice Hall PTR, Upper Saddle River, NJ, 1998

[13]L. McMillan and S. Gortler, 'Image-Based Rendering: A New Interface Between Computer Vision and Computer Graphics', *Computer Graphics*, vol. 33, pp. 61-64, 1999.

[14]D. Ragget, 'Extending WWW to support platform independent virtual reality', *Proc. Internet Society/European Networking*, pp. 242/1-242/3-6, Internet Soc. Press., 1995.

[15]R. Carey, G. Bell, 'The Annotated VRML 2.0 Reference Manual', *Addison-Wesley Pub Co*, (ISBN: 0201419742), 1997.

[16]R.Truyen, P.Lefere, S.Gryspeerdt, T.Deschamps, 'Speed and robustness of (semi-) automatic path tracking', *Second International Symposium on Virtual Colonoscopy*, October 16-17, 2000, Boston.

[17]Wegenkittl, R., Vilanova, A., Hegedüs, B., Wagner, D., Freund, M.C., Gröller, E.M., 'Mastering Interactive Virtual Bronchioscopy on a Low-End PC', in *Proc. IEEE Visualization Conf. 2000*, pp. 461-464, ACM Press, 2000.

Editors' Note: see Appendix, p. 344 for colored figure of this paper

Three-dimensional Reconstruction and Visualization of the Cerebral Cortex in Primates

Sergio Demelio[1], Fabio Bettio[1], Enrico Gobbetti[1], and Giuseppe Luppino[2]

[1] CRS4, VI Str. Ovest, Z.I. Macchiareddu, C.P. 94, I-09010 Uta (CA), Italy
http://www.crs4.it/vvr
{sergio, fabio, gobbetti}@crs4.it
[2] Institute of Human Physiology, University of Parma, I-43100 Parma, Italy
luppino@ipruniv.cce.unipr.it

Abstract. We present a prototype interactive application for the direct analysis in three dimensions of the cerebral cortex in primates. The paper provides an overview of the current prototype system and presents the techniques used for reconstructing the cortex shape from data derived from histological sections as well as for rendering it at interactive rates. Results are evaluated by discussing the analysis of the right hemisphere of the brain of a macaque monkey used for neuroanatomical tract-tracing experiments.

1 Introduction

One major field of interest in neuroscience is the study of the anatomical and functional organization of the cerebral cortex in primates. One widely accepted notion is that cortical information processing occurs in parallel, along different specialized cortical circuits, linking anatomical and functional cortical units, generally referred to as cortical areas (see, e.g., [7]). One major challenge is thus, at present, the definition of the exact number, extent and functional properties of the various cortical areas. Non invasive modern functional imaging techniques (PET, fMRI) allow addressing this issue in human subjects. These techniques, however, have a very poor temporal and a low spatial resolution, limited to the macrostructural level. On the other hand, a variety of more invasive methodological approaches may be used in experimental animals and, in this line of research, non human primates represent the experimental model closest to the human brain. These approaches allow acquisition of data with high spatial and temporal detail, so that the brain of non human primates is unanimously considered a crucial anatomical and functional frame of reference for studies on human subjects.

Basically, three main categories of experimental approaches may be used in non human primates, possibly combined together in the same animal: i) the *morphological approach* is based on the study of the regional variability of the cortical structure and is aimed to define the anatomical borders between different cortical areas in individual brains; given the interindividual variability no standard brain atlas may be used as anatomical frame of reference; ii) the *anatomical approach* is based on microinjections of substances transported by the neurons (neural tracers); the qualitative and quantitative distribution of neurons labeled with the tracers employed, allows the definition of the connections of a given cortical area with other cortical areas or subcortical structures; iii) the *electrophysiological approach* is based on the recording with micro-electrodes of the electrical activity of single neurons in different behavioral tasks and therefore, on the definition of their functional properties.

Each of these experimental procedures requires the analysis of the collected data at the microstructural level. Basically, the brain is cut in serial sections (thickness about $50\mu m$), oriented so as to have an optimal view of the region of interest. After appropriate histological processing, sections are mounted on slides and individually analyzed with low or high power microscopy. Typically, this step of the analysis relies on the manual selections in each sampled section of different sets of points, grouped in different classes, that represent the X-Y coordinates of the various data of interest. These data always include the outer and the inner border of the cortical mantle and may include the locations of borders between areas, locations of individual neurons labelled with each of neural tracers employed (see, e.g., figure 1) or the location of the tracks of microelectrode penetrations.

The typical final step in the analysis and interpretation of the data is, then, the analysis of static images of brain sections or, for an overall view of the collected data, of the 2D reconstruction of the cortical surface, made by flattening and aligning the cortical contours of the sampled sections [3]. However, the procedure of flattening a surface with a very complex geometry, due to the presence of deep cortical sulci with irregular shapes and spatial arrangement (see, e.g., figure 1), is an unavoidable source of distortions that can seriously affect a correct interpretation of the data. This problem becomes even greater in the comparison of data obtained in different animals in which, in order to have an optimal view of the regions of interest, brains are cut with different sectioning planes. In this case, sections and flattened views of the same region of brains cut with different angles can be very difficult to compare, since researchers are forced to mentally construct a model of 3D shapes, adding further complexity to what is an already challenging task. Given these problems, 3D reconstruction and, therefore, the creation of a virtual environment for the direct analysis in three dimensions of the cortical structure, would be of crucial value in the interpretation of these types of data.

In the remainder of this paper, we provide an overview of our prototype system and discuss experimental results obtained on the right hemisphere of the brain of a macaque monkey used for neuroanatomical tract-tracing experiments.

2   Methods and tools

Our interactive system takes as input data acquired by identifying contours and neuron positions on cryogenized brain sections and regards it as a three-dimensional environment to be interactively inspected.

The full potential of 3D rendering is harnessed when allowing operators to view and naturally interact in real-time with the reconstructed 3D data. In particular, the ability to render at interactive speeds dramatically improves depth perception, thanks to motion parallax effects. To support the required interactive performance, we have chosen to exploit the capabilities of current polygon-oriented graphics accelerators by first reconstructing the cortex boundary surfaces and then rendering them, using a multipass technique to also incorporate in the same image subsurface neurons. As illustrated in

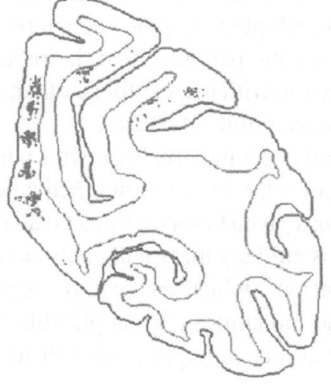

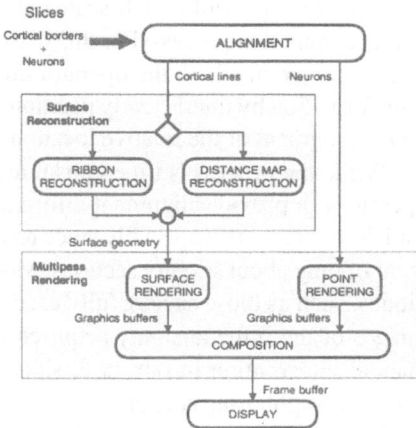

Fig. 1: **Typical brain section**. We can see the external and internal cortical lines, as well as different groups of neuronal cells

Fig. 2: **System overview.** Alignment and surface reconstruction are performed once for each data set in a pre-processing phase, while the the multipass rendering algorithm is embodied in a simple prototype analysis tool.

figure 2, the main operations performed by the system are thus *section alignment*, to position all data in the same frame of reference, *surface reconstruction*, to construct, using different algorithms, triangular meshes that approximate the cortex boundaries, and *multipass rendering*, to display at interactive rates images of the cortical surface and of subsurface neurons. Section alignment and surface reconstruction are performed once for each dataset in a preprocessing phase, while the multipass rendering algorithm is embodied in a prototype analysis tool that offers a simple user interface for performing standard viewing and inspection operations, including camera motion, control of rendering parameters (e.g. transparency), and of cutting planes.

The prototype has been developed with GLUT/OpenGL and is running both on Unix and Windows platform. We have implemented and tested it on a Silicon Graphics Onyx InfiniteReality, with 2 MIPS R10000 194 MHz CPU, and on a PC equipped with two PIII 600 MHz CPU and a GPU NVIDIA GeForce2. The results obtained on the PC have been comparable (and often superior!) to those obtained on the Onyx.

Details on each of the main operations are provided in the following sections.

2.1 Section alignment

In order to correctly reconstruct the brain, the different sections have to be aligned according to the same frame of reference. A variety of manual, semi-automatic, or automatic techniques have been presented in the literature. In our case, the system has to work without the use of fiducial markers, and the process is made more complex

150

because of the possible small-scale deformations (e.g. lobe motions) caused by the particular acquisition process. For this reason, we have adopted a simple manual retrospective technique in which the operator directly defines the parameters of the registration transformation by interactively positioning one section with respect to the other, relying on his judgment of the relative location of specific anatomical features.

While the process is time-consuming, and limited by the precision with which the operator can provide alignment information, we consider the solution practical enough for laboratory work, especially since reconstructing a given brain requires manual alignment of only about seventy sections. Moreover, it is unlikely that totally automatic techniques, such as those successfully used for data fusion applications (e.g. for registering images of the same anatomy acquired in different modalities), are applicable without manual intervention in our case, since semantic ability is required to "fill the gaps" between neighboring sections.

The major current limitation is coming from the constraint that we currently support only rigid transforms. We are currently considering the introduction of global stretching and second degree polynomial transforms (as in [1]) and of local transformations for correcting lobe displacements.

2.2 Surface reconstruction from aligned cross-sectional data

The goal of surface reconstruction is to to generate, from the set of contours present in the aligned brain sections, a triangular mesh of the cortex surfaces. Obtaining a triangular mesh for each boundary surface is extremely useful for high speed rendering on current graphics hardware.

A number of authors have presented techniques for reconstructing surfaces from the cross-sections of an object. All techniques can be classified in two distinct approaches: direct triangulation or shape based functional techniques [12]. The first approach directly triangulates the set of points making up each of these cross-sections, such that they become the vertices of the triangular mesh. While the technique is conceptually simple, many problems arise when the cross-sectional shape varies widely between planes (e.g. holes, branching structures), and much effort is required to detect and correct special cases where the triangulation of complex shapes might otherwise fail [10, 11]. The second approach is to estimate from the contours in the cross sections a 3D function, discretized on a grid, which represents the distance at any point in space to the nearest boundary surface. Numerous algorithms have been proposed to efficiently compute such a distance map (an excellent survey is provided in [5]). Once the function has been created, the iso-surface can be triangulated by a variety of algorithms, including marching cubes [13, 6].

It is important to note that it is intrinsically impossible to guarantee that a particular technique will reconstruct the actual anatomy from a particular set of cross-sections. For this reason, we have chosen to complement the reconstruction using a distance map technique with a discontinuous method which duplicates the profile of a contour along the thickness of the section, so that the representation we obtain is a surface made up of "ribbons". The second method makes it possible to evaluate the relative positioning of the different sections and judge possible misalignment problems. Figure 4 shows a typical exterior cortex surface reconstructed with both methods, which are described in more detail below.

Ribbon surface reconstruction The goal of the algorithm is to create well formed surfaces that represent the boundary of the solid constructed by stacking all the sections one on top of the other. For each of the boundaries, a triangular mesh is constructed using the following technique:

1. for each section, the contour is duplicated in the vertical direction at the distance given by the section thickness; the resulting strip is triangulated and the triangles are added to the mesh;
2. the gaps between adjacent sections are closed by triangulating the planar regions delimited by the boundaries of the sections (see figure 3); this can be efficiently done the following way:
   (a) the intersection points between two subsequent contours are computed (these are always in even number since the contours are closed);
   (b) the closed contours of each of the regions included between two subsequent intersection points are reconstructed;
   (c) a Delaunay triangulation of each new computed regions is carried out, and the computed triangles are added to the mesh.

The obtained reconstruction presents the advantage of not altering the initial dataset, and, while it is a quite rough representation of the most likely smooth surface, it still provides a good impression of the overall brain shape.

Fig. 3: **"Ribbon" surface reconstruction.** From the left: a pair of adjacent sections; detail of a region delimited by two points of intersection; detail of the region triangulation.

Distance map surface reconstruction The goal of the distance map algorithm is to construct a triangular mesh for each boundary surface that is a likely piecewise linear approximation of the corresponding real cortical boundary. Smoothed images can then be obtained at interactive speeds by exploiting the Gouraud shading algorithm. In our implementation, the reconstruction algorithm performs the following passes:

1. a planar distance function is created for every section; this distance transformation maps a binary image (consisting of boundary pixels on a zero background) into an image where all corresponding background pixels have a value proportional to the distance to the nearest boundary pixel. This mapping can be created analytically, but a global operation like this is prohibitively costly [4]. A more efficient way of calculating a distance transform is through a chamfer coding [2], which allows good approximations of the Euclidean distance with small masks applied sequentially on local neighborhoods. For our computations, we have used the $(5:7:11)\,3\times5$ chamfer mask;

2. a spatial distance function is created for the entire volume; the best results can be obtained by interpolating according to suitable directions among adjacent plane distance functions [13], in order to take into account the variations in the topology of the surface. In our case, the sections are not very sparse, and a 3D distance function obtained without interpolation has also provided good results;

3. the zero isosurface is extracted: the method applied is a marching cube algorithm [6].

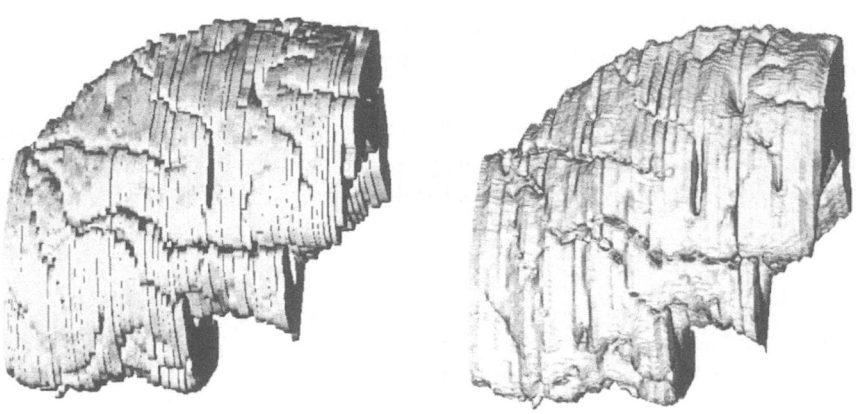

Fig. 4: **Surface reconstruction**. In the left, the cortex surface reconstructed with the "ribbon" method; in the right, the same surface reconstructed with the distance map method. The distance map surface is more "readable" because of the more likely distribution of curvatures.

2.3 Interactive rendering

The rendering system needs to present at interactive rates both surface elements (i.e. the cortex) and subsurface elements (i.e. the neurons). In order to present in the same image details of the interior of the brain that convey depth information, a volumetric approach has to be taken. To obtain interactive performance, we have decided to exploit the power of current graphics boards to accelerate rendering and compositing tasks. The implementation of advanced rendering algorithms using multipass techniques is becoming a standard approach in interactive applications, motivated by the increasing performance of graphics chips. Many special- or general-purpose techniques have been presented in the literature. An excellent survey is provided in [9].

In our case, we model neurons as emissive particles and the cortex material as an absorbing (semi-transparent) medium with constant absorption coefficient. Since the position of a given neuron is only known in its section's plane, we apply to each neuron a random shift in the thickness direction, to obtain a more likely neuron distribution. Rendering, using orthographic projections, is then performed using the following passes:

1. the cortical surface is rendered using lighting and Gouraud shading;
2. the Z-buffer and the color buffer are copied into memory;
3. neurons are rendered with Z-test enabled and no lighting; each neuron is assigned a color ID which identifies its type;
4. the Z-buffer and the color buffer are copied into memory;
5. the color buffer is updated by computing neuron colors as a function of their ID and of the difference between the two Z-buffer values;
6. the updated color buffer is written to frame memory for display; a convolution filter may be applied to perform image smoothing operations.

The attenuation computed by the algorithm is exact, to the limits of the Z-buffer precision, for convex surfaces, but is possibly overestimated for concave boundaries, since multiple intersections of the eye-neuron ray with boundary surfaces are ignored (see figure 5 left). We consider this error negligible for all practical applications, since the only visible neurons are those close to the exterior surface and the brain has a particular shape with very thin sulci. Figure 5 middle and right illustrate the effects of the emission-absorption model.

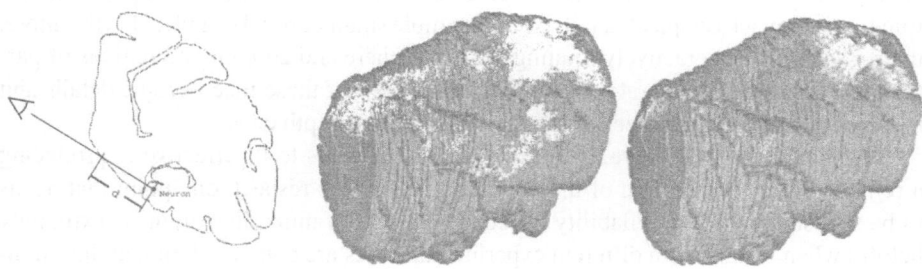

Fig. 5: **Neuron rendering.** Neurons are modeled as emissive particles, while the brain material is considered a semi-transparent medium with constant absorption coefficient. Absorption is computed based on the distance from the neuron to the surface closest to the eye (d in the left image). In the middle, neurons are rendered all with the same intensity. In the right, their color is attenuated by a factor which depends on the distance to the surface in the viewing direction, effectively providing depth information

3 Case study

Our system has been tested on medical datasets acquired at the Institute of Human Physiology of the University of Parma. The case discussed in this report is the right hemisphere of the brain of a macaque monkey used for neuroanatomical tract-tracing experiments. This experiment was made in the framework of a line of research devoted to the exact definition of the source of cortical projections to each of the various anatomical and functional areas that form the motor cortex in the monkey [8]. These data are of extreme importance for the interpretation of the possible functional role in motor control of each motor area and for the definition of the various cortical circuits involved in different types of information processing for voluntary action. In this particular case, in compliance with Italian and European laws on the care and use of laboratory animals, under aseptic condition and under microscopic guidance, three restricted injections of different fluorescent neural tracers (FB, DY, TB) were made in the two areas that form the so called "dorsal premotor cortex". After appropriate survival period the animal was sacrificed and the brain cut frozen in serial sections ($60\mu m$ thickness) in coronal plane. Under fluorescence microscopy each of the three fluorescent tracers emits light of different color so that neurons labeled with each tracer (and, therefore, neurons that are anatomically connected with the injected cortical site) can be easily identified at medium-power enlargement (200-400X). In one section each $600\mu m$ the outer and inner cortical contours were manually delineated and saved as different classes of points along with the location of individual labeled neurons plotted and grouped in three different classes of points according to the type of tracer for which they were positive. Figure 1 is an example of data collected from one individual brain section.

The source dataset is made up of about 70 sections that cover the whole extent of the brain regions labeled with the tracers. The number of points acquired in each section is about 10^3 per boundary, which means that the 3D model contains over 10^5 vertices. Figures 6 and 7 (color plates) show the results of the reconstruction procedure. In Figure 6 (left) is shown a lateral view of the hemisphere in which the frontal lobe is on the right and the occipital lobe on the left. As one can see from the figure, the result is very realistic. All the major cortical sulci are clearly distinguishable and the same is true for finer macroscopical details as for example small cortical dimples. Furthermore, the possibility of interactively rotating the hemisphere and zooming on regions of particular interest allows operators to have a closer view of these macroscopic details and to benefit from parallax effects that provide additional depth cues.

It is quite clear, therefore, that this procedure appears to be effective in producing a reliable 3D reconstruction of individual brains. In this respect, one point that needs to be stressed is that the reliability of reconstructions of individual brains is extremely helpful when results from different experimental cases are compared. In fact, interindividual variability in the gross morphology may account for apparent differences among different animals.

One problem faced in the alignment of the sections is represented by some discontinuities between profiles of adjacent sections. These discontinuities are actually due to some deformations of the sections that eventually occur during the histological processing or when mounted on slides. This problem can be overcome by improved alignment techniques (see section 2.1) or by an appropriate intervention before alignment. In the same figure is also shown the distribution of the labeled cortical neurons. Neurons labeled with different tracers are shown with dots of different colors and the color intensity is attenuated depending on the distance of the neurons from the surface in the viewing direction. In the upper part of the frontal lobe three densely aggregate of labeling surround the injections of the three tracers. The injected region can be better appreciated in Figure 7 that represents a dorsal view of the hemisphere. Because of the large and concentrated number of labeled neurons in these regions the dots appear to be fused together. However, when the region is zoomed in (Figure 7 right) the three clouds of labeling can be much better resolved. The sites of the injections are here provisionally represented as the regions empty of labeling within each of the three clouds of labeling. It is quite clear that this reconstruction procedure and appropriate reslicing could be a very powerful tool for the exact definition of the location and extent of the cortical region interested by the tracer injection. The differential distribution of the labeling in the various cortical region is clearly visible in all the figures presented. These presentations confirm, but also extend previous observations made in this case with conventional bidimensional flattening of the cortical surface. For example in the lateral view of the hemisphere, shown in Figure 6, left, it very clear the differential distribution of the labeling in the caudal part of the hemisphere, corresponding to the inferior parietal lobule. Although already noticed with conventional reconstruction techniques, the relative position of the positive neurons and their location relative to the sulci is now much clearer. The present reconstruction procedure appears therefore very helpful for the interpretation of the data and for the attribution of the labeling to particular cortical areas. Additional aspects that would be very difficult to appreciate with conventional bidimensional reconstruction can be clearly revealed by this procedure. For example, in the mesial view of the brain shown in the right part of Figure 6, it is possible to see, in the caudal part of the hemisphere, a differential distribution of various neuronal populations. Preliminary morphological data collected in the Institute of Human Physiology of the University of Parma indicate that several different anatomical areas form this region. Correlation of architectonic borders with the distribution of the labeling in this reconstruction would be a strong argument in favor of this hypothesis.

4 Conclusions and future work

We have presented a prototype interactive application for the direct analysis in three dimensions of primate cerebral structures. The system takes as input data acquired by identifying contours and neuron positions on cryogenized brain sections and regards it as a three-dimensional environment to be interactively inspected.

To support interactive performance, the system exploits the capabilities of current graphics accelerators by first reconstructing the cortex boundary surfaces and then rendering them, using a multipass technique to also show in the same image subsurface neurons.

While none of the techniques used are original from a graphics point of view, their combination in an interactive setting promises to be a valuable improvement over standard analysis tools used in neuroscience. In particular, the availability of low-cost, high performance graphics accelerators for PC platforms makes it possible to consider interactive 3D inspection of high resolution neuroanatomical data practical for everyday laboratory work. The system has been tested on actual datasets, and the feed back from end-users has been positive. In this work, we have discussed the inspection of the right hemisphere of the brain of a macaque monkey used for neuroanatomical tract-tracing experiments.

We are currently working on extending the system with improved reconstruction and visualization techniques. Particular areas of development are non-rigid section alignment and presentation of cytoarchitectural information about cortical tissue.

Acknowledgments

All the datasets used in the present work were supplied by the Institute of Human Physiology of the University of Parma. This work is partially funded by MURST under project: "Laboratorio Avanzato per la Progettazione e la Simulazione Assistita al Calcolatore". We also acknowledge the contribution of Sardinian regional authorities.

References

1. C. Bohm, T. Greitz, D. Kingsley, B. M. Berggren, and L. Olsson. Adjustable computerized stereotaxic brain atlas for transmission and emission tomography. *American Journal of Neuroradiology*, 4:731–733, 1983.
2. Gunilla Borgefors. Distance transformations in digital images. *Computer Vision, Graphics, and Image Processing*, 34(3):344–371, June 1986.
3. G. J. Carman, H. A.Drury, and D. C. van Essen. Computational methods for reconstructing and unfolding the cerebral cortex. *Cerebral Cortex*, 5:506–517, 1995.
4. O. Cuisenaire and B. Macq. Fast Euclidean distance transformation by propagation using multiple neighborhoods. *Computer Vision and Image Understanding: CVIU*, 76(2):163–172, November 1999.
5. Olivier Cuisenaire. *Distance Transformations: Fast Algorithms and Applications to Medical Image Processing*. PhD thesis, Université Catholique de Louvain, 1999.
6. W. E. Lorensen and H. E. Cline. Marching cubes: a high resolution 3D surface construction algorithm. In M. C. Stone, editor, *SIGGRAPH '87 Conference Proceedings (Anaheim, CA, July 27–31, 1987)*, pages 163–170. Computer Graphics, Volume 21, Number 4, July 1987.
7. G. Luppino and G. Rizzolatti. The organization of the frontal motor cortex. *News in Physiological Sciences*, 15:219–225, 2000.
8. M. Matelli, P. Govoni, C. Galletti, D.F. Kutz, and G. Luppino. Superior area 6 afferents from the superior parietal lobule in the macaque monkey. *The Journal of Comparative Neurology*, 402:327–352, 1998.
9. Tomas Moller and Eric Haines. *Real-Time Rendering*. A. K. Peters Limited, 1999.
10. J.-M. Oliva, M. Perrin, and S. Coquillart. 3D reconstruction of complex polyhedral shapes from contours using a simplified generalized voronoi diagram. *Computer Graphics Forum*, 15(3):307–408, August 1996. Proceedings of Eurographics '96. ISSN 1067-7055.
11. Bradley A. Payne and Arthur W. Toga. Surface mapping brain function on 3D models. *IEEE Computer Graphics and Applications*, 10(5):33–41, September 1990.
12. S. P. Raya and J. K. Udupa. Shape-based interpolation of multidimensional objects. *IEEE Transactions On Medical Imaging*, 9(1):32–43, March 1990.
13. G. M. Treece, R. W. Prager, A. H. Gee, and L. Berman. Surface interpolation for sparse cross sections using region correspondence. In *Medical Image Understanding and Analysis 1999*, July 1999.

Editors' Note: see Appendix, p. 345 for colored figures of this paper

Interactive and Multi-modal Visualization for Neuroendoscopic Interventions

Dirk Bartz[1], Wolfgang Straßer[1], Özlem Gürvit[2,3],
Dirk Freudenstein[3], and Martin Skalej[2]

[1] WSI/GRIS, University of Tübingen,
Auf der Morgenstelle 10/C9,
D72076 Tübingen, Germany
Email: {bartz,strasser}@gris.uni-tuebingen.de
[2] Department of Neuroradiology,
[3] Department of Neurosurgery,
University Hospital Tübingen
Hoppe-Seyler-Str. 3
D72076 Tübingen, Germany
Email: {oezlem.guervit,dkfreude,martin.skalej}@med.uni-tuebingen.de

Abstract. Based on the VIVENDI-framework for virtual endoscopy, we present a system for the interactive and multi-modal representation of important anatomical structures for neuroendoscopic interventions.

A serious problem of neuroendoscopic interventions is the possibility of injuring a blood vessel while performing endoscopic surgery inside the human brain. Besides the sudden loss of optical visibility due to the red-out of the injured vessel, a potential lethal mass bleeding can be the fatal outcome of the intervention. To avoid accidental lesions, we represent the relevant information using multiple volumetric MRI-based representations of the respective organs.

Keywords: Virtual Environments, Magnetic-Resonance-Imaging, MR Angiography, Virtual Neuroendoscopy, Computer Assisted Diagnosis.

1 Introduction

Minimal-invasive neurosurgical procedures are rapidly gaining importance in neurosurgery. Besides the reduced damage of brain tissue compared to commonly used surgical techniques, regions of the brain become accessible, which are not approachable for previous techniques. However, minimal-invasive neurosurgical procedures facilitate only limited access to the respective brain region. In case of a serious complication, such as a mass bleeding after injuring a major blood vessel, effective *stanching* of that injury is usually not possible.

Our focus is on minimal-invasive interventions in the ventricular system of the human brain, where the *cerebrospinal fluid (CSF)* is produced and distributed via the ventricles to the other cavities inside of the skull. Sometimes the cerebral aqueduct – a narrow passage between the third and fourth ventricle – is occluded due to a tumor, an

158

injury due to an accident, or a native defect, thus blocking the natural flow of the CSF. This blockage leads to a dangerous increase of brain pressure which can damage the brain severely. Other problems include the formation of a CSF-filled cyst which also introduces pressure on blood vessels, nerves, or the ventricular (cerebral) aqueduct.

To avoid these dangerous increases of pressure inside of the skull, the cyst is drained or a *ventriculostomy* is applied, where a new CSF drain is realized in the floor of the third ventricle using endoscopic tools. Unfortunately, the major basilar artery is located directly below the floor of the third ventricle without an optical visibility from the third ventricle. Lesions of these arteries result usually in a fatal outcome of the intervention. Even if only a small blood vessel is injured by one of the endoscopic tools, the sudden loss of optical visibility through the endoscope introduces severe difficulties for obtaining the desired results of the interventions.

To avoid traumas of such blood vessels, we modified the VIVENDI-framework for virtual endoscopy [2, 4] to represent multiple anatomical information of the patient data using several 3D scanning techniques.

In the remaining parts of the paper, we briefly discuss related work in the field of virtual endoscopy in Section 2, introduce the existing VIVENDI-framework and describe the additionally used techniques in Section 3. In Section 4, we discuss the learned lessons and finally, we summarize our paper in Section 5.

2 Related Work

Research on virtual endoscopy is one of the most active areas in virtual medicine. The developed methods have been applied to virtual colonoscopy [10, 16], bronchoscopy [7], ventriculoscopy [1, 3], and angioscopy [6, 5, 8, 4].

Different rendering techniques are used to provide sufficient visual quality and/or interactivity. Standard graphics hardware is used to render surface models [17, 12, 10, 2], extracted with the Marching Cubes algorithm [11]. In contrast, volume-rendering techniques are used, partially for better visual quality, partially for interactive speed [15, 20, 8, 1]. Unfortunately, interactive speed was always compromising visual quality, general applicability, or flexibility. In [15] and [5], key-framed animations are generated offline, which frequently leads to the time-intense refinement of the key-framed animation. You et al. used a 16 processor SGI Challenge for parallel volume-rendering of isosurfaces [20]. In contrast, Gobetti et al. used the 3D texture mapping hardware abilities of high-end graphics systems for volume rendering. However, the lack of shading reduced the visual quality significantly [8][1]. Furthermore, the size of the texture memory limits the size of datasets severely, while swapping techniques like bricking reduce the framerate. The Navigator software of General Electric uses isosurface ray casting with approximately one frame per second. Even if the performance of the 1996 results has significantly improved, it hardly can be viewed as interactive [6].

[1] In 1998, Westermann and Ertl presented 3D texture mapping-based volume-rendering with isosurface shading [19]. However, this approach does not provide sufficient performance for interactive endoscopy applications.

Besides rendering, the used navigation paradigm determines the usability of a virtual endoscopy system. Many systems [12, 5, 16, 13] use a planned or automatic navigation, which generates an offline animation of a fly-through after specifying a camera path. This simple scheme reduces the interaction to a VCR-like functionality, requiring a costly refinement of the camera path (and of the animation), if the structure of interest is not well covered. A variation of the planned navigation is the "reliable navigation" [9], in which a complete "visit" of all structures of the organ is guaranteed. However, this also means that user interaction is limited and that not relevant regions cannot easily be skipped.

A free navigation approach is followed by [6, 7, 1]. Unfortunately, the complexity of the anatomical structures commonly found in the datasets is very high. Even for a specifically trained physician, it can be difficult to navigate to the target. Furthermore, collision avoidance is a costly operation which is frequently not available in these systems. In [10, 2, 4], a guided navigation paradigm was adopted in order to provide full navigation flexibility, combined with user guidance and an efficient collision avoidance scheme.

Considering the pros and cons of these approaches, the VIVENDI-framework adopted a surface rendering approach based on [2] and a guided-navigation approach.

3 The VIVENDI-framework for Virtual Endoscopy

The first step after data acquisition of the volume dataset – based on the image stack from rotational angiography, MRI, or Computed Tomography (CT) –, is the segmentation of the respective organs. Standard 3D region growing-based segmentation algorithms [2] can be used as well as advanced segmentation methods [18]. Subsequently, an octree-hierarchical, MarchingCubes-based isosurface algorithm is applied to extract the isosurface representing the inner surface of the respective organ system. Finally, distance fields are computed to provide a guided-navigation system for the virtual camera [10].

VIVENDI uses the octree-oriented subdivision of the isosurfaces geometry to apply an OpenGL-based view-frustum culling and Hewlett-Packard occlusion culling flag [14] scheme to cull on average 90% of the geometry [2–4], resulting in more than 25 fps on an HP J7000/fx6 graphics workstation, and about 20 fps on an HP X-class PC with a fx6 graphics card running LINUX. The distance fields are employed to implement a collision avoidance system and a guided-navigation scheme, which directs the user to a specified target point, while it still guarantees full navigation flexibility [10]. If a 3D, intra-operative navigation system is present, the position and orientation information of the optical endoscope can be fed into VIVENDI to synchronize the virtual and optical endoscopic exploration throughout the neuroendoscopic intervention[2].

---

[2] At this point no commercially available navigation system provides this information to an external entity. VIVENDI however, can process information of this kind.

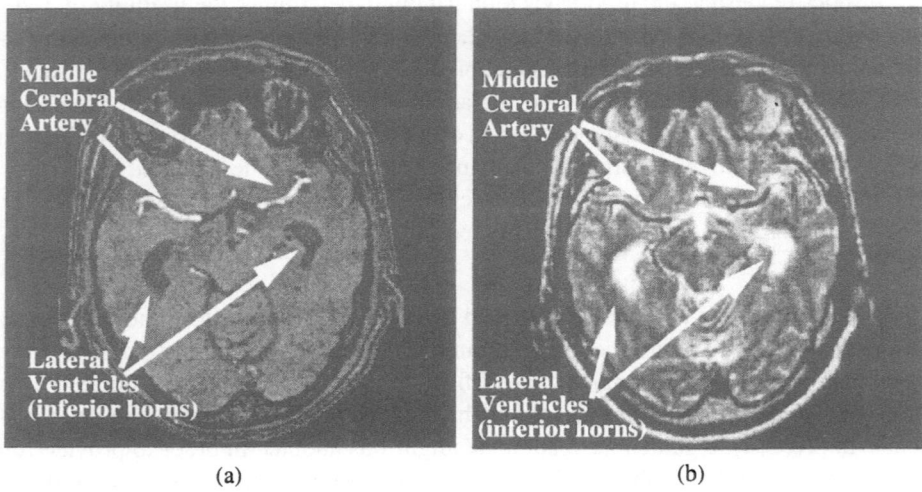

Fig. 1. MRI axial slice with the inferior horns of the lateral ventricles and middle cerebral artery (left and right). (a) MRI Angiography sequence, (b) MRI TSE sequence.

Based on techniques developed for combining multiple target points for guided-navigation and disconnected anatomical structures [2], we integrate the concurrent rendering of information from two data modalities. We use an MRI TSE (Turbo Spin Echo) sequence to generate the inner surface of the CSF-filled cavities (Fig. 1b) – such as the ventricular system or cysts –, and an MRI Angiography (Time of Flight sequence) to derive information on the vascular system (Fig. 1a) in the area of interest. Both sequences were performed subsequently, without changing the position and orientation of the patient. It later turned out in our experiments that patient movement during both scans is negligible. Although the resolution within the axial slices is twice as large in the MRI Angiography as in the MRI TSE sequence, the scans generate two well-aligned data volumes. However, the number of axial slices in the MRI data is different, and hence so is the covered scanning area. This difference requires a manual slice matching step that is performed by a neuroradiologist or neurosurgeon. The resulting axial translation generates an error which is at most the distance between two slices in the data volumes. A manifestation of this error can be found in Figures 3 and 4 (see Appendix), where the red/dark artery geometry reconstructed from the MRI Angiography sequence penetrates the geometry of the transparent CSF-filled cavity geometry[3]. Especially in Figure 4 (see Appendix), the "original" position of the blood vessel is also visible in the geometry extracted from MRI TSE sequence. Fortunately, the maximum error (if the matching step is correct) is always sufficiently below a critical threshold, where the "clearance area" would also cover the proposed target area of the endoscope.

---

[3] The depth sorting of the geometry for correct transparent rendering is obtained by the view-frustum culling. All subdivision entities are sorted according to their closest depth values.

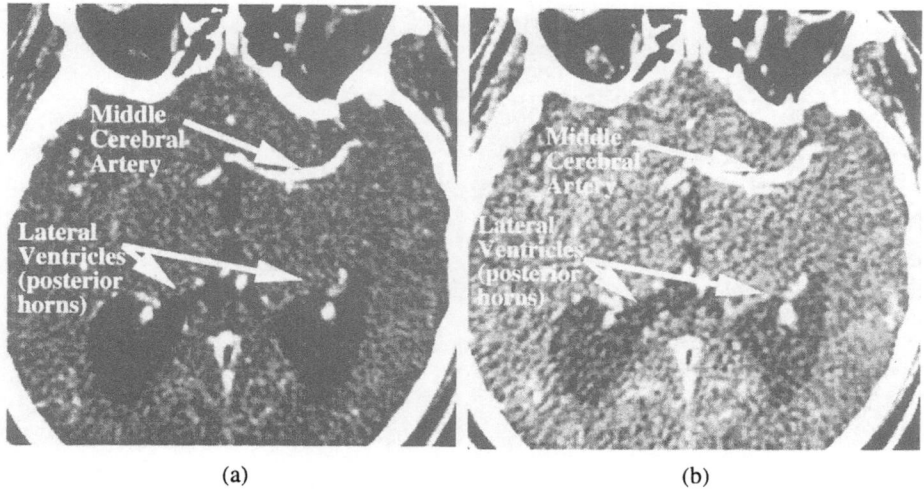

(a)   (b)

Fig. 2. CT axial slice with the posterior horns of the lateral ventricles and middle cerebral artery. (a) Contrast window one, (b) Contrast window two.

4   Lessons Learned

We applied our methods to a set of different patient datasets. Most patients received a ventriculostomy to treat the aqueductal stenosis (see Figures 6 and 5 in Appendix). Two other patients suffered from a large cyst in the third ventricle and near the left temporal lobe (see Figs. 3 and 4 in Appendix) and were treated with a drainage of the cyst. As mentioned earlier, the neurosurgeons performed these interventions using minimal-invasive procedures that introduce a non-negligible risk of damaging a major artery close to the new drain.

We conducted several experiments to determine an appropriate scanning protocol which produces volume datasets which can be matched using CT- and MRI scanners. For this purpose, two anatomical structures need to be identified by the scanning procedure; the CSF-filled ventricular system and cysts, and the blood filled major arterial blood vessels in proximity to the CSF-filled target areas.

The (contrast agent-enhanced) CT scan provided a good contrast and a high resolution for the vascular system within the region of interest. However, this sequence did not produce a sufficient contrast between the brain tissue and the CSF-filled cavities, while still preserving the complete inner surface of the cavities (Fig. 2a/b). Furthermore, Computed Tomography inherently introduces radiation, an additional drawback compared to MRI.

Blood-flow induced MRI Angiography (Fig. 1a) also reconstructs the vascular system with good quality, although the resolution is slightly lower than with a CT scan. However, it is not usable for the segmentation of CSF-filled cavities. Therefore, we perform a second sequence that focuses on these cavities right after the MRI Angiography.

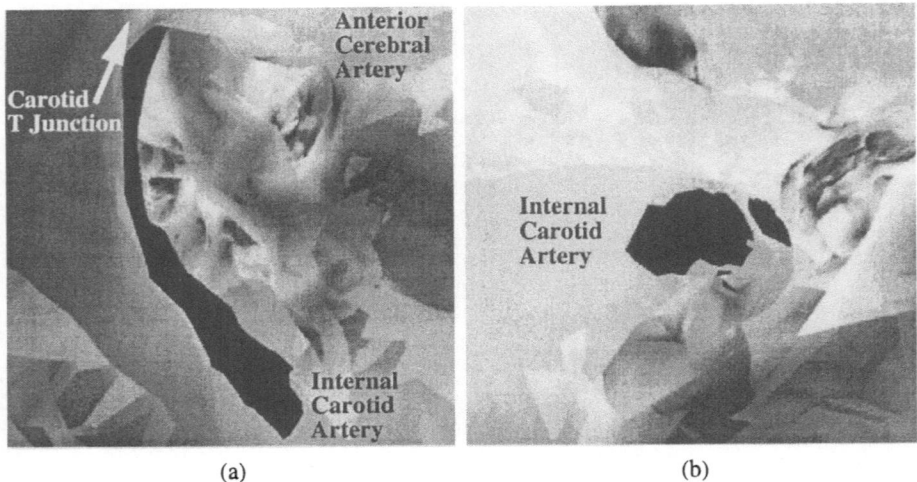

(a)
(b)

Fig. 3. (a) Close-up to internal carotid artery, (b) View downwards on the "carotid siphon", below the carotid T junction.

We previously used an MRI 3D CISS (Constructive Interference in Steady States) sequence to reconstruct the ventricular system in patient datasets. Unfortunately, the different scanning orientation (sagittal and axial) introduced a surprisingly difficult match procedure, which qualified this sequence as impractical. Consequently, we modified the 3D CISS sequence to an MRI TSE sequence (Fig. 1b) which provides the same orientation as the angiography sequence and unfortunately, also a smaller slice range than the 3D CISS. However, it turned out that the resolution is sufficient for our purposes. Finally, the combination of angiography and MRI TSE data delivered a satisfying matching and image quality and was henceforth used in all later experiments. Furthermore, MRI does not expose radiation, in contrast to a CT scan.

The current system is now providing the vascular topography combined with the (already previously available) information of the anatomical structure of the CSF-filled ventricular cavities. This information is successfully used to represent the location of the blood vessels to carefully plan the neuroendoscopic intervention. Lesions of the respective arteries can be avoided, resulting in a substantial reduction of the risk of serious complications. Up to now, five patients were scanned using the described protocol and provided a useful planning aid for the neuroendoscopic interventions.

5 Conclusion

In this paper, we presented a virtual neuroendoscopy system, based on the VIVENDI framework for virtual endoscopy. Interactive performance on a graphics workstation (or PCs) and intuitive handling was achieved using a visibility driven rendering and by adopting the guided-navigation paradigm.

A reliable MRI-based scanning protocol was established which provides the necessary information for the successful planning of neuroendoscopic procedures which significantly reduce the risk of serious complications. Note that the problems described in this paper are not strictly specific to neuroendoscopy, but are also present in other medical application fields, such as oral-maxillofacial or cardio-thoracic surgery, which will be the focus of future work.

Acknowledgments

This work has been supported by the Hewlett-Packard Workstation System Lab, Ft. Collins, USA, by the Ministry of Science, Research, and Arts of the State of Baden-Württemberg, and by the Deutsche Telekom AG. Datasets were provided by the Department of Neuroradiology. Especially, we would like to thank Sabine Joachim, Violeta Adis, and Bernd Kardatzki of the Department of Neuroradiology for their help obtaining and transferring the various patient datasets, and Mike Doggett for proof-reading.

References

1. D. Auer and L. Auer. Virtual Endoscopy - A New Tool for Teaching and Training in Neuroimaging. *International Journal of Neuroradiology*, 4:3–14, 1998.
2. D. Bartz and M. Skalej. VIVENDI - A Virtual Ventricle Endoscopy System for Virtual Medicine. In *Proc. of Symposium on Visualization*, pages 155–166,324, 1999.
3. D. Bartz, M. Skalej, D. Welte, W. Straßer, and F. Duffner. A Virtual Endoscopy System for the Planning of Endoscopic Interventions in the Ventricle System of the Human Brain. In *Proc. of BiOS'99: Biomedical Diagnostics, Guidance and Surgical Assist Systems*, volume 3514, pages 91–100, 1999.
4. D. Bartz, W. Straßer, M. Skalej, and D. Welte. Interactive Exploration of Extra- and Intracranial Blood Vessels. In *Proc. of IEEE Visualization*, pages 389–392,547, 1999.
5. J. Beier, T. Diebold, H. Vehse, G. Biamino, E. Fleck, and R. Felix. Virtual Endoscopy in the Assessment of Implanted Aortic Stents. In *Proc. of Computer Assisted Radiology*, pages 183–188, 1997.
6. C. Davis, M. Ladds, B. Romanowski, S. Wildermuth, J. Knoplioch, and J. Debatin. Human Aorta: Preliminary Results with Virtual Endoscopy Based on Three-dimensional MR Imaging Data Sets. *Radiology*, 199:37–40, 1996.
7. G. Ferretti, D. Vining, J. Knoplioch, and M. Coulomb. Tracheobronchial Tree: Three-Dimensional Spiral CT with Bronchoscopic Perspective. *Journal of Computer Assisted Tomography*, 20(5):777–781, 1996.
8. E. Gobbetti, P. Pili, A. Zorcolo, and M. Tuveri. Interactive Virtual Angioscopy. In *Proc. of IEEE Visualization*, pages 435–438, 1998.
9. T. He and L. Hong. Reliable Navigation for Virtual Endoscopy. In *Proc. of IEEE Medical Imaging*, 1999.
10. L. Hong, S. Muraki, A. Kaufman, D. Bartz, and T. He. Virtual Voyage: Interactive Navigation in the Human Colon. In *Proc. of ACM SIGGRAPH*, pages 27–34, 1997.
11. W. Lorensen and H. Cline. Marching Cubes: A High Resolution 3D Surface Construction Algorithm. In *Proc. of ACM SIGGRAPH*, pages 163–169, 1987.

164

12. W. Lorensen, F. Jolesz, and R. Kikinis. The Exploration of Cross-Sectional Data with a Virtual Endoscope. In R. Satava and K. Morgan, editors, *Interactive Technology and New Medical Paradigms for Health Care*, pages 221–230. 1995.

13. G. Rubin, C. Beaulieu, V. Argiro, H. Ringl, A. Norbash, J. Feller, M. Dake, R. Jeffrey, and S. Napel. Perspective Volume Rendering of CT and MR Images: Application for Endoscopic Imaging. In *Radiology*, volume 199, pages 321–330, 1994.

14. N. Scott, D. Olsen, and E. Gannett. An Overview of the VISUALIZE fx Graphics Accelerator Hardware. *The Hewlett-Packard Journal*, (May):28–34, 1998.

15. R. Shadidi, V. Argiro, S. Napel, L. Gray, H. McAdams, G. Rubin, C. Beaulieu, R. Jeffrey, and A. Johnson. Assessment of Several Virtual Endoscopy Techniques Using Computed Tomography and Perspective Volume Rendering. In *Proc. of Visualization in Biomedical Computing*, volume LNCS 1131, pages 521–528, 1996.

16. D. Vining, R. Shifrin, E. Grishaw, K. Liu, and R. Choplin. Virtual Colonoscopy (abstract). In *Radiology*, volume 193(P), page 446, 1994.

17. D. Vining, D. Stelts, D. Ahn, P. Hemler, Y. Ge, G. Hunt, C. Siege, D. McCorquodale, M. Sarojak, and G. Ferretti. FreeFlight: A Virtual Endoscopy System. In *First Joint Conference, Computer Vision, Virtual Reality and Robotics in Medicine and Medical Robotics and Computer-Assisted Surgery*, volume LNCS 1205, pages 413–416, 1997.

18. D. Welte and U. Klose. Segmentation and Selective Imaging of Arteries and Veins from Contrast-Enhanced MRA Data. In *Proc. of European Congress of Radiology*, 1999.

19. R. Westermann and T. Ertl. Efficiently Using Graphics Hardware in Volume Rendering Applications. In *Proc. of ACM SIGGRAPH*, pages 169–177, 1998.

20. S. You, L. Hong, M. Wan, K. Junyapreasert, A. Kaufman, S. Muraki, Y. Zhou, M. Wax, and Z. Liang. Interactive Volume Rendering for Virtual Colonoscopy. In *Proc. of IEEE Visualization*, pages 343–346, 1997.

Editors' Note: see Appendix, p. 346 for colored figures of this paper

Visualization of Generalized Voronoi Diagrams

Alexandru Telea, Jarke J. van Wijk

Eindhoven University of Technology,
Den Dolech 2,Eindhoven 5600 MB, The Netherlands,
{alext,vanwijk}@win.tue.nl

Abstract. Voronoi diagrams are an important data structure in computer science. However well studied mathematically, understanding such diagrams for different metrics, orders, and site shapes is a complex task. We propose a new method to visualize k-order diagrams and give an efficient adaptive implementation for this method. The algorithm is easy to customize for different metrics and site shapes. Its real-time performance makes it suitable for interactive planning and analysis of complex Voronoi configurations in 2D. We illustrate the method for different combinations of metrics and site shapes.

1   Introduction

Voronoi diagrams are a fundamental data structure in computer science. Mostly used in computational geometry, Voronoi diagrams have found their way in many application areas, such as computer graphics (collision detection, motion planning), optimization theory (associative file searching, clustering, scheduling), and physics (crystal and cell growth studies).

Voronoi diagrams based on Euclidean distance are the best known. Such diagrams partition a 2D plane in regions such that all points within a region are closest to one site from a given site set. Visualizing such diagrams is straightforwardly done by drawing the set of disjoint, adjacent planar polygons that represent the diagram.

We present a new visualization method for two generalizations of the Voronoi diagrams. The first generalization regards *k-order diagrams*, which partition the plane in cells such that all points in a given cell have the same k closest sites. Although many studies cover the mathematics of k-order Voronoi diagrams [2, 1], getting an intuitive understanding of such diagrams is a difficult task. Specifically, we would like to answer questions such as 'which are the k sites that influence a given partition' and 'which are all points under the k-order influence of a given site' in a simple, visual manner. The second generalization concerns using different metrics besides the Euclidean distance and different site shapes besides points. Diagrams for higher orders, different site shapes and metrics lead to complex shape-site relationships that require a more elaborate visualization than a straightforward polygonal drawing.

In Section 2, we give a mathematical overview of Voronoi diagrams and outline the difficulties inherent to their visualization. In Section 3, we introduce our method for visualizing generalized Voronoi diagrams and illustrate it with several examples. Section 4 presents an efficient implementation of the method. We conclude in Section 5 with future research directions.

2   Background

We begin with a description of the elementary properties of 2D Voronoi diagrams. More on the mathematical aspects is available from several surveys [2, 4, 5].

Let $P = p_1, ..., p_n$ be a set of n distinct points in the plane, called sites, and $d(p, q)$ the Euclidean distance between points p and q. The first order Voronoi diagram of P is a subdivision of the plane in n cells, one for each site in P, such that a point q lies in the cell of a site p_i if and only if $d(q, p_i) < d(q, p_j)$ for all sites $p_j \in P$ with $j \neq i$. The cell boundaries lie thus on the perpendicular bisectors of the line segments $p_i p_j$ (Fig. 1 a). First order Voronoi diagrams are used, for example, to partition a city map into regions (cells), given a set of fire station positions (sites), such that any city location is assigned to the closest fire station.

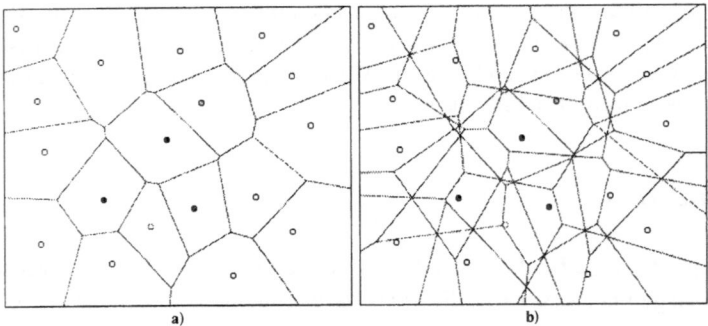

Fig. 1. Order-1 (a) and order-2 Voronoi diagram of a point set

A *k-order* Voronoi diagram subdivides the plane in cells such that all points in a cell have the same ordered set of k closest sites from the site set P. In our example, a k-order diagram would indicate which are the second, third, and k-th fire stations that serve a given city location if the closest (first order) fire station is unavailable for some reason. Conversely, one can visualize the regions served by a given fire station when some neighboring stations fail to work. In an interactive city planning setup, one could add, delete, or move the locations of the fire stations to optimize the area coverage and redundancy in case of fire station failure.

However, visualizing k-order diagrams by simply drawing the involved cells produces hardly readable drawings (Fig. 1 b). It is hard to tell from such a drawing which is the area served by a fire station if other stations fail or, conversely, which are the k stations serving a given city location. These questions could be answered by interactively selecting a site and highlighting its influence region or, conversely, by highlighting all stations that serve a given city location. However, such a method is not capable of producing an overview of the influence of *all* fire stations on *all* city locations simultaneously.

A second generalization of the Voronoi diagrams involves the distance function used. The L_1 (Manhattan distance) metric $d(p, q) = |x_p - x_q| + |y_p - y_q|$ is used to model access times to strategic locations in a city where the streets form an orthogonal grid, or the access time to given records in mass storage systems where the read/write head can only move in orthogonal directions [3]. All edges in such a Voronoi diagram are vertical, horizontal, or diagonal at 45 degree angles. Weighted Voronoi diagrams assign a weight w_i to each site p_i and define the distance $\delta(q, p_i)$ by adding or multiplying the Euclidean distance with the weight w_i. Additive weights were used by Johnson and Mehl [2, 4] to model the growth of crystals from a given seed set. Multiplicative weights lead to the so-called Apollonius model that describes the growth of plant cells, coverage areas of trees, or areas of best received transmitters [6]. Finally, distances can be computed from other site shapes than points, such as lines and curves.

So far, visualizing Voronoi diagrams for higher orders and/or different metrics has been limited to drawing the cell boundaries, such as in Fig. 1. As mentioned, such drawings communicate little or no insight in the k-level hierarchy of subregions generated by the sites. In the following, we shall exploit other graphical dimensions, such as color and shading, to communicate more insight into the complexity of Voronoi diagrams.

3 Shaded Voronoi Diagrams

Our basic idea is to use the natural ability of the human visual system to interpret shade and color cues as boundaries of illuminated objects. We construct two types of such graphical objects, i.e. cushions and bevels, as follows.

3.1 Cushions

Suppose first we have a first order Voronoi diagram. For each site $p_i = (x_i, y_i)$ of the diagram, consider a curved surface $z(x, y)$ given by:

$$z(x, y) = h_1 \min_{i=1}^{n} [(x - x_i)^2 + (y - y_i)^2], \tag{1}$$

where h_1 is a (usually negative) height scale factor. In other words, we build a parabolic cushion under each site p_i. To shade the cushion, we use a diffuse shading model. The surface normal $\mathbf{n}$ is given by:

$$\begin{aligned}
\mathbf{n} &= [-\partial z/\partial x, -\partial z/\partial y, 1] \\
&= [2h_1(x_i - x), 2h_1(y_i - y), 1]
\end{aligned} \tag{2}$$

The final pixel intensity I is given by:

$$I = I_a + I_l \max(0, \frac{\mathbf{n} \cdot \mathbf{l}}{|\mathbf{n}||\mathbf{l}|}) \tag{3}$$

where I_a is the ambient light intensity and $\mathbf{l}$ and I_l are the direction and the intensity of a directional light source. This shading scheme (Fig. 2 a) is similar with the rectangular treemap cushions described by Van Wijk et al [9]. The shading of the cushions gives a visual cue for the distance of any point to the closest site, as any pixel is ultimately shaded in function of the closest site p_i.

For a k-order Voronoi diagram, we superimpose extra cushions on top of the k cushions generated by the k closest sites $p_1, ..., p_k$ for every pixel, to visualize the higher order structure. The center of the order-1 cushion is p_1, the center of the order-2 cushion is between p_1 and p_2, and the center of the order-i cushion is $c_i = \frac{1}{i} \sum_1^i p_i$. To emphasize the

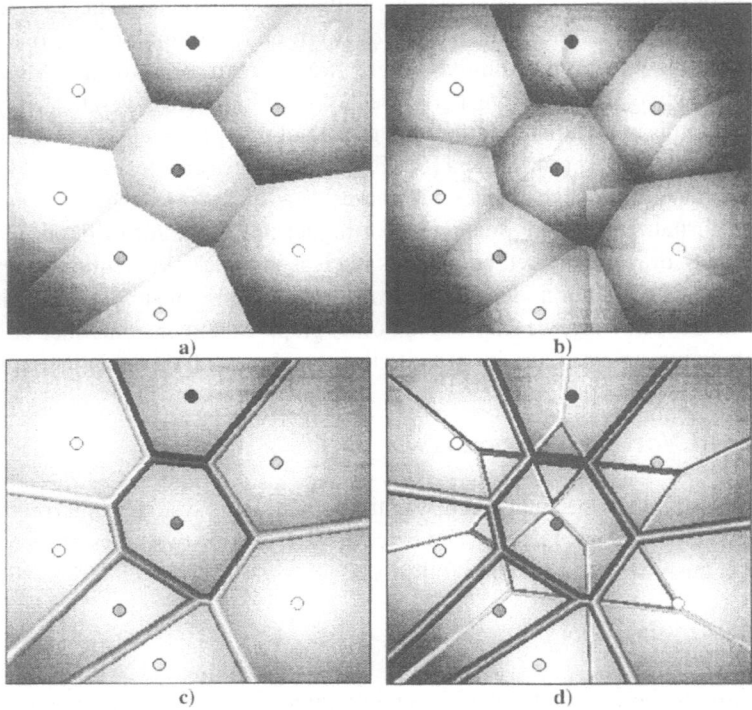

a)   b)

c)   d)

Fig. 2. Cushions (a,b) and cushions and bevels (c,d) visualization

cell nesting in the diagram, the height h_i of the i th level cushion is scaled to $h_i = s^{i-1}h_1$, where the scale factor s lies between 0 and 1. In this way, closer sites have a stronger effect on the cushions than farther ones. Figure 2 b shows the 2-order diagram rendering for the same sites as in Fig. 2 a. To enhance the nesting effect, we assign a hue to every site p_i and color all cushions influenced by that site with that hue (see Appendix). If desired, one could easily automate the assignment of hues to sites such that no two sites with neighboring first order diagram cells have the same hue.

3.2 Bevels

The shading described so far emphasizes the discontinuities between the cells of a Voronoi diagram as well as the nesting of higher order cells within the lower order ones. The shade of a pixel gives a cue to its distance to the sites that influence it. Although better than drawing cell contours only, shaded cushions still cannot indicate clearly the sites that influence a given pixel and the influence region of a given site. To emphasize these aspects, we augment our visualization by drawing *colored bevels* along the cell edges. The basic bevel idea is similar to the concept introduced by Bruls et al. for visualizing squarified treemaps [10].

A bevel is defined here as a parabolic surface of given height h_{bevel} and width w_{bevel} whose medial axis follows the Voronoi cell boundaries. Just as cushions, bevels are scaled to reflect the hierarchy level. For a bevel of level i we have:

$$w_{beveli} = t^{i-1} w_{bevel0} \quad \text{and} \quad h_{beveli} = t^{i-1} h_{bevel0} \tag{4}$$

where the bevel scaling factor t is between 0 and 1. Figures 2 c and d show the order-1 and order-2 bevels corresponding to the cushion diagrams in Figs. 2 a and b respectively.

While the bevel size depends on the order of the Voronoi cells that meet at that edge, its color indicates the two sites that are at equal distance from that edge. For two sites p_i and p_j, we color the two halves of the bevel equidistant to p_i and p_j with the hues of p_i and p_j.

The size and color coding of bevels is a non-ambiguous, intuitive way to interpret a k-order Voronoi diagram. To identify the k sites closest to a point p in the plane, we first look at the cushion containing p (see Appendix). The cushion's hue indicates the closest site, i.e. the site that has the strongest influence on p. The cell edge's bevels describe the other (at most $k - 1$) sites influencing the cell as follows:

– the *inner bevel hues* identify the sites
– the *bevel widths* are proportional with the distances to the sites

Conversely, to identify the regions influenced by a given site p_i, we first look at the cell in which p_i is found, i.e. the cell bearing p_i's hue. This is the region to which p_i is the closest site. To identify the regions to which p_i is the second,third,...k th closest site, we look at the regions bordered by bevels of p_i's hue of decreasing bevel width. These regions are nested (order $k + 1$ region contains order k region). Since the bevels of a region *never* overlap with the ones of a region of different order of the same site, it is rather easy to visually follow a bevel of a given width and color.

3.3 Interactive visualization

We have built an interactive application in which the user can control several aspects of the Voronoi diagram visualization orthogonally. First, the diagram order k, saturation, hue, height h, and scale factors s, t of both cushions and bevels can be chosen independently. This allows viewing a k-order Voronoi diagram in a multitude of ways, e.g. cushions or bevels only, or a gradual blend of the two, depending on the height and saturation factors. An effective combination is for example displaying only the first order cushions with low hue saturation and height values, together with the 1 to k order bevels with high hue and height values. In this way, the background is split into a tiling of non distracting cushions, whereas the bevels form the visualization foreground. A small bevel scale factor $t \leq 0.4$ produces an effective display of the diagram hierarchy: thicker bevels show the first order cells whereas thinner bevels depict the higher order cells.

Secondly, one can interactively add, remove, or drag the sites with the mouse. This gives a direct insight in the way the 2D plane is subjected to the influence of the sites and assists the user effectively in finding an optimal site distribution to a given situation. Problem-specific configurations are very easy to implement e.g. by constraining the sites to stay within or outside prescribed areas on a city or geographical map, such as markets, urban areas, lakes, forests, etc. Following this idea, we have constructed a user interface for a musical synthesizer in which the sites represent musical instruments and the Voronoi cells their influence areas (Fig. 4). To create a new sound, one drags a marker and/or the

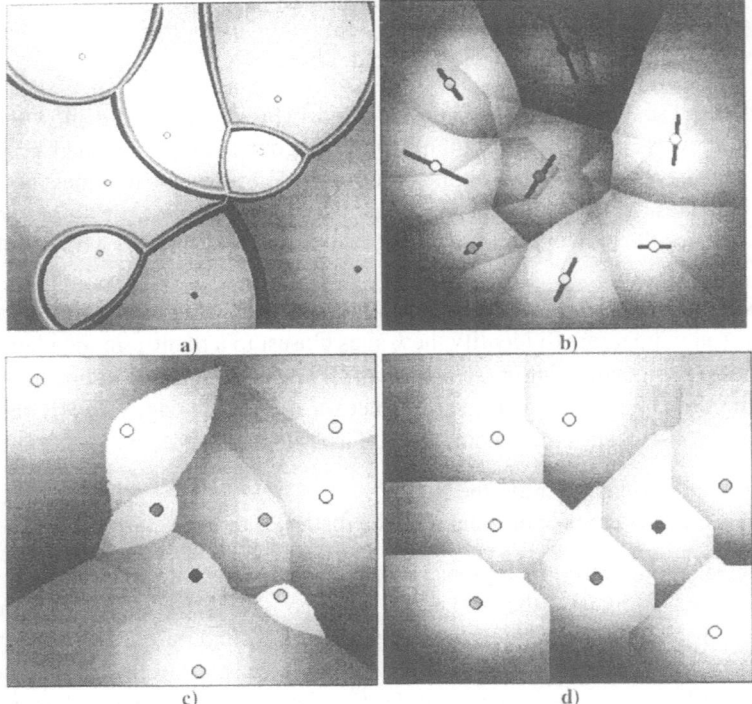

Fig. 3. Voronoi diagrams for different metrics

sites interactively on the Voronoi diagram. The sound is synthesized based on the marker-instrument sites distances.

Thirdly, the distance metric and site shape can be specified by the user. Figure 3 shows renderings for the multiplicative-weighted L_2 (Apollonius model) metric (a), an order-2 diagram for line sites (b), the additive-weighted L_2 (Johnson-Mehl model) metric (c), the L_1 (Manhattan distance) metric (d) (see also Appendix). In Fig. 5 a, a diagram defined by a discrete distance metric on a hexagonal board is shown. The map and cushion rendering are combined into a single image that shows the influence areas of several strategic sites in a computer military simulation [11] (Fig. 5 b).

4   Algorithm

A brute force implementation of the diagram rendering would compute $O(kNS)$ distances for a k-order diagram rendering of N pixels and S sites. Dragging a site with the mouse stops being interactive for $N > 200^2$ pixels and $S > 3$ sites. However, we can exploit the fact that all pixels of a given Voronoi cell share the same k closest sites. One idea would be to compute the k-order Voronoi diagram geometrically, which is $O(S log^3 S + k(S-k))$ complex [1], and then to use it as a spatial search structure to locate the k closest sites for each pixel. However, implementing a robust k-order analytic Voronoi diagram algorithm for any distance metric is a very complex task.

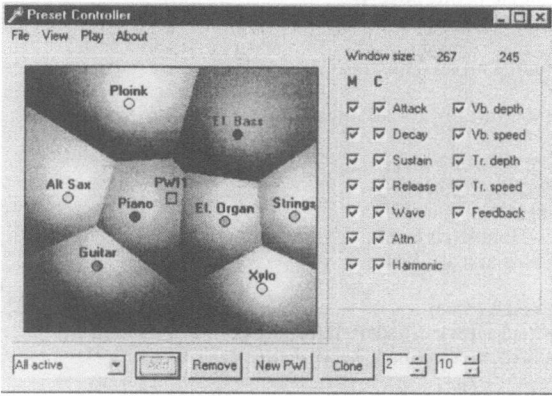

Fig. 4. Voronoi diagram used as input interface for a musical synthesizer

We present a simpler, yet very efficient pixel-based rendering approach that accepts different distance metrics and site shapes in a generic fashion for a k-order diagram.

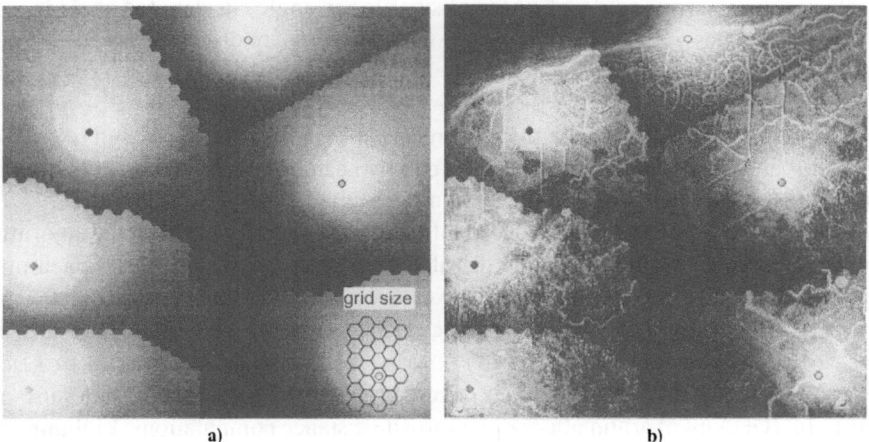

Fig. 5. a) Cushions for discrete distance metric b) Cushions layed over geographical map

4.1 Rendering first order diagrams

We use a recursive quadtree approach to reduce the number of sites considered per pixel. The recursion termination is based on the following observation. Consider an axis-aligned pixel rectangle to be shaded under the influence of two sites p_1 and p_2 (Fig. 6 b). Denote the minima and maxima of the distances of the sites i to the rectangle by $\min(d_i)$, $\max(d_i)$. If $\max(d_2) < \min(d_1)$, then all pixels p in the rectangle are influenced only by the site p_2.

```
ShadeRect(Rectangle r, Site sites[])
   {
1      for i in sites
2         compute min(d(i,r)) and max(d(i,r));

3      dlim := min(max(d(i,r))) for all i in sites;

4      if (s = {s1}) or (r is of 1 pixel size)
5            ShadeRectSingle(r,s1);
6      else if (s = {s1,s2})
7            ShadeRectDouble(r,s1,s2);
8      else  // s contains at least 3 sites
       {
9          split r in r1,r2,r3,r4;
10         ShadeRect(r1,sites);  ShadeRect(r2,sites);
11         ShadeRect(r3,sites);  ShadeRect(r4,sites);
       }

   }
```

a)   b)

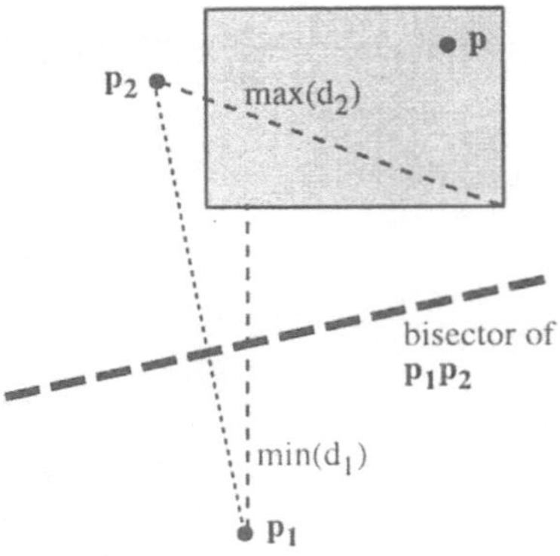

Fig. 6. a) Recursive algorithm b) Algorithm principle

We start with the complete pixel rectangle to shade r and a list of all sites *sites*. The recursive algorithm (Fig. 6 a) proceeds as follows. First, we determine $\min(d_i)$ and $\max(d_i)$ for all sites i to the rectangle r (distance of site i to rectangle r is denoted as d(i,r) in Fig 6 a). Next, sites j for which $\min(d_j) > \min(\max(d_i))$ are discarded since they don't influence the rectangle, according to the previous observation. If one or two sites remain, the rectangle is scan converted by the ShadeRectSingle, respectively ShadeRect-Double optimized routines, else the rectangle is split in four and processed recursively. As a result, the recursion arrives at pixel level only near points that are at equal distances from at least three sites. For S sites, there are less than $2S - 4$ such points [5, 2]. Figure 7 a shows the subdivision for an order-1 diagram of 8 sites. To estimate the algorithm's complexity, assume S sites randomly distributed over an area of N pixels. Each of the $2S - 4$ points where subdivision reaches pixel level influences thus $\frac{N}{S}$ pixels on the average. On the other hand, a region of n pixels needs $\log_2 \sqrt{n}$ steps to be subdivided until pixel level. Therefore we need $O(S\log_2 \sqrt{N/S})$ ShadeRectSingle and ShadeRectDouble calls to render the whole N pixels covered by S sites. These two routines use only fixed-point (DDA-like) arithmetic for pixel-to-site distance computations. Lighting (3) is accelerated via table lookup. For each discrete gray level $\frac{l_i}{|n||l|}$, we store the corresponding value of d^2 in a table. As starting point for looking up shade as function of distance, the gray level of the previous pixel is used. Overall, we achieve 16 frames per second in software rendering for a 400x400 image with 20 sites on a Pentium II 350 MHz processor.

4.2   Rendering k-order diagrams, bevels, and different metrics

To render k-order diagrams, we modify the line 3 of the algorithm such that $dlim$ is the kth smallest $\max(d)$ instead of the first. To render bevels of a given image-size width w, we adapt the ShadeRectSingle and ShadeRectDouble routines to compute the *Euclidian* distance from any pixel p to the closest Voronoi cell edge. This edge is found by searching for the closest pixel q where $f(q) = d(s_i, q) - d(s_j, q) = 0$, for any sites

s_i, s_j. The search is implemented independently on the user-chosen Voronoi metric d by following f's steepest descent (gradient) from the current pixel p.

An important result is that the subdivision algorithm is independent on the distance metric and site shape. The only operations needed are the distance between a site and a point and the minimum and maximum distances between a site and an axis-aligned rectangle. Overall, rendering Voronoi diagrams with circular or hyperbolic cell edges (the Apollonius, respectively Johnson-Mehl models) and diagrams for line sites (Fig. 3) is as fast as rendering the L_2 norm- and point-based diagrams.

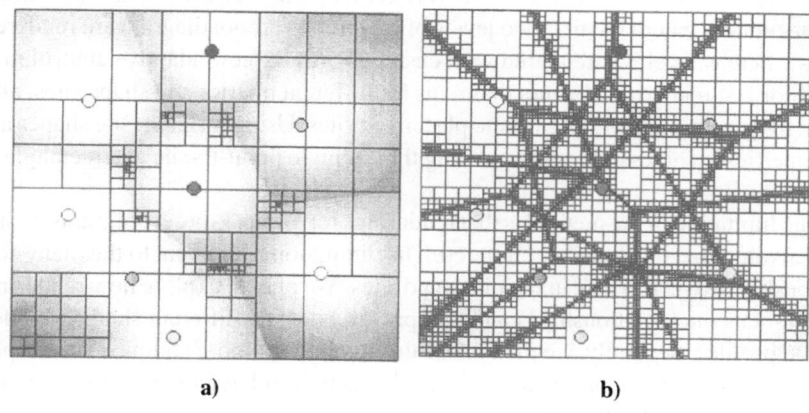

a)   b)

Fig. 7. Subdivision for order-1 (a) and order-2 (b) diagram rendering

For $k > 1$, rendering k-order diagrams is slower than for first-order diagrams, since we haven't implemented a k-order `ShadeRectDouble` procedure. Pixel-level subdivision calling the `ShadeRectSingle` routine occurs now on all pixels along the cell edges. Figure 7 b shows the subdivision for the 2-order diagram of the sites in Fig. 7 a. To estimate the complexity, consider first an order-1 diagram of S sites spread over N pixels. In the worst case, a site causes an edge of $n = O(\sqrt{N})$ pixels length. To subdivide such an edge to pixel level, $O(n)$ steps are needed. For S sites we perform thus at most $O(S\sqrt{N})$ `ShadeRectSingle` calls. An order-k diagram has a total edge length at most k times larger than the order-1 diagram, so its rendering complexity is $O(S\sqrt{kN})$, which is still better than the $O(kNS)$ brute-force approach. The method still performs in near real-time for 2-order diagrams on the above metioned platform.

Other pixel-level Voronoi diagram algorithms exist. A similar algorithm presented by Vleugels [7] uses a top-down rectangle subdivision based on the distances between the rectangle corners and the sites. However, this algorithm proceeds till pixel level along *all* Voronoi boundaries even for order-1 diagrams, and computes distances to *all* sites at any subdivision level. The authors reported around 5 seconds per frame for configurations similar to the ones we compute in real time. Other fast pixel-level algorithms use OpenGL hardware to compute Voronoi diagrams by rendering polygonal approximations of the distance functions [8]. Although impressive as performance, it is not clear how easy is to extend such algorithms to arbitrary distance functions, k-order diagrams, and maintain pixel-level distance computing accuracy.

174

5 Discussion and Future Work

We have presented a new method to visualize k-order Voronoi diagrams generated by different metrics and site shapes. A simple graphical element - a shaded, colored parabolic cushion - is used to visualize both edge and surface information. Spatial nesting is exploited in two different ways to depict the k levels of a k-order diagram in a single image. This visualization answers two questions in a compact way, namely: "which are the k sites influencing any given point?" and "which are the k hierarchical influence areas of a given site?". By varying the cushions' hue, height, and saturation one can intuitively emphasize the different aspects of the Voronoi diagrams. The above are best illustrated by Appendix b, where the first two levels of a 3-order Voronoi diagram are rendered with bevels whereas level 3 is rendered with cushions only. A new adaptive algorithm is presented for fast rendering Voronoi diagrams for different metrics and shape sites, allowing interactive exploration of optimal site placement sites. Using different site shapes and distance metrics involves only implementing three simple point-to-site and rectangle-to-site distance functions.

One limitation of the method is that renderings for S-site, k-order diagrams with higher k and S values become quickly cluttered. This limitation is inherent to the many-to-many relationship between points in the plane and sites. We plan to explore how rendering such higly-dimensional relationships can be improved by using different shading models and cushion profiles. Secondly, we plan to use this method for more complex site shapes than lines, for visualizing several mathematical abstractions related to Voronoi diagrams such as medial axes and skeletons [2].

References

1. P. AGARWAL, M. DE BERG, J. MATOUSEK, AND O. SCHWARZKOPF, *Constructing levels in arrangements and higher order Voronoi diagrams*, SIAM J. Comp., no. 27, 1998, pp. 654-667.
2. F. AURENHAMMER, *Voronoi Diagrams: A Survey of a Fundamental Geometric Data Structure*, ACM Computing Surveys, no. 23, 1991, pp. 345-405.
3. D. LEE AND C. WONG, *Voronoi diagrams in the L_1 and L_{inf} metrics with 2-dimensional storage applications*, SIAM J. Computing, no. 9, 1980, pp. 200-211.
4. A. OKABE, B. BOOTS, AND K. SUGIHARA *Spatial Tessellations: Concepts and Applications of Voronoi Diagrams*, John Wiley & Sons, Chichester, UK, 1992.
5. M. DE BERG, M. VAN KREFELD, M. OVERMARS, O. SCHWARZKOPF, *Computational Geometry - Algorithms and Applications*, Springer, 1998.
6. M. SAKAMOTO AND M. TAKAGI, *Patterns of weighted Voronoi tessellations*, Sci. Forum, no. 3, 1988, pp. 103-111.
7. J. VLEUGELS AND M. OVERMARS *Approximating Generalized Voronoi Diagrams in Any Dimension*, Technical Report UU-CS-1995-14, Utrecht University, 1995.
8. K. HOFF, T. CULVER, J. KEYSER, M, LIN, D. MANOCHA, *Fast Computation of Generalized Voronoi Diagrams Using Graphics Hardware*, Proc. SIGGRAPH '99, ACM Press, 1999, pp. 277-286.
9. J. J. VAN WIJK, H. VAN DE WETERING, *Cushion Treemaps: Visualization of Hierarchical Information*, Proc. Information Visualization '99, IEEE Press, pp. 73-78.
10. M. BRULS, J. J. VAN WIJK, K. HUIZING, *Squarified Treemaps*, Proc. IEEE VisSym 2000, Springer, 2000, pp. 33-42
11. STRATEGIC SIMULATIONS, INC., *Panzer General 3D*, http://www.ssionline.com/, 1999.

Editors' Note: see Appendix, p. 347 for colored figure of this paper

Preserving the Mental Map
using Foresighted Layout

Stephan Diehl, Carsten Görg and Andreas Kerren

University of Saarland, FR 6.2 Informatik,
PO Box 15 11 50, D-66041 Saarbrücken, Germany
{diehl,goerg,kerren}@cs.uni-sb.de

Abstract. First we introduce the concept of graph animations as a sequence of evolving graphs and a generic algorithm which computes a Foresighted Layout for dynamically drawing these graphs while preserving the mental map. The algorithm is generic in the sense that it takes a static graph drawing algorithm as a parameter. In other words, trees can be animated with a static tree layouter, graphs with a static Sugiyama-style layouter or a spring embedder, etc. Second we discuss applications of Foresighted Layout in algorithm animation and visualization of navigation behaviour.

1   Introduction

Most work on graph drawing addresses the problem of layouting a single, static graph. Algorithms have been developed for different classes of graphs (trees, dags, digraphs, ...) and different aesthetic criteria, like minimizing crossings and bends or maximizing symmetries [1, 7]. But the world is full of dynamic graphs, e.g. animations of graph algorithms or algorithms which work on pointered data structures, dynamic visualisations of resource allocation in operating systems and project management, network connectivity and the constantly changing hyperlink structure of the web.

Dynamic graph drawing addresses the problem of layouting graphs which evolve over time by adding and deleting edges and nodes. This results in an additional aesthetic criterium known as "preserving the mental map" [8].

The ad-hoc approach is to compute a new layout for the whole graph after each update using those algorithms developed for static graph layout. In most cases this approach produces layouts which do not preserve the mental map. The common solution is to apply a technique known from key-frame animations called inbetweening to achieve "smooth" transitions between subsequent graphs, i.e. animations show how nodes are moved to their new positions. This approach yields decent results if only a few nodes change their position or whole clusters are moved without substantially changing their inner layout. But in most cases the animations are just nice and do neither convey much information nor help to preserve the mental map. Incremental algorithms try to change the layout just as far as to accomodate the update. Unfortunately, in the worst case they have to compute the layout of the whole graph.

In this paper we present a totally different approach. Given a sequence of n graphs we compute a global layout which induces a layout for each of the n graphs. A unique features of this approach is that once they are drawn on the screen neither nodes nor the bends of edges change their positions in graphs subsequently drawn. Using static graph layouters, which accepts fixed node positions as an additional input, it is also possible that only the bends change their positions. We call the algorithm *Foresighted Layout* as it knows the future of the graph, i.e. the next $n - 1$ modifications.

In particular applications can be visualized post mortem using information stored in log files. In Sections 5 and 6 we discuss two applications of Foresighted Layout. First in the area of algorithm animation we used it to visualize the generation of finite state automata which are drawn as state transition diagrams. Second, in the area of web visualization, we have implemented a portal which logs the web pages and links visited by a user. We visualize these logs later to analyse the navigation behavior of different users.

2 Graph Animations

In the following we consider graphs with multi-edges. For this we add unique identifiers to each edge.

Definition 1 (Graph). *A* **graph** $g = (V, E)$ *consists of a set of nodes V, a set of edges $E \subseteq V \times V \times \mathrm{Id}$ and for all $(v_1, v_2, n), (v_1', v_2', m) \in E : n = m \Rightarrow v_1 = v_1', v_2 = v_2'$.*

We define a graph animation as a sequence of graphs. A graph results from modifications (adding or deleting nodes and edges) of its preceding graph. Usually subsequent graphs in a graph animation share some nodes and edges. But in the worst case each graph can consist of totally different nodes and edges.

Definition 2 (Graph Animation). *A* **graph animation** G *is a sequence $G = [g_1, \ldots, g_n]$ of graphs with $G_i = (V_i, E_i)$ and for all $(v_1, v_2, n) \in E_p, (v_1', v_2', m) \in E_r$ with $1 \le p, r \le n : n = m \Rightarrow v_1 = v_1', v_2 = v_2'$.*

The restriction in this definition ensures that edge identifiers are used consistently in all graphs, i.e. for edges between the same nodes.

3 Foresighted Layout

A first approach to layout a graph animation is to compute its super graph and to reuse its layout information for the layout of the individual graphs in the animation.

Definition 3 (Super Graph). *Let G be a graph animation $G = [g_1, \ldots, g_n]$ with $g_i = (V_i, E_i)$, then the* **super graph** $\widehat{G}$ *of G is defined as $\widehat{G} = (\widehat{V}, \widehat{E})$ with $\widehat{V} = \bigcup_{i=1}^{n} V_i$ and $\widehat{E} = \bigcup_{i=1}^{n} E_i$.*

In general the super graph will be large and there will be much unused space in the layout of each individual graph. To avoid this *Foresighted Layout* constructs on the basis of the super graph a smaller graph by taking into account the live times of the nodes and edges in the graph animation.

Definition 4 (Live Time). *Let* $G = [g_1, \ldots, g_n]$ *be a graph animation and* $\widehat{G} = (\widehat{V}, \widehat{E})$ *its super graph where* $g_i = (V_i, E_i)$. *Then* $T(v) = \{i | v \in V_i\}$ *are the* **live times** *of the node* $v \in \widehat{V}$ *and* $T(n) = \{i | (v, w, n) \in E_i\}$ *are the live times of the edge identified by* n.

3.1 Graph Animation Partitionings

Definition 5 (Graph Partitioning). *Let* $g = (V, E)$ *be a graph and* $\widetilde{V} \subseteq \mathcal{P}(V)$ *and* $\widetilde{E} \subseteq \widetilde{V} \times \widetilde{V} \times \mathrm{Id}$. *A graph* $\widetilde{g} = (\widetilde{V}, \widetilde{E})$ *is a* **graph partitioning** *of* g *iff the nodes in* $\widetilde{V}$ *are disjoint,* $\bigcup_{v \in \widetilde{V}} v = V$ *and* $(\widetilde{v_1}, \widetilde{v_2}, n) \in \widetilde{E} \Leftrightarrow \exists v_1 \in \widetilde{v_1}$ *and* $v_2 \in \widetilde{v_2} : (v_1, v_2, n) \in E$. *We call* $\widetilde{E}$ *the set of edges induced by* $\widetilde{V}$.

In other words, $\widetilde{V}$ is a partitioning of V. Each node in $\widetilde{V}$ represents one or more nodes from V and all edges between two nodes in V are converted into edges between the representatives of the two nodes.

Definition 6 (Graph Animation Partitioning GAP). *Let* $G = [g_1, \ldots, g_n]$ *with* $g_i = (V_i, E_i)$ *be a graph animation and* $\widehat{G} = (\widehat{V}, \widehat{E})$ *be the super graph of* G. *A graph partitioning* $\widetilde{g} = (\widetilde{V}, \widetilde{E})$ *of* $\widehat{G}$ *where* $\widetilde{V} = \{P_1, \ldots, P_k\}$ *is a* **graph animation partitioning** *of* G *iff* $v, v' \in P_r \Rightarrow T(v) \cap T(v') = \emptyset$.
We call $\widetilde{g}$ *a minimal GAP of* G, *if there exists no GAP of* G *with less nodes.*

In a GAP nodes with disjoint live times are grouped together. Unfortunately, the problem of computing a minimal GAP (hence mGAPP) is $\mathcal{NP}$-complete. We have proven the $\mathcal{NP}$-completeness of mGAPP and mRGAPP (see Section 3.3) by reduction on the minimal graph coloring problem [2,6]. Now we present an algorithm which computes a GAP in $O(n^2)$ where n is the number of nodes of the super graph.

Algorithm 1 (Computing a GAP).
$W := \widehat{V}, P := [\,], p := 0$
While $v \in W$ do
　If $\exists j : T(v) \cap T(P_j) = \emptyset$ then
　　　$P_j := P_j \cup \{v\}, T(P_j) := T(P_j) \cup T(v)$
　else
　　　$p := p + 1, P_p := \{v\}, T(P_p) := T(v)$
　$W := W - \{v\}$

3.2 Strategies for Computing a GAP

From an aesthetical point of view it is not too bad that we do not compute minimal GAP's. A minimal GAP is often not the best choice as we pay for the minimal number of nodes by an increased number of edge crossings. In Algorithm 1 we have not specified in which order the life times of the node v and the already computed partitions P_j are compared, i.e. how to find a j such that $T(v) \cap T(P_j) = \emptyset$. In our implementation we can choose one of the following strategies which in general yield different GAPs:

1. Search the list from P_1 to P_p.
2. Search the list from P_p to P_1.
3. Add v to the partition with the smallest number of nodes.
4. Only allow a limited number of nodes in a partition. If there is no partition with less nodes, then create a new partition.
5. Only allow a limited number of edges in a partition.
6. Give priority to nodes with induced edges to the same already computed partitions.

3.3 Reduced Graph Animation Partitionings

In a GAP the number of nodes of the super graph of a graph animation is reduced. In a similar way, the number of edges can be reduced.

Definition 7 (Reduced Graph Animation Partitioning RGAP). *Let $G = [g_1, \ldots, g_n]$ with $g_i = (V_i, E_i)$ be a graph animation and $\widetilde{g} = (\widetilde{V}, \widetilde{E})$ be a GAP of G. The graph $\overline{g} = (\widetilde{V}, \overline{E})$, where $\overline{E} \subseteq \widetilde{V} \times \widetilde{V} \times \mathcal{P}(\mathrm{Id})$, is a reduced GAP, iff $\forall (\widetilde{v}_1, \widetilde{v}_2, \{m_1, \ldots, m_k\}) \in \overline{E}$ the following holds:*
$(\widetilde{v}_1, \widetilde{v}_2, m_i), (\widetilde{v}_1, \widetilde{v}_2, m_j) \in \overline{E} : T(m_i) \cap T(m_j) = \emptyset$ *for* $1 \leq i < j \leq k$
We call $\overline{g}$ a minimal RGAP of G, if there exists no RGAP of G with less edges.

An edge $(\widetilde{v}_1, \widetilde{v}_2, \{m_1, \ldots, m_k\})$ of the RGAP represents k edges which exist at different times, i.e. in different graphs of the graph animation, between a node in $\widetilde{v}_1$ and $\widetilde{v}_2$. But it does not represent two or more multi-edges which exist at the same time; they can not be represented by a single edge in the RGAP. Also the problem of computing a minimal RGAP (hence mRGAPP) is $\mathcal{NP}$-complete.

As computing minimal RGAPs is $\mathcal{NP}$-complete, we present a faster algorithm ($O(m^2)$ where $m = |\widehat{E}|$) which does not compute minimal RGAPs, but yields good results in practice, i.e. RGAPs with small numbers of edges. The algorithm actually computes the partitioning of the edge identifiers for an RGAP.

Algorithm 2 (Computing a RGAP).
$W := \{m_1, \ldots, m_k\}$, i.e. the set of all identifiers occuring in $\widetilde{E}$
$P := [], p := 0$
While $n \in W$ do
  Let $(\widetilde{v}, \widetilde{w}, n)$ be the edge identified by n.
  $p := p + 1, P_p := \{n\}, T(P_p) := T(n)$

While $\exists m \in W$ with $(\widetilde{v}, \widetilde{w}, m)$ and $T(P_p) \cap T(m) = \emptyset$ then
$\qquad P_p := P_p \cup \{m\}, T(P_p) := T(P_p) \cup T(m), W := W - \{m\}$
$W := W - \{n\}$

3.4 Algorithm

After we have seen how to compute RGAPs, we now show how they can be used in combination with a static graph layouter to draw a sequence of graphs while preserving the mental map.

Algorithm 3 (Forsighted Layout).
```
foresightedLayout([g_1,...,g_k], staticLayouter())
  { g =computeGAP(g_1,...,g_k)
    g̅ =computeRGAP(g)
    layout=staticLayouter(g̅)
    for i = 1 to k
        drawGraph(g_i, layout)
  }
```

We call the static layouter to compute a layout of the RGAP of the graph animation. We assume that the static layouter returns a layout, i.e. a data structure containing the positions of each node and polylines (or bends) for each edge. The function drawGraph() gets this data structure and a graph of the graph animation. For each node in the graph it uses the layout information of its super node, i.e. the node in the RGAP it is a member of. For each edge it uses the layout information of the bends of the edge in the RGAP which contains its identifier.

4 Implementation

We have implemented Foresighted Layout in Java as part of an API which we use for algorithm animations [5]. The class AnimatedGraph of this API has the following interface:

```
class AnimatedGraph {
  public AnimatedGraph(GView view)
  public void insertNode(Node n)
  public void insertEdge(Edge e)
  public void deleteNode(Node n)
  public void deleteEdge(Edge e)

  public void snapshot()

  public void play()
  public void next()
  public void back()
```

```
public void perform(Object target, String methodname, Object arg)
        throws NoSuchMethodException
public void perform(Object target, String methodname, Object arg,
              Object reverseTarget, String reverseMethodname,
              Object reverseArg)
        throws NoSuchMethodException
}
```

The class provides methods to build and modify a graph, to record a graph animation by doing snapshots of individual graphs and replay the animation afterwards.

A node can be a specialization of any AWT component which has to implement a certain interface (a few additional methods). Thus it is also possible to draw a graph in a node of a graph again. As the nodes can be AWT components, one can also destructively change attributes of these objects during the recording sessions. To defer these changes until the animation is replayed, such changes must be done using the method perform(), which puts the method calls into a data structure and invokes them later using Java Reflection.

The basic idea of the static layout algorithm used in our examples is to divide the nodes into several levels. Then the algorithm computes the relative positions of the nodes within these levels, so that edge crossings are minimized [11, 9]. Ideally this method can be used for directed graphs, because the direction of the edges can be used for the layouting process.

4.1 Drawing Graphs in 4 Dimensions

In addition to showing the different graphs in a graph animation one after another in 2D, we have implemented a 3D viewer in Java3D which uses the third dimension as a time axis. As a result we can show several graphs simultaneously in a history view. In addition the 3D viewer can show the supergraph in the background, see Figure 3. Finally it allows to interact and customize the view in various ways including recolouring, translating and rotating the graph.

5 Algorithm Animation

Algorithm animation is one of the most prominent areas of software visualization. The GaniFA applet visualizes and animates several generation algorithms from automata theory including the generation of a non-deterministic finite automaton (NFA) from a regular expression RE [12]. We have included GaniFA into an electronic textbook on automata theory to allow interactive exercises [3–5].

In case of visualizing transition diagrams of finite automata our static layout algorithm is a good choice, but the algorithm $RE \rightarrow NFA$ changes the graph successively. Animations of algorithms which change graphs, i.e. add or delete nodes and edges, are often very confusing, because after each change a new layout of the current graph is computed. In this new layout nodes are moved to

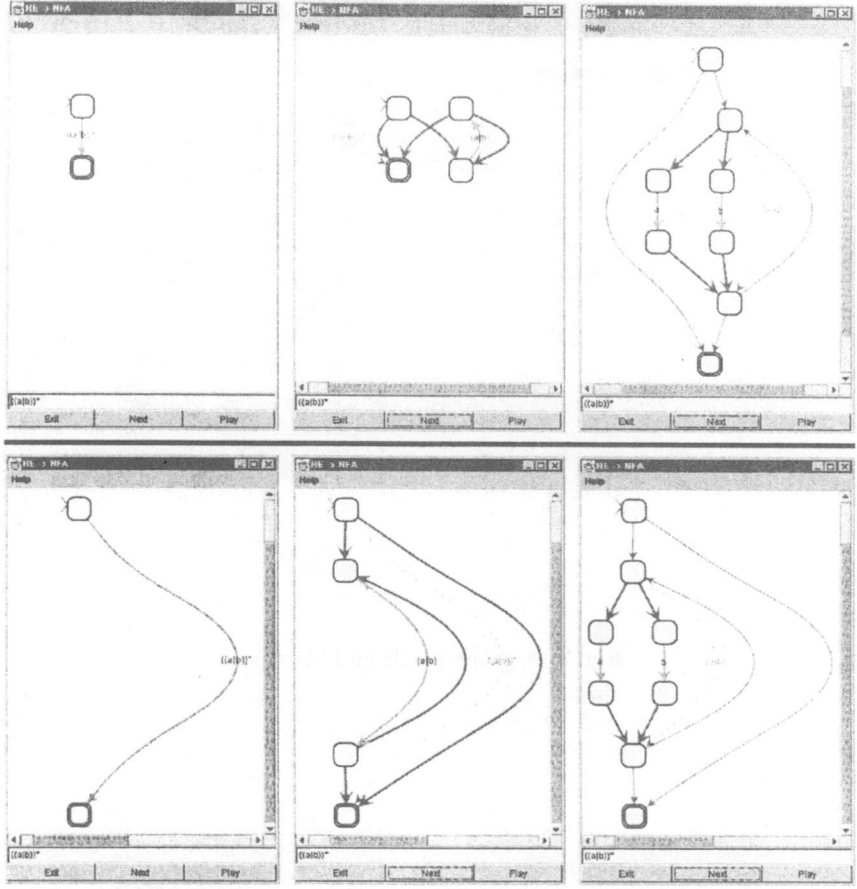

Fig. 1. Ad-hoc and foresighted layout of the intermediate and final NFA for $(a|b)^*$.

different places although the algorithm didn't actually change these nodes. As a result it is not clear to the user what changes of the graph are due to the graph algorithm and what changes are due to the layout algorithm.

The lower part of Figure 1 shows how Foresighted Layout can be used to animate the conversion of a regular expression $(a|b)^*$ into an appropriate nondeterministic finite state automaton ($RE \rightarrow NFA$). In contrast to the upper part of Figure 1, which shows the same conversion, this visualization is significantly more clear because once created, a node doesn't change its position.

6 Visualization of Navigation Behavior

As a user browses through web pages he unfolds a subgraph of the web. As he moves from one page to another, new pages become directly accessible as hy-

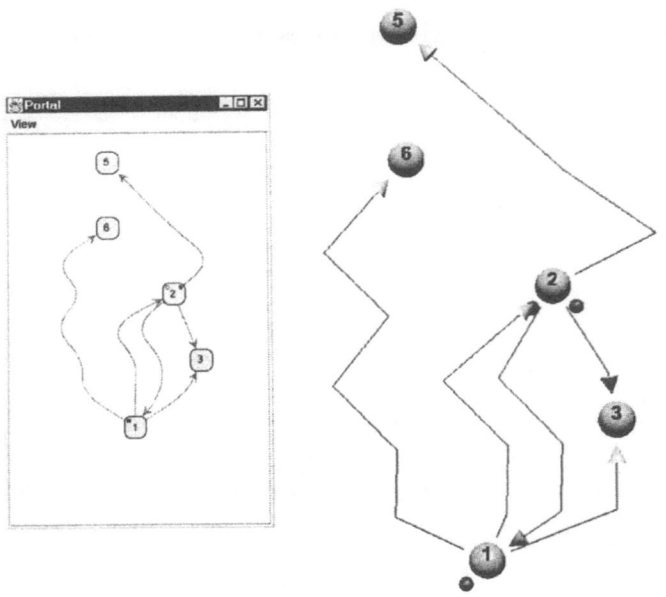

Fig. 2. A graph in 2D and 3D view.

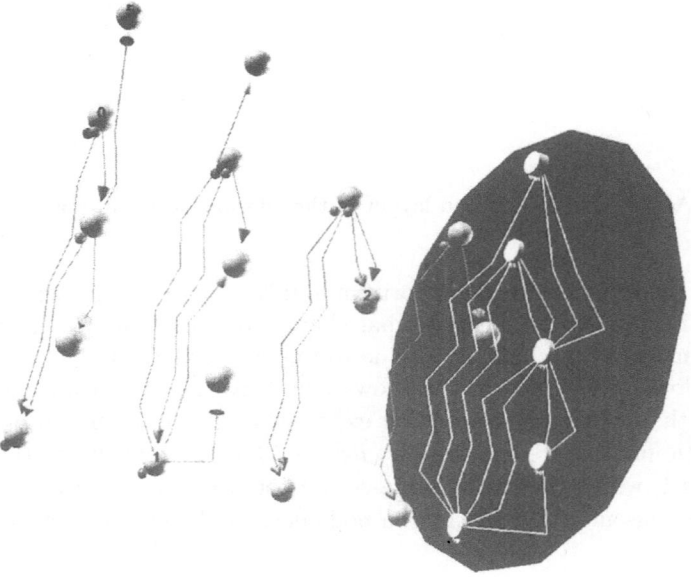

Fig. 3. A history view of navigation by two users.

perlinks, whereas those which have been directly accessible before, might not be accessible from the current page. To analyse the browsing or navigation behavior of one or several users we visualize the subgraphs which they are currently aware of. We can visualize the subgraphs of different users at the same time sharing nodes, if these represent the same web page. The edges in each subgraph are drawn in the users color. Thus we see when they visit the same page or have a links to the same pages. In Figure 2 we see, that both users are currently on web page 2 as indicated by the two circles with the colors of the users in node 2.

There are various ways to acquire the necessary information in log files. We have implemented a web page, which we call portal, which logs the web pages and links visited by a user who enters our web server through this portal. We visualize these logs later to analyse the navigation behavior of different users.

With the help of the portal and a tool which uses foresighted layout it is possible to analyse the learner navigation through the electronic textbook discussed above. Thus, we can see what pages a student looks at or that a student frequently uses the glossary. If we visualize the behavior of several students we might find that different students catch up (synchronize) on certain pages, although their navigation behavior differs considerably on intermediate pages.

These analyses are facilitated by the history view of the 3D version of our foresighted graphlayout, see Figure 3.

7 Interactive Graph Animations

In most applications the future of a graph depends on user input. Nevertheless between such points in time when the user interacts with the application, the program can perform several "foreseeable" changes of the graph. Thus the execution of such an interactive application can be modeled as a sequence of graph animations. When we draw a graph animation of such a sequence on the screen, we do not know the next animation in the sequence but we know the one before. As a "smooth" transition between the previous and the actual graph animation we can use the traditional morphing approach. More precisely: Let $G = [g_1, \ldots, g_n]$ be the previously drawn graph animation. Then graph g_n was drawn on the screen using the Foresighted Layout for an RGAP $\bar{g}$ of G. Now the user does some input and triggers the graph animation $G' = [g_1', \ldots, g_k']$. To draw this animation the application computes an RGAP $\bar{g'}$ of G' and uses morphing between the graph g_n with node and edge positions as in $\bar{g'}$ and g_1' with node and edge positions as in $\bar{g'}$.

8 Conclusion

We have presented the motivation and theory behind Foresighted Layout. Using our generic algorithm existing static graph drawing algorithms can be used for graph animations which preserve the mental map. The algorithm has been implemented in Java and in particular used for algorithm animations and visualization of navigation behaviour. For these kinds of application it provides

184

better results than traditional approaches which use smooth transitions and/or incremental changes of the layout, e.g. using the VCG tool [10]. For the analysis of navigation behaviour the history of our 3D viewer turned out to be very beneficial.

Acknowledgement This research has been partially supported by the German Research Council (DFG) under grant WI 576/8-1 and WI 576/8-3. For helping with the implementation of the 3D viewer the authors thank Peter Blanchebarbe.

References

1. G. Di Battista, P. Eades, R. Tamassia, and I. G. Tollis. Algorithms for Drawing Graphs: an Annotated Bibliography. *Computational Geometry: Theory and Applications*, 4:235–282, 1994.
2. S. Diehl, C. Görg, and A. Kerren. Foresighted Graphlayout. Technical Report A/02/2000, FR 6.2 - Informatik, University of Saarland, December 2000. http://www.cs.uni-sb.de/tr/FB14.
3. S. Diehl and A. Kerren. Increasing Explorativity by Generation. In *Proceedings of World Conference on Educational Multimedia, Hypermedia and Telecommunications, EDMEDIA-2000*. AACE, 2000.
4. S. Diehl and A. Kerren. Levels of Exploration. In *Proceedings of the 32nd Technical Symposium on Computer Science Education, SIGCSE 2001*. ACM, 2001.
5. GANIMAL. Project Homepage. http://www.cs.uni-sb.de/GANIMAL.
6. M. R. Garey and D. S. Johnson. *Computers and Intractability. A Guide to the Theory of $\mathcal{NP}$-Completeness*. Freeman and Company, 1979.
7. I. Herman, G. Melancon, and M. S. Marshall. Graph Visualization and Navigation in Information Visualization: A Survey. *IEEE Transactions on Visualization and Computer Graphics*, 6(1):24–43, 2000.
8. K. Misue, P. Eades, W. Lai, and K. Sugiyama. Layout Adjustment and the Mental Map. *Journal of Visual Languages and Computing*, 6(2):183–210, 1995.
9. G. Sander. *Visualization Techniques for Compiler Construction*. Dissertation (in german), University of Saarland, Saarbrücken (Germany), 1996.
10. G. Sander, M. Alt, C. Ferdinand, and R. Wilhelm. CLaX - a Visualized Compiler. In F. J. Brandenburg, editor, *Graph Drawing (Proc. GD '95)*, volume 1027 of *Lecture Notes Computer Science*. Springer-Verlag, 1996.
11. K. Sugiyama, S. Tagawa, and M. Toda. Methods for Visual Understanding of Hierarchical Systems. *IEEE Transactions on Systems, Man and Cybernetics*, SMC-11(2):109–125, 1981.
12. Reinhard Wilhelm and Dieter Maurer. *Compiler Design: Theory, Construction, Generation*. Addison-Wesley, 2nd printing edition, 1996.

Editors' Note: see Appendix, p. 348 for colored figures of this paper

Visualization of directed associations in e-commerce transaction data

Ming C. Hao, Umeshwar Dayal, Meichun Hsu,
Thomas Sprenger*, Markus H. Gross*
Hewlett Packard Research Laboratories, Palo Alto, CA.
(ming_hao, dayal, mhsu)@hpl.hp.com

Abstract. Many real-world e-commerce applications require the mining of large volumes of transaction data to extract marketing and sales information. This paper describes the Directed Association Visualization (DAV) system that visually associates product affinities and relationships for large volumes of e-commerce transaction data. DAV maps transaction data items and their relationships to vertices, edges, and positions on a visual spherical surface. DAV encompasses several innovative techniques (1) items are positioned according to their associations to show the strength of their relationships; (2) edges with arrows are used to represent the implication directions; (3) a mass-spring engine is integrated into a visual data mining platform to provide a self-organized graph. We have applied this system successfully to market basket analysis and e-customer profiling Internet applications.

1 Introduction

Market basket analysis has become a key success factor in e-commerce. Effective market basket analysis methods employ association [1, 4] and clustering [6] as methods of analyzing such data. E-commerce transactions often are comprised of several products (items) that are purchased together. An example of an association is that 85% of the people who buy a printer also buy paper. Understanding these relationships across hundreds of product lines and among millions of transactions provides visibility and predictability into product affinity purchasing behavior.

To date, there are many technologies that allow the visualization of associations for retail stores to make business decisions such as product recommendations, cross selling, and store shelf arrangements. As illustrated in Figure 1A, a common technique [7, 8], for visualizing associations is the matrix display. The matrix technique positions pairs of items on separate axes to visualize the strength of their relationships. The association visualizer from SGI MineSet [8] lays out the rules on a 3D grid landscape. Visual filtering and querying allow users to focus in on selected rules. However, to visualize millions of association rules, we have found that association matrixes are too restrictive. The number of rules shown at the same time needs to be pre-decided and can only be a small range of rules (10-20).

An alternative to the association matrix, as illustrated in Figure 1B, is to lay out associations on a graph. For example, LikeMinds [11] uses an individual purchase history to make suggestions to shoppers based on a graph. However, when the number of items grows large, the graph can quickly become cluttered with many interactions. Also, associated items may not be placed close together. The market analysis graph from Insights' Advizor [12] has achieved dramatic improvements by utilizing dynamic

*Presently with Department of Computer Science, Swiss Federal Institute of Technology, Zurich, Switzerland
sprenger@inf.ethz.ch

queries and presentations.

Besides the use for data associations, graph visualization methods have been very popular in information visualization. For instance, cone trees [17] and their hyperbolic projections [18, 19] are prominent examples for web and file system visualization. Eric [15] used fast graph layout to display various types of statistical data. A central approach to graph visualization are physics-based paradigms being exploited in [2], [9], or [10]. Recently, clustering algorithms improved performance and scalability of physics-based methods [20, 21].

In spite of the advances in the field, it is still difficult to mine and visualize customer's purchasing behavior from millions of Internet transactions. As the volume of e-commerce data grows and as the transaction data is integrated into offline data, new data visualization associations are required to extract useful and relevant information.

In this paper, we describe a system for visualizing associations of purchasing transactions for two and more items.

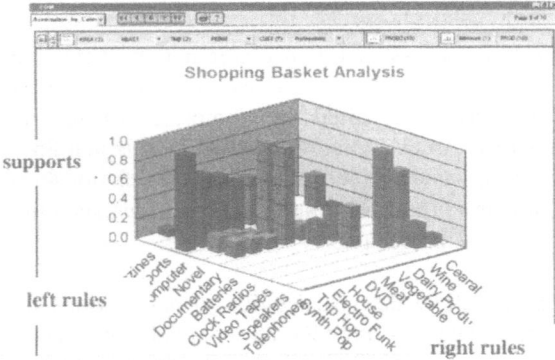

Figure 1A: A Matrix Technique

Figure 1B: A Graph – Based Technique

2 Our approach

At HP Laboratories, we have devised a "Directed Association Visualization (DAV)" 3D system. To meet the needs mentioned previously, we attempt to visualize the internal relationships and implications between large volumes of transaction data. DAV maps the transaction items and relationships to vertices and positions on a visual spherical surface. DAV uses weighted edges with arrows to represent association directions and levels. In addition, DAV employs dynamic aggregation and hierarchical link lists to enhance the scalability.

DAV integrates a well-known physics-based "Mass Spring" visualization system [2,5,9,10,14] into a visual mining platform [3]. DAV uses a sphere layout to place the most tightly related item in the center and all others around. The most tightly related item is the item with the highest correlation with other item. With various association algorithms, DAV provides a self-organized graph. It consists of the following:

1. The "distance" between each pair of items represents *support*.
2. A "directed edge" represents the direction of the association. The color of the edge is used to represent the *confidence level*.
3. A "cluster" is used to wrap around highly related items using an ellipsoidal surface.

3 Component architecture

DAV is built on a Java-based client-server model. Its architecture contains four basic components - initialization, relaxation, direction, and clustering.

Figure 2 illustrates the overall architecture of DAV. Each of the above components is described in the following sections.

3.1 Directed association definition
In order to better understand what follows, we start with a few definitions:

An association rule is of the form $X \rightarrow Y$ where X and Y are sets of items. X is called the antecedent and Y the consequence of the rule. The strength of a rule is expressed by two factors: *support and confidence*. The support of rule $X \rightarrow Y$ is the frequency of occurrence of $X \cup Y$ in all transactions, i.e. support of $X \cup Y$ is defined as the ratio of the number of transactions in which X and Y occurs to the total number of transactions. The confidence of rule $X \rightarrow Y$ is the probability that if a transaction contains the antecedent, then it also contains the consequent, i.e. the ratio of the number of transactions that contain $X \cup Y$ to the number of transactions that contain X. Thus if 85% of the customers who bought printer also bought paper, and only 10% of all the customers bought both, then the association rule has confidence 85% and support 10%. The association direction is from the printer to the paper.

3.2 Initialization
DAV arranges items extracted from the web transaction data in a spherical surface. Items are represented as vertices. The transaction data is described as the following:

Transactions $\{T_1, T_2, ..., T_n\}$
Products $\{P_1, ... P_m\}$
Transaction $T_i = \{P_1, ... , P_{mi}\}$ i = [1..n]

The initial positions of items on the spherical surface could be at random. To avoid random pre-clustering, DAV distributed items equally on a sphere. The computation of equally spaced positions is based on a Poisson Disc Sampling [13] for approximation. After the computation of those positions, the most tightly related item is in the center and others are evenly distributed around. The tightness of an item is the sum of all supports to its directly adjacent items.

3.3 Relaxation
DAV constructs a frequency (support) matrix F. This matrix defines the stiffness of the spring attached to a pair of items. The strength of the relationship between items is represented by the stiffness of the spring. Each element contains the frequency of

occurrence of the association in all transactions (normalized relative to the most frequent item).

$$F = \begin{bmatrix} f_{11} & & & \\ f_{1i} & f_{2i} & f_{ii} & \\ \cdots & & & \\ f_{1n} & & \cdots & f_{nn} \end{bmatrix}$$

$f_{ij} = \#trans\ (Pi,\ Pj)/max\{PTk \mid k=[1..m]\}$
Where trans (Pi, Pj) is the set of transactions that contain Pi and Pj.

DAV transforms the spring stiffness to the distance in a 3D sphere after the graph has relaxed and converged to a state of local minimal energy.

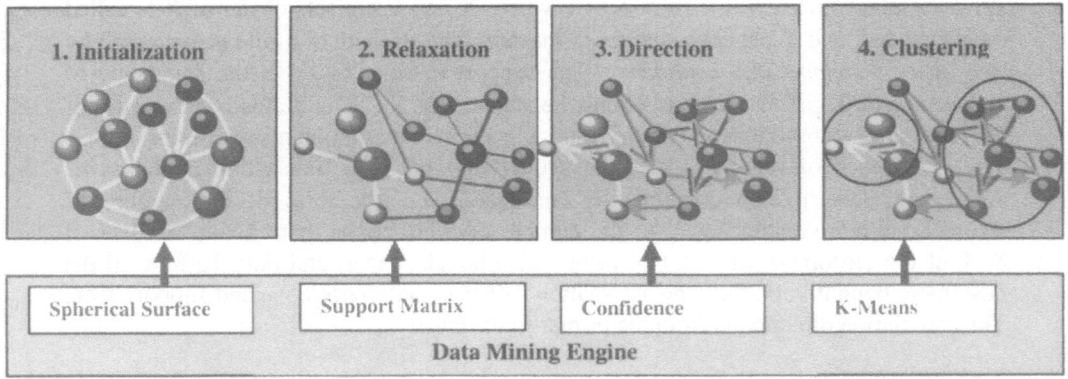

Figure 2: A Directed Association Visual Mining Component Architecture

3.4 Direction of association
DAV joins the antecedent of a rule with the consequence using a directed edge (arrow) to represent the direction of the association. The confidence levels are given in a matrix D. It is obtained by dividing the support of the item set by the support of the antecedent of the rule.

$$D = \begin{bmatrix} d_{11} & d_{12} & \cdots & d_{1n} \\ \cdots & & & \cdots \\ d_{1i} & d_{2i} & d_{ii} & \\ \cdots & & & \\ d_{1n} & & \cdots & d_{nn} \end{bmatrix}$$

$d(Pi,\ Pj) = \#trans\ (Pi,\ Pj)\ /\ \#trans\ (Pi)$

dij = association confidence direction and level

Pi → Pj

To identify rules with sufficient predictive power, DAV allows users to specify a minimum confidence level (threshold). The asymmetry of D between Pi→ Pj and Pj→ Pi is defined as the following:

(1) Draw an arrow from Pi to Pj
    If confidence level (Pi ⇒ Pj) exceeds threshold but confidence level (Pj ⇒ Pi) does not.
(2) Draw a double arrow between Pi and Pj
    If both confidence levels (Pi ⇒ Pj) and (Pj ⇒ Pi) exceed the thresholds

Figure 3 only draws the items above a minimum confidence value. The others are hidden. The user can easily follow the edges and directions to discover implications between items. For example, the user is able to find all antecedents that have "paper" as consequence. This visualization may help plan what the store should do to promote the sales of "paper".

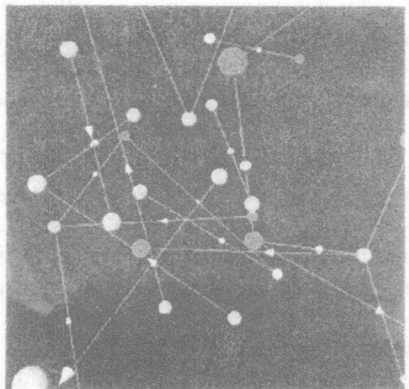

Figure 3. Directed Items with Minimum Confidence Level

3.5 Clustering
In addition to visualizing associations, DAV automatically clusters related items into groups using the "k-means" algorithm [16]. DAV wraps ellipsoids around each cluster. The number of items in a cluster and their positions control the shape of an ellipsoid.

4 Applications

DAV is a java-based client-server model. It is built on a VisMine [3] platform. VisMine uses a web browser with the Java activator that allows the user to mine large volumes of transaction data. The Web interfaces are based on standard HTML and the

190

use of Java applets. The client can run on a notebook. The user at the client side visually mines the knowledge results. The server is integrated with the data warehouse and mining engines.

We have applied DAV to several data mining visualization applications. In this paper, we illustrate its use in market basket analysis and customer profiling applications.

4.1 A Market basket analysis

One of the common problems electronic store managers want to solve is how to use e-customer purchase history for cross selling and up selling. They want to understand which products are purchased together and when to make real-time recommendations. Using our "directed association" system, we prototyped a market basket analysis visualization application to discover product affinities and relationships from transaction data.

An e-commerce manager can navigate a DAV-generated product sales graph and answer questions on which product groups are frequently bought together, how strong the correlation is, and in which direction. For instance, from the previous example where 85% of the people who buy a printer also buy paper, this visualization may help determine what products should be sold together with printers. Also, it helps to find what products may be impacted if the store discontinues selling printers.
Figure 4 illustrates a series of market basket analysis graphs to visualize data taken from one of the Hewlett-Packard sample shopping web site. The vertices (balls) represent products. The distances represent the support of items bought together.

Figure 4 (A) illustrates the initial layout of the graph generated from a web log. In this sample dataset, there are 182 different products (represented as balls), 250,000 transactions, and 1,383 edges. The color of the ball is used to show how often the product appears in the transaction database over a period of time. The most tightly related product is in the center and all others are evenly distributed around.

Figure 4 (B) shows the graph after it has been relaxed with 212 iterations and reached the local minima. The relaxation is based on the support/product affinities. The highly related products are self-organized into individual groups. The user can select an area to zoom in for further analysis shown in Figure 4(C). Figure 4(D) represents the graph after zoom-in. Figure 4(E) represents the graph after automatic clustering. Each highly related group is wrapped with an ellipsoidal surface for visibility. DAV allows users to interact with the graph. When the user clicks on a cluster, such as the large cluster at the upper right side in Figure 4(E), the detail information (i.e. associated product names, prices...) are displayed in a separate window as the user clicks on a selected cluster. For rapid discovery of patterns, DAV is able to monitor multiple simultaneous views of associated products.

4.2 Customer profiling

We applied this technology to analyze customer profiles. As illustrated in Figure 5, we use balls to represent customers making transactions on the web. DAV places customers with similar purchasing behaviors (i.e. product type, $ amount, geographical location, and income) near to each other. The store manager can rapidly discover patterns and issue coupons to the right customers for promotions.

5 Conclusion

Information visualization of e-commerce applications is an emerging technology. It needs new techniques to visualize large volumes of massive transaction data. At Hewlett-Packard Laboratories, we have integrated a mass-spring system into a visual mining platform. We have used the system to visually mine over a dataset containing 500,000 transactions covering 600 different products for market basket analysis. DAV provides a useful, fast, and interactive way for e-commerce managers to easily navigate through large-volume purchasing data to find product affinities for cross selling and up selling. Further research is continuing on scalability issues.

Acknowledgements

Thanks to Sharon Beach from HP Research Laboratories for her encouragement; to Prof. Daniel A. Keim from Halle University, to Graham Pollock from Agilent Laboratories for suggestions and comments; to Patrick Barthelemy -- "Template Graphics Software" for technical supports.

References

[1] Rakesh Agrawal, Tomasz Imielinski, Arun Seamil, "Mining Association Rules Between Sets of Items in Large Databases", Sigmod 5/93, Washington.
[2] T.C.Sprenger, M.H. Gross, "Ivory – An Object Oriented Framework for Physics-Based Information Visualization in Java", IEEE InfoVis98, North Carolina.
[3] Ming Hao, Umesh Dayal, Meichun Hsu, etc. "A Java- based Visual Mining Infrastructure and Applications", IEEE InfoVis99, CA.
[4] Pak Chung Wong, Paul Whitney, Jim Thomas, "Visualizing Association Rules for Text Mining", IEEE InfoVis99, CA.
[5] Giuseppe Di Battista, Peter Eades, "Graph Drawing Algorithms for the Visualization of Graph", Prentice Hall, 1999.
[6] Mihael Ankerst, Stefan Berchtold, Daniel A Keim, "Similarity Clustering of Dimensions for an Visualization of Multidimensional Data", IEEE InfoVis99, CA.
[7] "Quest": IBM Data Mining Technologies.
[8] "MineSet": SGI MineSet 3.0 Enterprise Edition.
[9] T.C.Sprenger, M.H. Gross, A. Eggenberger, M.Kaufmann:"A Framework for Physically-based Information Visualization". Eight Euro Graphics-Workshop on Visualization in Scientific Computing, France, 1997.
[10] M.H. Gross, T.C. Spenger, J.Finger: "Visualizing Information on a Sphere", IEEE VisInfo97.
[11] LikeMinds: LikeMinds Partner Program.
[12] Adviszor: Visual Insights data visualization.

[13] A.S.Glassner: Principles of Digital Image Synthesis, Morgan Kaufmann Publishers, San Francisco, 1995.

[14] R.J.Hendley, N.S.Drew, A.M.Wood & R.Beale, "Case Study Narcissus: Visualizing Information", IEEE InfoVis95.

[15] Stephen G. Eick, Joseph L. Steffen, Eric E, Sumner, Jr.: "SeeSoft-A Tool for Visualizing Line Oriented Software Statistics", IEEE Transactions on Software Engineering, 1992

[16] J. MacQueen "Some Methods for Classification and Analysis of Multivariate Observations", The 5th Berkeley symposium on mathematical statistics and probability, Berkeley, CA. 1967.

[17] G.G. Robertson , J.D. Mackinlay, S. K. Card, " Cone Tree: Animated 3D visualizations of hierarchical information. SIGCHI'91, 1991.

[18] John Lamping and Ramana Rao, "Laying out and Visualizing Large Trees Using a Hyperbolic Space". ACM /UIST'94, 1994

[19] Tamara Munster, "Exploring Large Graphs in 3D Hyperbolic Space" IEEE Computer Graphics. Vol. 18, Number 4. 1998

[20] Matthias Kreuseler, Norma Lopez, Heidrun Schumann, "A Scalable Framework for Information Visualization" InforVis 2000, 2000, Utah.

[21] T.C.Sprenger, R. Brunella, M.H. Gross, "H-BLOB: A Hierarchical Visual Clustering Method Using Implicit Surfaces", IEEE/VIS2000.

Editors' Note: see Appendix, p. 349 for colored figures of this paper

Space-Efficient Boundary Representation
of Volumetric Objects

Lukas Mroz and Helwig Hauser

VRVis Research Center, Vienna, Austria
http://www.vrvis.at/vis/
{mroz|hauser}@vrvis.at

Abstract. In this paper we present a compression technique for efficiently representing boundary objects from volumetric data-sets. Exploiting spatial coherency within object contours, we are able to reduce the size of the volumetric boundary down to the size of just a few images. Allowing for direct volume rendering of the down-scaled data in addition to compression ratios up to 250:1, interactive volume visualization becomes possible, even over the Internet and on low-end hardware.

1 Introduction

One major challenge of visualization in general is to deal with a whole lot of data. Especially in volume visualization, common data-sets range between several hundreds of Kilobytes, at the minimum, up to Gigabytes of uncompressed size. In medical visualization, for example, volumetric data-sets of size $256^3 \times 16$ Bit, i.e., 32 MBytes in total, are quite usual. If standard compression like gzip [7], for example, is applied, data-sets usually shrink to about 30–60 percent of the original size – still MBytes.

Processing huge data-sets itself poses high-performance requirements on the visualization software, but also storage and transmission of volumetric data-sets easily get into bandwidth problems, especially if multiple data-sets are to be treated. From medical applications, for example, we know that archiving 3D data-sets, which accompany diagnosis data, significantly stresses storage devices currently available in common clinical setups.

Even more critical, concerning the size of volumetric data-sets, and compared to storage problems, is visualization over the Internet. Web applications like remote diagnosis, for example, suffer from low transmission rates, even over local networks. In general, client-server solutions in the field of visualization usually are classified by the point, at which the visualization pipeline [8] is cut into a server-part and a client-part. Doing most of the visualization job at the client, for example, usually is referred to being a fat-client solution [10]. Thin clients, on the other hand, just display results of the visualization process, namely images, which entirely have been computed at the server beforehand. The trade-off between thin- and fat-client solutions is driven by the fact, that cutting the visualization pipeline at an earlier stage (fat-client solution) allows for more flexibility at the client's side (without any need to reload data). However, this advantage is gained at the expense of large-sized (volumetric) data to be downloaded, whenever necessary (initial download, changes to parameters of the preprocess). Respectively, thin clients deal with smaller data – just result images, for example – but need to download new data, whenever any of the parameters, even just viewing parameters, are changed.

The applicability of the more flexible fat-client solution to volume visualization strongly depends on the effectivity of the compression techniques used for transmission of the data-set. Lossless compression techniques – general purpose [7] as well as volumetric data specific [6] – usually achieve rather low compression ratios (around 2), which is not sufficient to significantly widen the bandwidth bottleneck. Using lossy compression [17, 2, 12] ratios in the range of 5 to 50 can be achieved while maintaining acceptable quality of the visualization results. On the other hand, medical applications, for example, prohibit changes to the accuracy of the data, as induced by lossy compression methods. Hierarchical methods, like wavelet compression [12] combine advantages of lossy and lossless compression. By transmitting and considering just a small fraction of the coefficients (around 5%) images of acceptable quality can be generated, data values of the original volume can be reconstructed if all coefficients are considered. A useful property of wavelet compression and many lossy compression techniques is the ability to render compressed data directly, without prior expansion and decompression.

Polygonal representations of structures within the volume (e.g. of iso-surfaces) can be used to realize solutions which are compromises between a pure thin and fat client approach. The volume is maintained at the server, just the polygonal model is transmitted and rendered at the client. Changes of viewing parameters require local rendering only, just changes affecting the shape of the model require a recomputation at the server and transmission of surface data over the network. To reduce the bandwidth required to transmit the model and to improve the interactivity of rendering at low-end clients, progressive refinement as well as focus and context techniques can be used [5], trading quality of representation (in less relevant regions of the volume) for speed. A combination of server-side and client-side approaches for direct volume rendering has been presented by Engel et al. [4]. They transmit a sub-sampled volume to the client and use it for local rendering during interactions. The original volume at the server is used to create and transmit a high-quality image whenever the interaction is finished/paused.

Pure thin-client solutions on the other hand, allow to perform visualization using low-end clients making at the same time shared use of special purpose hardware at the server (multiple CPUs, VolumePro board [18] for example).

One approach to determine the effectiveness of compression techniques for volumetric data-sets and their suitability for Internet-based visualization is to compare the size of compressed volumes versus the size of images of the same data. This comparison is useful as it directly corresponds to the trade-off between thin and fat-client solutions. If sizes of compressed volume data-sets range in the same magnitude as sizes of images thereof, and given the client to provide sufficient computational performance to carry out most of the visualization steps itself, then fat-client solutions become feasible even via the Internet. In our case, we achieve compression rates such that, given a 256^3 data-set as well as 512^2 images (24 Bits per pixel) in compressed GIF-format, about 2–5 images already are bigger in size than the compressed volume data-set.

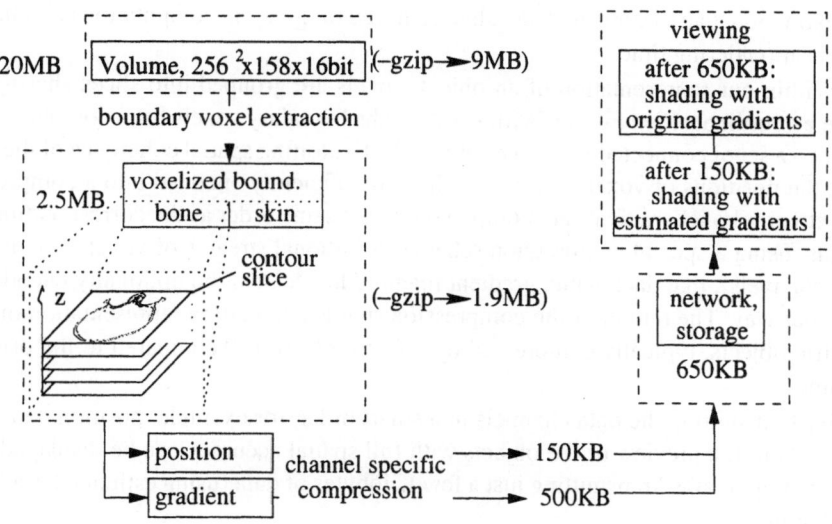

Fig. 1. Boundary extraction, compression and visualization pipeline

2 The Basic Idea

The effectivity of our approach is based on the observation that for the vast majority of applications, especially in medical visualization, volumetric data is rendered by displaying either iso-surfaces [14] or surface-like structures defined by areas of high gradient magnitude [13]. In both cases, the result of the visualization is determined by contributions of just a small fraction of all volume samples. By just coding those voxels of an object, which actually contribute to its visual appearance, the size of the data-set is greatly reduced. Thereby, a small-scale boundary representation of volumetric objects is generated (Fig. 1, Sect. 3). Compression of the boundary representation, which exploits spatial coherence among neighboring voxels, produces an extremely compact object representation (Sect. 4) which is well-suited for network transmission (Sect. 5). The information contained within this representation of objects allows interactive rendering at a client without any dependency on hardware-support, and with more flexibility regarding visualization parameters than polygonal surface representations (Sect. 6).[1]

The first step to obtain an efficient representation of bounded objects within a volumetric data-set is the identification and extraction of voxels which contribute to the object's visual representation, i.e., the boundary of the object. In our case, boundary voxels are data samples located within the object and have at least one neighboring voxel outside the object. The extraction process generates a separate boundary representation for each object within the volume. Usually 5–10% of all voxels belong to the boundary representation.

Typically, gradient information is required to evaluate a shading equation at each voxel during rendering. It is more efficient to precompute voxel gradients during bound-

---

[1] A demonstration applet is available from
http://bandviz.cg.tuwien.ac.at/basinviz/compression/

ary extraction than to store all data values required for gradient computation at boundary voxels at rendering time.

Within our representation of an object, voxels are grouped into slices sharing the same z coordinate (See Fig. 1). Within a slice, the boundary voxels form contours of the object – a set of connected sequences of voxels. Exploiting spatial coherence of the contour, the positions of voxels within the slice are efficiently encoded into a compressed data stream. Voxel gradients are compressed in the same order as the corresponding positions, using a special compression scheme. Additional streams of voxel attributes (= data channels), like data value, gradient magnitude, etc., can be optionally encoded in a similar way. The output of the compression step is a boundary representation of volumetric objects, typically compressed by a factor of 10–100 compared to the original volume.

By transmitting the data channels in a smart order, for example, position data first, gradients last, a preview of the objects with full spatial accuracy can be displayed (appendix, Fig. 3) after transmitting just a few Kilobytes of data (using estimated gradients for shading).

The decompressed boundary representation can be rendered directly [15, 9], without prior reconstruction of a full-sized volume. Compared with a polygonal representation of the boundary surfaces, our approach preserves the full accuracy of the data-set at much lower memory cost, allows interactive rendering on low-end hardware and provides more flexibility with respect to rendering parameters. Transparency, non-photorealistic shading and the fusion with truly volumetric objects (which can be compressed using the same method) are easily possible without performance degradation.

3   Extraction of Boundary Voxels

In our approach we either use the iso-surface metaphor to specify boundary voxels, or use a predefined and explicit segmentation mask for this purpose. In the first case, voxels with a data value $\geq$ iso-value and at least one 26-connected neighbor with a value smaller than the iso-value are considered to be part of the boundary. This definition results in 6-connected sets of boundary voxels, a property useful for exploiting coherence during compression of the contours. Boundaries of objects defined using a segmentation mask, can be extracted in a similar way and also result in 6-connected sets of voxels. As voxel identification accounts only for a small part of extraction time (gradient computation is most expensive), a simple sweep method is used for this purpose. The extraction of object boundaries is performed during an interactive volume visualization session. The resulting object representation can be rendered immediately, the compressed boundary can be stored for later viewing and presentation of visualization results.

Although best compression efficiency is achieved for surface-like voxel sets, truly volumetric objects can be extracted and compressed in the same way. This is especially useful for the visualization of spatially complex structures, like vessels in medical angiography data-sets or complex chaotic attractors in the field of dynamical systems [1].

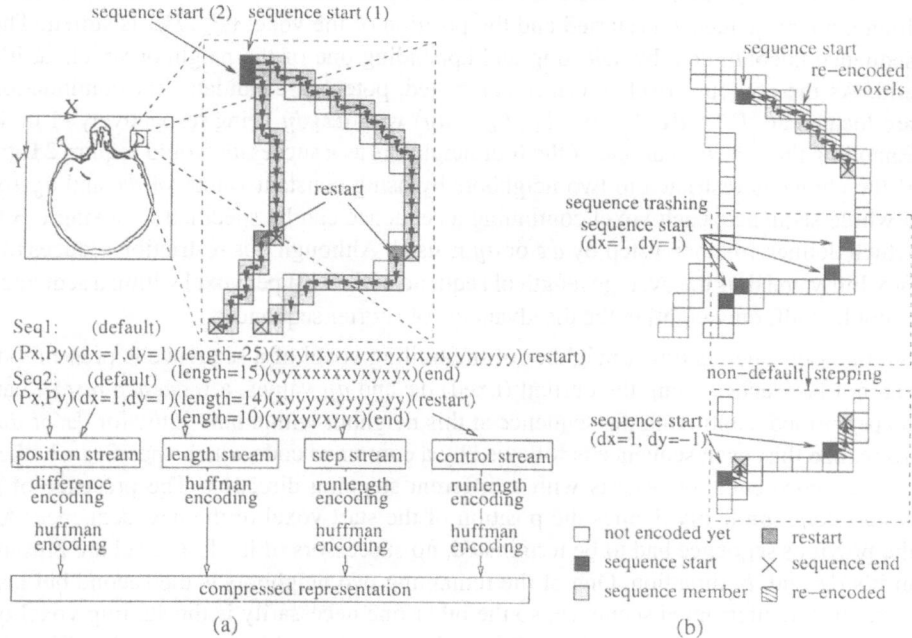

Fig. 2. Encoding of voxel positions: slice scanned from top left to bottom right a) long sequences and sequence continuations b) re-encoding of voxels and non-default stepping direction to reduce number of sequence starts and thus position specifications

For the extracted voxels, attributes (data channels) like voxel position, data value, gradient direction and magnitude, and application specific attributes are stored. When only the display of shaded surfaces is desired, storing voxel position and gradient direction is sufficient.

4 Data Compression

Individual objects within the volume are compressed separately. Voxels of each object are grouped into slices of voxels with the same z coordinate which are processed sequentially (See Fig. 1). To ensure effectivity, different data channels have to be compressed using specialized compression methods.

4.1 Position Data Channel

Boundary voxels within a single z-slice form object contours which consist of face-connected voxels (See Fig. 2). Exploiting spatial coherence and connectivity, voxels can be grouped into "*sequences*" which spatially follow the object contour. The approach is similar to "chain coding" used in binary image (text) compression [11] for contour encoding. In contrast contours of 2D objects, the boundary of a volumetric object may be thicker than a single layer of voxels, and thus requires a modified approach.

During compression, the slice is scanned for non-encoded voxels. Whenever one is found, a new sequence is started and the position of the voxel (P_x, P_y) is stored. The sequence is continued, by selecting and appending one of the neighbor voxels at it's end. As the contour voxels are face-connected, potential candidates for continuation are located at $(P_x + dx, P_y)$ or $(P_x, P_y + dy)$ with dx, dy being respectively -1 or 1. Encoding the selection of one of the four neighbors as a successor would require 2 Bits. If the choice is restricted to two neighbors by using constant values of dx and dy for a whole sequence, each voxel continuing a sequence can be specified by a single Bit, which defines whether a step by dx or dy is used. Although this restriction reduces the flexibility and thus the average length of sequences, the cost per voxel within a sequence is cut by half, outweighting the disadvantage of shorter sequences.

In cases where a direct neighbor of the trailing voxel of a sequence is present, but can not be reached using the current (fixed) dx and dy values, a *sequence restart* can be performed, continuing the sequence at this neighbor with a new value for dx or dy. To realize this, each sequence is followed by a command code which specifies whether the sequence ends, or restarts with a different stepping direction. The presence of a restart code implicitly defines the position of the start voxel of the new sequence. As the previous sequence had to be terminated, no successors of it's last voxel are present in it's dx and dy direction. One of the remaining two neighbors is the second but last voxel of the interrupted sequence, so the other one necessarily is the starting voxel of the new sequence. The new values of dx and dy are derived from dx and dy of the old sequence. Depending on whether the last step of the sequence was dx or dy either dx or dy is inverted. Although being more restrictive than with an explicit specification of dx and dy, this strategy still allows encoding of cyclic structures with a single position specification and restart commands within a *chain of sequences*.

For each combination of dx and dy values, one of the possible stepping directions is preferred, whenever both ways can be taken. The preference is chosen in a way, that a clockwise processing of closed objects will stay as close to the outer border as possible. For example, for $dx = 1$ and $dy = 1$ like in the first sequence of Fig. 2a, steps by dx are preferred.

After the creation of long sequences, usually groups of short sequences or even non-connected voxels remain. Starting a new sequence for each of these voxels is expensive. Usually most of these voxels can be encoded at a lower cost by joining them into sequences re-using voxels already encoded earlier in the process (Fig. 2b). In general, a sequence has to be continued, reusing already encoded voxels, if this allows to reach non-encoded voxels at a cost which is lower than a "sequence end" and the start of a new sequence.

As the scan for non-encoded voxels within a slice is performed in ascending x and y direction, using $dx = 1$ and $dy = 1$ as a default stepping direction for newly started sequences is usually a good solution – voxels with smaller x and y coordinates compared to the current one are already encoded in this case. In some cases however, keeping $dx = 1$ and $dy = 1$ as default directions tends to generate a lot of short sequences "sequence trashing" (Fig. 2b). Instead it is better to first search "backwards" (using $dx = -1$, $dy = 1$) and to start the new sequence using $dx = 1$ and $dy = -1$ at the last voxel found (for reasons of simplicity no backward scan was performed for the sequences of Fig. 2a). At each sequence start 1 Bit is used to store, whether the default stepping direction $(1, 1)$, or the direction of the backward scan $(1, -1)$ is used.

For further compression, the sequence data is separated into four streams. The *position stream* stores starting positions and stepping directions. Positions are stored using Huffman encoded differences between successive coordinate values (Typically 12 Bits per starting code). The *length stream* stores information about sequence lengths (Huffman encoded, 5 Bits per sequence). The *step stream* stores the information for building up sequences (1 Bit per voxel). As dx and dy steps tend to cluster due to the presence of a preferred stepping direction, this information is run-length encoded, using again Huffman encoding for the run-lengths. The *control stream* is used for the sequence control information (end/restart, 1 Bit per sequence). As many restarts at the beginning of encoding a slice are followed by short sequences collecting isolated voxels towards the end of encoding, which leads to clustering of restart and end commands, run-length encoding combined with Huffman encoding is also used here. Combining all those streams, an average of 2 Bits is required to encode the position of a single voxel.

Within all other data channels, voxels are encoded in the same order as their position data. This ordering allows to exploit spatial coherence within voxel sequences also for attribute encoding. For subsequent occurrences of re-encoded voxels, no attribute information is stored.

4.2 Gradient Direction Channel

As a first step, gradient vectors are normalized, transformed to polar coordinates and quantized to 2x6 Bits. This gradient representation is also used by our rendering algorithm for interactive shading. By exploiting spatial coherence within the encoded stream of voxels the gradient information is reduced to 3–8 Bit per voxel, depending on the smoothness of the boundary. Both polar coordinates are encoded into separate streams, storing differences between coordinates of successive voxels. As most of the difference data consists of sequences of values in the range of $[-1, 1]$ which are occasionally interrupted by larger values or clusters thereof, the encoder switches between two different coding schemes. The first scheme is used to encode sequences of differences in the range of $[-1, 1]$ using 1 Bit for 0 (most common), 2 and 3 Bits for -1 and 1, and a 3 Bit code to switch to the other encoding scheme. Larger differences are encoded using Huffman coding with an extra symbol to switch to the encoding scheme for small values. A switch to the code for small values is only performed to encode sufficiently long sequences of small values (cost of switching).

The use of prediction techniques for estimating gradients and the encoding of the prediction error instead of encoding gradient differences seems to promise good results at the first glance. Nevertheless, tests performed using linear regression [16] with a diameter of 3 and 5 for gradient estimation, indicate that compression rates obtained using this technique are worse than our current approach.

4.3 Other Data Channels

Additional data channels, like gradient magnitude, data value, etc., are compressed in the same order as the positions of the voxels to exploit spatial coherency also. Huffman encoding of differences of successive values and additional `zlib` compression (for further reduction of uniform areas) is used.

5 Data Transmission and Decompression

The compressed data-set consists of two parts: a header, which contains control-information about the objects and their position within the data, information about additional data channels and how to use them for rendering. The body contains voxel positions and other data channels for all objects. The data within the body is arranged in a way which allows to obtain a view on the data as early as possible during loading. Objects and data channels which are more significant for the preset visualization mappings are stored and transferred earlier than less significant data. Data channels are subdivided into blocks of a few Kilobytes each. As soon as an entire block has arrived, it can be decompressed and displayed while the following data is arriving. This allows voxel data to be rapidly updated, without having to wait for the arrival of the entire channel. Finally, as gradient information usually accounts for most of the data to be transmitted (See table 1), for boundary objects a locally computed gradient approximation (linear regression [16] with a filter size of 5 while interpreting the data as a binary object) can be displayed before the original gradient data arrives (appendix, Fig. 3). For inherently binary objects, like basins of attraction within the phase space of a dynamical system [1] the locally computed gradients can entirely replace the transmission of gradients, significantly decreasing the amount of transmitted data.

6 Rendering

In our test application, tendering of the data is performed by a Java applet at the client. A fast shear-warp-based method previously described by the Authors [15, 9] is used and extended to provide more flexible influence of data channels on the results of rendering. The 12 Bit gradient representation is used to directly index a look-up table containing shading information for interactive lighting (appendix, Fig. 4a). Using a shading table filled according to a non-photorealistic shading equation [3] and using the result to modulate voxel opacity, interactive non-photorealistic rendering can be realized. Using gradient-based shading and an additional gradient magnitude channel, the classical gradient-modulated transfer functions of Levoy [13] can be realized. Using one additional data channel (containing distance information) to modulate either color or opacity (appendix, Fig. 4b) the visualization of contacts between objects can be enhanced.

7 Results

Table 1 presents the compression rates obtained by applying our technique to a collection of data-sets from different application fields. The head and hand data-sets are CT scans containing objects typical for medical applications. Bone and skin surfaces extracted from the data are usually made up from 1–4% of all voxels. Using our compression scheme the boundary data is compressed by a factor of 20–90 compared to

data-set	volume size	obj. voxels	Bit/pos	Bit/grad	Bit/voxel	file(w/o grad.)	ratio to gzipped vol.
head-bone	$256^2 * 158$	378k	2.0	7.0	9.0	430k(95k)	1:22(1:97)
head-skin	$256^2 * 158$	231k	2.1	5.8	7.9	229k(60k)	1:40(1:154)
hand-bone	$256^2 * 232$	191k	2.5	7.8	10.3	246k(60k)	1:45(1:186)
hand-skin	$256^2 * 232$	170k	2.0	4.0	6.0	126k(41k)	1:89(1:273)
engine	$256^2 * 110$	298k	1.7	5.1	6.8	253k(64k)	1:13(1:51)
teapot	256^3	152k	1.7	3.4	5.1	80k(28k)	1:4(1:11)
attractor	256^3	769k	1.8	4.9*	6.7	639k(170k)	–**
basin	256^3	292k	2.2	0.6*	2.8	104k(80k)	–**

Table 1. Compression survey. * Scalar value channel instead of gradients. ** The attractor and basin data-sets have been extracted from a volume with a vector of several scalar values at each voxel directly within the simulation application. No volumetric representation was available.

the original volume when compressed with gzip. If gradient information is not stored but approximated at the client the compression factor increases to 100–270. The cost of compressing voxel positions within such data-sets is relatively independent of the surface shape 2–2.5 Bit/voxel. The cost for storing gradients depends on the smoothness and curvature of the surface and varies between 4 and 8 Bit/voxel. For objects with artificial, "well-behaved" surfaces like the CT scan of an engine block or the voxelized teapot, better compression is achieved for both voxel position and gradient data. The attractor and basin of attraction data, obtained from a simulation of a dynamical system, is also effectively compressed – especially as the basin boundary is derived from a binary classification of space and no gradient information has to be stored – it can be reconstructed from the surface shape at the client. Compression for each of the examples mentioned above takes approximately one second on a PIII/733 PC. Decompression timings for locally stored data are similar on the same PC. An applet which implements the techniques described in this paper and all compressed data-sets discussed and depicted here are available at http://bandviz.cg.tuwien.ac.at/basinviz/compression/.

8 Conclusions

Many applications of volume visualization require the display of objects boundaries. Using our compact volume representation, volume visualization becomes feasible even over the Internet, while still providing full spatial accuracy. Representing just the boundary voxels of objects reduces the amount of data to be transmitted or stored dramatically. By exploiting known properties of the boundary voxels (like spatial coherence and intervoxel connectivity) the data is further compressed. The resulting data representation is smaller by a factor of 20-250 than the volume compressed with gzip. The location of voxels within the volume is compressed very efficiently to about 2 Bit/voxel. The compression rates for gradient data are lower, in the range of 3-8 Bit/voxel, as gradient data is derivative information compared to the original data, containing less spatial coherence. Using a proper gradient reconstruction scheme, gradients can be estimated from voxel positions only, allowing to display objects just after the well-compressed position data has arrived, instead of waiting for the original gradient information. The display of

202

the boundary data can be performed in pure software (Java) at interactive frame rates without the need for any hardware support .

Acknowledgements

The work presented in this paper has been financed within the Kplus program of the Austrian government, as well as the BandViz project (http://bandviz.cg.tuwien.ac.at/) which is supported by the FWF under project number P 12811. The medical data-sets are courtesy of Tiani MedGraph.

References

1. G.-I. Bischi, L. Mroz, and H. Hauser. Studying basin bifurcations in nonlinear triopoly games by using 3D visualization. *accepted for publication in Journal of Nonlinear Analysis.*
2. T. Chiueh, C. Yang, T. He, H. Pfister, and A. Kaufman. Integrated volume compression and visualization. In *Proceedings IEEE Visualization '97*, pages 329–336, 1997.
3. D. Ebert and P. Rheingans. Volume illustration: non-photographic rendering of volume models. In *Proceedings IEEE Visualization 2000*, pages 195–202, 2000.
4. K. Engel, P. Hastreiter, B. Tomandl, K. Eberhardt, and T. Ertl. Combining local and remote visualization techniques for interactive volume rendering in medical applications. In *Proceedings IEEE Visualization 2000*, pages 449–452, 2000.
5. K. Engel, R. Westermann, and T. Ertl. Isosurface extraction techniques for web-based volume visualization. In *Proceedings IEEE Visualization '99*, pages 139–146, 1999.
6. J. Fowler and R. Yagel. Lossless compression of volume data. In *Proceedings IEEE Volume Visualization Symposium '94*, pages 43–50, 1994.
7. J. Gailly and M. Adler. gzip. URL: http://www.gzip.org.
8. R. Haber and D. McNabb. *Visualization idioms: A conceptual model for scientific visualization systems, visualization in scientific computing*, pages 74–93. 1996.
9. H. Hauser, L. Mroz, G.-I. Bischi, and E. Gröller. Two-level volume rendering - fusing MIP and DVR. In *Proceedings IEEE Visualization 2000*, pages 211–218, 2000.
10. M. Jern. Information drill-down using web tools. In *Proceedings of the 8th EUROGRAPHICS Workshop on Visualization in Scientific Computing*, pages 1–12, 1997.
11. R. Estes JR and V. Algazi. Efficient error free chain coding of binary documents. In *Proceedings of the Data Compression Conference*, pages 122–132, 1995.
12. C. Kurmann L. Lippert, M. Gross. Compression domain volume rendering for distributed environments. In *Proceedings of Eurographics '97*, pages C95–C107, 1997.
13. M. Levoy. Display of surfaces from volume data. *IEEE Computer Graphics & Applications*, 8(5):29–37, May 1988.
14. W. Lorensen and H. Cline. Marching cubes: A high resolution 3D surface construction algorithm. In *Proceedings of ACM SIGGRAPH '87*, pages 163–189, 1987.
15. L. Mroz, R. Wegenkittl, and E. Gröller. Mastering interactive surface rendering for java-based diagnostic applications. In *Proceedings IEEE Visualization 2000*, pages 437–440.
16. L. Neumann, B. Csébfalvi, A. König, and E. Gröller. Gradient estimation in volume data using 4D linear regression. In *Proceedings of Eurographics 2000*, pages C–351–C–357.
17. P. Ning and L. Hesselink. Fast volume rendering of compressed data. In *Proceedings IEEE Visualization '93*, pages 11–18, 1993.
18. H. Pfister, J. Hardenbergh, J. Knittel, H. Lauer, and L. Seiler. The VolumePro real-time ray-casting system. In *Proceedings of ACM SIGGRAPH '99*, pages 251–260, 1999.

Editors' Note: see Appendix, p. 350 for colored figures of this paper

Salient Representation of Volume Data

Jiří Hladůvka, Andreas König, and Eduard Gröller

Institute of Computer Graphics and Algorithms
Vienna University of Technology

Abstract. We introduce a novel method for identification of objects of interest in volume data. Our approach conveys the information contained in two essentially different concepts, the object's boundaries and the narrow solid structures, in an easy and uniform way. The second order derivative operators in directions reaching minimal response are employed for this task. To show the superior performance of our method, we provide a comparison with its main competitor – surface extraction from areas of maximal gradient magnitude. We show that our approach provides the possibility to represent volume data by a subset of a nominal size.

1   Introduction

The importance of edge information for machine vision is usually motivated from the observation that under rather general assumptions about the image formation process, a discontinuity in image brightness can be assumed to correspond to a discontinuity in either depth, surface orientation, reflectance, or illumination [8]. A different type of discontinuity – a line is also a structure of particular interest. While in a 2D image the representatives of narrow solid structures are spots and lines, in volume data this is more general. Identification of blobs, cylinders, and sheet–like structures plays a crucial role in medical visualization [16].

To represent volume data by just a small subset of important voxels is desirable for and addressed by a number of applications. Interactive volume visualization over the internet based on a client/server architecture profits from elaborated strategies for progressive data transmission. Here it is desirable that the content of a volume is visually interpretable already in the early stages of transmission to and visualization by a client. To achieve this, the server may start transmitting salient features earlier than the rest of the data. Non-distributed visualization may benefit from storing a small, representative subset of the data to disk. Such a representation can be reused later for a quick preview display using, e.g., non-photo realistic techniques [14] or algorithms yielding more realism [9, 11, 13].

For these applications, the subset of high salience has to be identified. The best established algorithms involve isosurface extraction [17], boundary identification or emphasis [7], or narrow solid structures identification or emphasis [1, 15, 16]. While the usual paradigm to identify object boundaries is to evaluate the magnitude of the gradient, the identification of narrow solid structures requires the use of either special filters or 2nd order derivative filters.

In this work we propose a filtering technique for the identification of both boundaries and narrow structures. Our algorithm is based on identification of areas with large negative second derivatives, and handles both of the cases in a uniform way. Defining a salience function based on this quantity allows us to identify those voxels of the input volume which provide a significant content necessary for visualization.

In the following section we discuss the theoretical background for our method. In section 3 its complexity and implementation issues are discussed. Results and a comparison to the gradient method are presented in section 4.

2 The proposed method

2.1 Edge detectors and line detectors revisited

For edge detection two kinds of filters have been designed – those based on looking for maxima of the first derivative and those based on looking for zero–crossings of the second derivative.

While these concepts are intuitive for 1D signals, the situation in higher dimensions gets more complicated. We assume that the volume is given as a density function I. To use the extrema of 1st derivatives we need to know the directions in which they occur. From calculus it is known that for the first derivative this direction is the gradient vector ∇I and the magnitude of the derivative in this direction is the magnitude of the gradient: $I'_{max} = I'_{\nabla} = \|\nabla I\| = (\sum_{i=1}^{3}(\partial I/\partial x_i)^2)^{1/2}$. Looking for maxima of gradient magnitudes yields an isotropic edge detector which responds both to outer and inner side of the object equally (Fig. 1a). For the second derivative approach it is necessary to check the neighborhood of a voxel for zero–crossings, i.e., for areas where the 2nd derivative changes its sign. The 2nd derivative is usually estimated by the rotationally invariant Laplacian $\triangle I = \sum_{i=1}^{3} \partial^2 I/\partial x_i^2$. A less referred feature of the Laplacian operator is that it responds with negative values at the inner part and by positive values at the outer part of the object's edge[1] (Fig. 1b). We put this into contrast to the gradient–based edge detection, which yields an equal response on both sides of an edge and exploit this fact to represent the objects' edges only by their internal side. Such a representation requires, compared to the gradient method, a smaller amount of voxels, or in other words it provides a better distribution over the surface in early stages of the progressive transmission for a limited bandwidth.

Considering the density profile, it is evident that the concepts of 1st derivative maxima can not be directly applied for spot and line detection (or, more generally speaking, for detection of narrow areas which in 3D correspond to blobs, lines, and sheet–like structures). The response of a 1st order derivative filter to a line, for instance, results in two lines, which would require a special, nontrivial

---

[1] Assuming the objects are of a higher density than background and not vice versa

mechanism for detection of the area in-between (Fig. 1a). A 2nd order derivative filter, on the other hand, responds to a line by negative values at its interior (Fig. 1b).

As a result we get a twofold interpretation of areas where the 2nd order derivative operator responds with negative values. Firstly such areas correspond to internal parts of a boundary, secondly they identify narrow structures. To make the search for negative areas more feasible for separation by thresholding, we are interested in the directions where the 2nd order derivatives reach the minima.

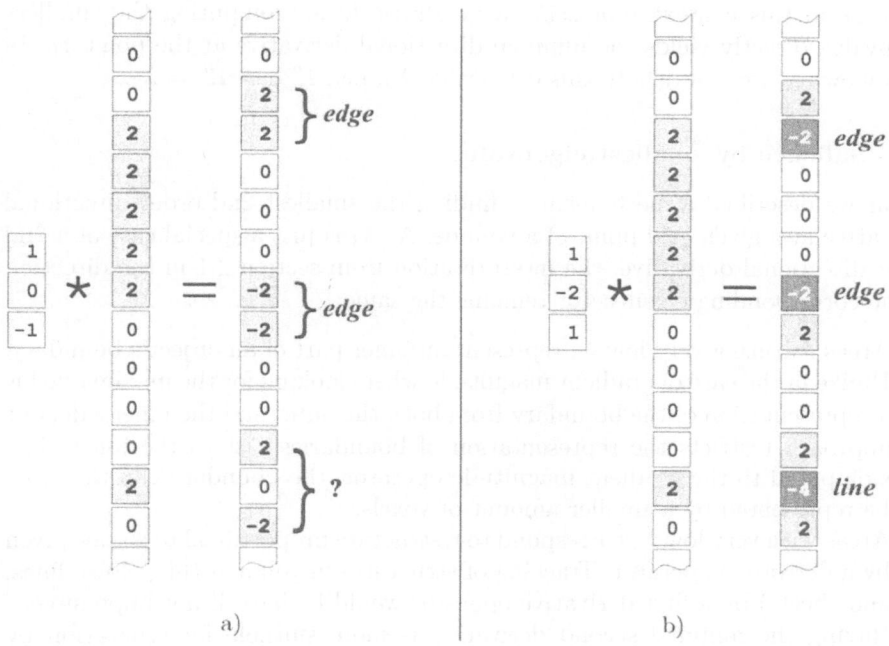

Fig. 1. Examples for 1D density profile. Responses of a 1st order derivative filter (a) and a 2nd order derivative filter (b) to an edge and to a line.

2.2 The smallest 2nd order directional derivative

Stating the Taylor expansion of a 3D density function I for the first three terms in the vicinity, spanned by vectors Δx, of a 3D point x_0

$$I(x_0 + \Delta x) \approx I(x_0) + \Delta x^T \nabla I(x_0) + \Delta x^T H(x_0) \Delta x \qquad (1)$$

it is evident, that the 2nd order information is entirely expressed by the symmetric Hessian matrix:

$$H = \begin{pmatrix} I_{xx} & I_{xy} & I_{xz} \\ I_{yx} & I_{yy} & I_{yz} \\ I_{zx} & I_{zy} & I_{zz} \end{pmatrix} \qquad \text{where the terms} \quad I_{ab} = \frac{\partial^2 I}{\partial a \partial b} \qquad (2)$$

denote the 2nd order mixed partial derivatives. For a fixed x_0, the term $\Delta x^T H \Delta x$ gives the second derivative of the density in the direction of Δx. The Hessian matrix, as a real–valued and symmetric matrix, has real–valued eigenvalues λ_i. From the definition of the eigenvalues, $H v_i = \lambda_i v_i$, follows that the eigenvalues λ_i give the second derivatives in the direction of the eigenvectors v_i: $\lambda_i = v_i^T H v_i$. Since H in this context represents a quadratic form, computing the smallest eigenvalue directly yields the minimal directional derivative at the point x_0. In the following, we will denote this eigenvalue λ_3, i.e., $I''_{min} = I''_{v_3} = \lambda_3$.

2.3 Salience by smallest eigenvalue

So far we described a mechanism to finding the smallest 2nd order directional derivative at a given grid point of a volume. As λ_3 is just a special case of a 2nd order directional derivative, the interpretation from section 2.1 in the direction of the corresponding eigenvector remains the same:

1. Areas featuring very low λ_3 represent an inner part of an object's boundary. Unlike in the case of gradient magnitude where looking for the maxima yields a representation of the boundary from both the outer and the inner side, our approach restricts the representation of boundaries just to the inner side. Compared to the gradient magnitude operator, the boundary can therefore be represented by a smaller amount of voxels.
2. Areas with very low λ_3 correspond to a structure proportional to a scale given by a derivative operator. Tracking of structures in volumes (like blobs, lines, and sheets) by a first derivative operator would be hard if not impossible.
3. Having the minimal second derivative is more suitable for separation by thresholding than having a second derivative in an arbitrary direction.

Due to these reasons, areas featuring low negative eigenvalues λ_3 yield a better representation of a volume then those with high positive values of the gradient magnitude. In order to provide a comparative study between these two approaches, we define the two following salience functions S_Γ, S_Λ of a voxel v and the two corresponding $p\%$-subsets of an input volume V they determine:

$$\begin{aligned} S_\Gamma[v] &= ||\nabla|| \, [v] & \Gamma[p] &= \{p\,\% \text{ of } V \text{ with the highest } S_\Gamma\} \\ S_\Lambda[v] &= -\lambda_3[v] & \Lambda[p] &= \{p\,\% \text{ of } V \text{ with the highest } S_\Lambda\} \end{aligned} \qquad (3)$$

For a given percentage p, functions S_Γ, S_Λ determine the $p\%$ of 'top salient' voxels which will represent the volume. For a progressive transmission of data through a network, these functions determine the priority of transmission: the voxels with higher salience will be transmitted earlier.

Obviously, there are also other candidates which might succeed well in the task of volume representation by a fraction of the data. In the following we give an overview of possible competitors and argue why we do not compare them to our method:

Isosurface methods require a user input to specify the density which determines an isosurface. The result is dependent on and yields only structures defined by this choice. Our method processes data automatically and delivers surfaces of more than just one isolevel.

The "distance to closest boundary" method, as introduced by Kindlmann and Durkin [6] yields an opacity transfer function for boundary emphasis. The algorithm performs a statistical analysis of the zero, 1st, and 2nd order derivatives in the direction of the gradient, providing information on which densities contribute most to boundaries. Defining a salience function based on density is essentially inconsistent with our approach, which is position based, and a comparison would be hardly possible.

Density distribution analysis based on *all* eigenvalues of the Hessian as proposed by Frangi et al. [1] or Sato et al. [15, 16] restricts the search space just to structures of a particular shape and a certain scale, and excludes boundaries of objects. In contrast, our approach handles both boundaries and structures in a uniform way.

In our previous work [2] we suggested taking into account also areas featuring high magnitudes of the largest eigenvalue λ_1. The maxima of this eigenvalue correspond to outer parts of objects' boundaries and do not contribute to the output significantly.

3 Implementation and complexity

3.1 Hessian matrix versus gradient vector

Computation of both the gradient vector and the Hessian matrix at grid points involves an approximation of the first and the second partial derivatives, respectively. For this task, convolution of the data with kernels designed for a particular derivative in a specific direction is usually employed.

For the first derivatives, kernels of size up to three are usually found in the textbooks: Roberts, Prewitt and Sobel filters are feasible for fast computation.

The Hessian matrix requires an estimation of 2nd order derivatives which is, especially for small kernels, much more sensitive to noise. The usual practice is to pre-smooth the input data with a Gaussian filter. Due to the associativity of convolution, the smoothing step and the derivation can be combined, resulting in a convolution of the data with a derivative of the Gaussian filter of a bigger size. The second reason to use the derivatives of the Gaussian filter is that we want to detect features represented at a certain *scale*. The Gaussian filter is the *only* filter which meets both the *minimum-maximum principle* and *scale invariance* necessary for such a representation. For more details on scale spaces we refer the reader for instance to Jähne [5].

To remain consistent for comparison of both the quality of results and the computational costs we used filters of the same size both for 1st and 2nd derivatives. Using the Gaussian filter requires that its size k is proportional to the standard deviation, so the kernels usually involved are 5, 7, or 9 voxels wide. Convolution with moderately–sized kernels is usually a computationally expensive process. To speed it up, hardware features can be used for specific platforms [3, 4]. For a software implementation, the separability of the Gaussian derivative kernels can be exploited:

$$\overbrace{\left(\frac{\partial^o}{\partial x^a \partial y^b \partial z^c} G_\sigma(x,y,z)\right)}^{k\times k\times k} * I = \overbrace{\frac{d^a}{dx^a} G_\sigma(x)}^{k\times 1\times 1} * \left(\overbrace{\frac{d^b}{dy^b} G_\sigma(y)}^{1\times k\times 1} * \left(\overbrace{\frac{d^c}{dz^c} G_\sigma(z)}^{1\times 1\times k} * I\right)\right)$$

$$(4)$$

where nonnegative integers $a + b + c = o \in \{1,2\}$ determine the order of derivation, and σ is the standard deviation of the Gaussian filter $G_\sigma(x) = exp(-\frac{x^2}{2\sigma^2})/\sqrt{2\pi}\sigma$. The decomposition according equation (4) reduces the overhead, for a partial derivative at a grid point, from convolution with a 3D kernel (complexity $O(k^3)$) to three convolutions with a 1D kernel (complexity $O(3k)$).

A direct application of equation (4) would require 18 1D convolutions for Hessian elements as compared to 9 1D convolutions for the gradient vector. Further speed-up can be achieved by appropriate reorganization and caching. Three 1D convolutions can be saved for the computation of the Hessian matrix, (e.g., $G_\sigma(x) * I$ can be reused three times and $G'_\sigma(x) * I$ twice) and one convolution can be saved for the gradient (e.g., $G_\sigma(x) * I$ can be reused twice). This reduces the number of required 1D convolutions to 15 for the Hessian and 8 for the gradient.

3.2 Eigenvalues of the Hessian versus magnitude of the gradient

While computing the gradient magnitude by the Euclidean norm requires three multiplications, two additions and one square root, the computation of eigenvalues of the Hessian matrix is more complex. The explicit formula would require solving cubic polynomials. In our implementation we used a numerical solution – the fast converging Jacobi's method as recommended by Press et al. [12] for real-valued, symmetric matrices.

Table 1 summarizes the overall costs concluding that the computation of eigenvalues is, as compared to the computation of the gradient magnitude, in average 2.7 times more expensive.

3.3 Construction of representative subsets

To build the subsets $\Gamma(p)$ and $\Lambda(p)$, we firstly construct cumulative histograms of quantities $||\nabla||$ and $-\lambda_3$, respectively, in one pass through the volume in linear time. The required percentage p controls the number of voxels to be included into the respective subset. The search for adjacent histogram bins straddling this number is logarithmic. The indices of bins correspond to a threshold which is used as a decision function.

Input Volume		$T_{\|\nabla\|}$		$T_{(\lambda_1, \lambda_2, \lambda_3)}$		factor
Data set	Dimensions	8 *	+norm	15 *	+Jacobi	
Lobster	$120 \times 120 \times 34$	3.37	3.43	6.30	8.69	2.53
Vertebra 1	$128 \times 128 \times 74$	8.52	8.69	15.91	24.58	2.83
CT Head	$128 \times 128 \times 113$	12.92	13.17	24.20	36.87	2.80
MRI Head	$256 \times 256 \times 109$	50.64	51.60	93.36	141.98	2.75
Engine Block	$256 \times 256 \times 110$	50.36	51.29	94.05	139.49	2.72
Tooth	$256 \times 256 \times 161$	73.49	74.71	137.49	184.35	2.47
Vertebra 2	$256 \times 256 \times 241$	109.89	111.96	206.58	275.45	2.46

Table 1. Time in seconds for computing the magnitude of the gradient and the eigenvalues of the Hessian as measured on a Pentium III, 800 MHz. 1D cyclic convolutions (eq. 4) with kernel of size $k = 7$ have been used. The meaning of columns from left to right: name of the data set and its dimensions; time for the computation of all partial derivatives for the gradient vector, and after Euclidean norm; time for the computation of all partial derivatives for the Hessian, and after eigenvalues search; the ratio of overall times for eigenvalues and gradient magnitude.

4 Results

To compare the quality of a volume representation by subsets Γ and Λ from equation (3), we generated sparse volumes where the density of voxels not presented in either of the subsets have been set to zero. Such volumes have been rendered by direct volume rendering provided by the VolumePro architecture [10]. In the following we refer to Figure 2 and to the project's web site[2].

Lobster: We compared 2, 4, 6, 8, and 10% representations of this data set. While the legs of the lobster are, due to the line filter, visible and good recognizable already in $\Lambda(2\%)$, in $\Gamma(6\%)$ they just start to appear. Representation by $\Gamma(2\%)$ is insufficient. The differences between Λ and Γ vanish with increasing percentage. Nevertheless, they are still noticeable between $\Lambda(10\%)$ and $\Gamma(10\%)$.

Vertebra 1: Neither $\Gamma(2\%)$ nor $\Lambda(2\%)$ provide a good representation, though there is much more content visible in $\Lambda(2\%)$. $\Gamma(4\%)$ features broken contours and is approximately on a level of $\Lambda(2\%)$. $\Lambda(4\%)$ and $\Gamma(6\%)$ represent approximately the same level, but $\Lambda(4\%)$ provides more details and more closed contours. $\Lambda(6\%)$ is already close to a good representation of the original data set.

Vertebra 2: Subset $\Gamma(2\%)$ features only high density screws. While the ribs only begin to appear in $\Gamma(4\%)$, their are better visible already in $\Lambda(2\%)$ due to a more even distribution of boundary voxels. $\Gamma(6\%)$ yields even less information than $\Lambda(4\%)$. $\Lambda(6\%)$ is a good approximation of the original data set.

---

[2] accessible via http://www.cg.tuwien.ac.at/research/vis/vismed/

210

Tooth: There are two significant features identified in this data set: a tooth inside a surrounding cylinder. The ratio between the number of voxels belonging to either of the features is quite big, and the subsets Γ and Λ contribute to a large extent to the wall of the cylinder. For this reasons we have noticed an obvious difference in the appearance of the tooth just by a very small percentages. $\Lambda(0.65\%)$ yields more content than the corresponding $\Gamma(0.65\%)$. Rendering this data set we also have noticed a suppression of the partial volume effect in Λ data sets. We explain this suppression as a consequence of representing object boundaries by their inner parts.

5 Conclusion

We introduced a novel approach for the representation of volume data sets by a subset which contains the salient features. Our attempt is to convey information contained in two essentially different modalities, the object's boundaries and the narrow structures in an easy and uniform way. For this task we employ second order derivative operators in the directions reaching minimal response.

Compared to the methods based on gradient maxima, our method represents objects only by the internal side of their boundaries, reducing thus the amount of necessary voxels.

Looking for the minima of second order derivative yields also a structure detector proportional to the scale of the derivation operator. In contrast to the concepts of Frangi [1] and Sato [15, 16] we enforce no shape restrictions, making no distinction among blob, tubular, and sheet–like structures.

The drawback of our method is a higher computational cost. Computation of the Hessian's eigenvalues is approximately 2.7 times more expensive than the computation of the gradient magnitude.

We evaluated our method and compared it to the gradient method for several data sets. Due to the results we conclude that our method performs better than methods based on gradient magnitude. For the same level of quality of visualization it allows to represent a data set by a reasonably smaller subset. The possible applications of such an advantageous representation are, e.g., progressive transmission over the internet and the generation of preview data sets.

Acknowledgements

The work presented in this publication has been funded by the V^{is}M^{ed} project, http://www.vismed.at. V^{is}M^{ed} is supported by *Tiani Medgraph*, Vienna, http://www.tiani.com, and the *Forschungsförderungsfonds für die gewerbliche Wirtschaft*, Austria, http://www.fff.co.at. The Vertebra1 and Vertebra2 data sets are courtesy of *Tiani Medgraph*, Vienna.

References

1. A. F. Frangi, W. J. Niessen, K. L. Vincken, and M. A. Viergever. Multiscale vessel enhancement filtering. *Lecture Notes in Computer Science*, 1496:130–137, 1998.
2. J. Hladůvka, A. König, and E. Gröller. Exploiting eigenvalues of the Hessian matrix for volume decimation. In *Proceedings of 9th International Conference in Central Europe on Computer Graphics, Visualization, and Computer Vision (WSCG 2001)*, pages 124–129, 2001.
3. M. Hopf and T. Ertl. Accelerating 3D convolution using graphics hardware. In *Proceedings of the 1999 IEEE Conference on Visualization (VIS-99)*, pages 471–474, 1999.
4. Intel Corporation. IPL–Intel Image Processing Library, v2.5, 2000.
5. B. Jähne. *Digital Image Processing*, chapter 5: Multiscale Representation, pages 121–138. Springer–Verlag Berlin Heidelberg, 4th edition, 1997.
6. G. Kindlmann and J. W. Durkin. Semi-automatic generation of transfer functions for direct volume rendering. In *Proceedings of IEEE Volume Visualization*, pages 79–86, 1998.
7. M. Levoy. Display of surfaces from volume data. *IEEE Computer Graphics and Applications*, 8(3):29–37, May 1988.
8. T. Lindeberg. Edge detection and ridge detection with automatic scale selection. In *Proceedings of IEEE Computer Vision and Pattern Recognition*, pages 465–470, 1996.
9. L. Mroz, H. Hauser, and E. Gröller. Interactive high-quality maximum intensity projection. *Computer Graphics Forum*, 19(3):341–350, 2000.
10. H. Pfister, J. Hardenbergh, J. Knittel, H. Lauer, and L. Seiler. The VolumePro real-time ray-casting system. In *Proceedings of ACM SIGGRAPH*, pages 251–260, 1999.
11. H. Pfister, M. Zwicker, J. van Baar, and M. Gross. Surfels: Surface elements as rendering primitives. In *SIGGRAPH 2000, Computer Graphics Proceedings*, pages 335–342, 2000.
12. W. H. Press, S. A. Teukolsky, W. T. Vetterling, and B. P. Flannery. *Numerical Recipes in C*, chapter 11: Eigensystems, pages 456–469. Cambridge University Press, 2 edition, 1992.
13. S. Rusinkiewicz and M. Levoy. QSplat: A multiresolution point rendering system for large meshes. In *SIGGRAPH 2000, Computer Graphics Proceedings*, pages 343–352, 2000.
14. T. Saito. Real-time previewing for volume visualization. In *Symposium on Volume Visualization*, pages 79–106, 1994.
15. Y. Sato, S. Nakajima, N. Shiraga, H. Atsumi, S. Yoshida, T. Koller, G. Gerig, and R. Kikinis. 3D multi–scale line filter for segmentation and visualization of curvilinear structures in medical images. *Medical Image Analysis*, 2(2):143–168, 1998.
16. Y. Sato, C.-F. Westin, A. Bhalerao, S. Nakajima, N. Shiraga, S. Tamura, and R. Kikinis. Tissue classification based on 3D local intensity structures for volume rendering. *IEEE Transactions on Visualization and Computer Graphics*, 6(2):160–180, 2000.
17. P. M. Sutton, C. D. Hansen, and H.-W. S. D. Schikore. A case study of isosurface extraction algorithm performance. In *Data Visualization 2000*, 2000.

Editors' Note: see Appendix, p. 351 for colored figure of this paper

A Selective Refinement Approach for Computing the Distance Functions of Curves

Daniel E. Laney[1], Mark A. Duchaineau[2], Nelson L. Max[1],[2]

[1] Department of Applied Science, University of California at Davis
[2] Lawrence Livermore National Laboratory* * *

Abstract. We present an adaptive signed distance transform algorithm for curves in the plane. A hierarchy of bounding boxes is required for the input curves. We demonstrate the algorithm on the isocontours of a turbulence simulation. The algorithm provides guaranteed error bounds with a selective refinement approach. The domain over which the signed distance function is desired is adaptively triangulated and piecewise discontinuous linear approximations are constructed within each triangle. The resulting transform performs work only were requested and does not rely on a preset sampling rate or other constraints.

1 Introduction

The distance function of a shape encodes the minimum distance to the shape at every point in space. The encoded shape can be extracted by taking the isocontour of the distance function with an isovalue of zero. Distance functions have proven useful in modeling deformations of solid objects [11], representation of medical data [4], modeling swept surfaces and volumes [9], and increasing the performance of ray-casting via space leaping [12]. A closed shape allows a signed distance function to be defined that encodes inside/outside information and enables boolean operations and object morphing. Signed distance representations vary smoothly across shape boundaries, enabling accurate reconstruction of surface properties from the distance function. As an implicit representation, distance functions have the potential to simplify the storage and visualization of time varying surfaces with changing topology because they do not explicitly store the topology.

Distance functions are usually represented as sampled scalar fields with values between samples generated by interpolation. A propagation algorithm such as that of Breen [1] is the most commonly used method for obtaining an approximate distance function of a shape. These algorithms initialize a sampled scalar field with closest point information near the boundary of a shape and propagate this information throughout the volume. For large problems it would be advantageous to limit the amount of computation spent on areas with less detail. The adaptive sampling technique of Frisken [3] provides many of the benefits of distance representations without the excessive memory requirements. However, adaptive sampling implies that a distance function exists in closed form, or an approximate distance function exists at a high resolution. Our

* * * {laney1, duchaineau1, max2}@llnl.gov

214

contribution is to provide adaptivity in the distance function itself through a distance transform based on selective refinement.

In this paper we present an adaptive signed distance transform for curves in two dimensions. We demonstrate the algorithm on isocontours of a turbulence simulation [7]. We are investigating distance functions in the plane as a precursor to a full three dimensional method. In two dimensions, the error analysis is simplified, and the behavior of the algorithm and data structures can be clearly visualized. Our algorithm utilizes a selective refinement approach which concentrates computation where a more accurate approximation is required.

The algorithm operates on an adaptive distance approximation represented by a triangulation of the desired 2D domain. Figure 1a shows the adaptive triangulation of the distance function for a small piece of isocontour from the test data. Figure 1b shows the bounding boxes which contain curve dependency information for the highlighted distance triangle. The dependency information reduces the work required to compute distance approximations within each triangle. We require the input curve to provide a bounding box hierarchy so that distance triangles can track potential contributing points efficiently. Figure 1c shows a continuous reconstruction of the distance approximation which satisfies the error bounds of every triangle.

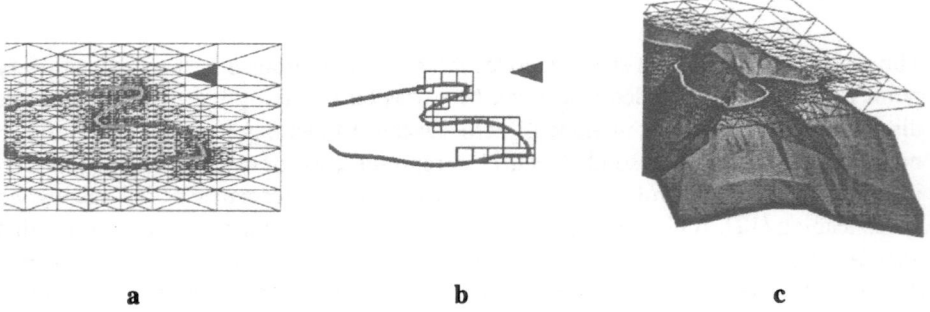

a   b   c

Fig. 1. A node in the distance hierarchy (a) contains a list of nodes in the curve hierarchy (b)

The adaptive distance approximation has the following properties:

1. **Triangle bintree representation:** The adaptive distance is defined over a triangle bintree [2, 6]. The refinement rules for triangle bintrees insure that all adjacent triangles meet at a common vertex or a common complete edge. This property allows the construction of continuous functions on the triangle mesh.
2. **Per-triangle linear approximations with guaranteed error bounds:** Selective refinement operations must be as local as possible. Our approach uses per-triangle linear approximations which may be discontinuous with respect to the linear approximations in neighboring triangles. Each triangle also maintains an error bound for the linear approximation which is guaranteed to be correct with respect to the

exact distance function. The per-triangle approach requires an extra step to construct a continuous distance approximation, but this is always possible.

3. **Per-triangle tracking of dependencies on the input curve:** In general the distance function within a triangle depends on a subset of the input curve. Each triangle tracks the subset of the curve that contains every point which potentially contributes to the distance function. Linear approximations are computed with respect to the per-triangle dependency information. The selective refinement approach is based on the observation that the curve dependencies are typically reduced when a triangle is refined.

4. **Easy reconstruction of a continuous per-vertex distance approximation:** The per-triangle error bounds are guaranteed to overlap at the vertices and edges. A simple algorithm is presented which produces distance values at the vertices of the triangle bintree that satisfy the error bounds of the per-triangle approximations. The triangle bintree automatically insures that no cracks are present in the continuous distance approximation.

The remainder of the paper first presents related work in distance transforms. We then describe the curves on which the algorithm was tested, including our method of obtaining a bounding box hierarchy. Finally, we detail the distance transform algorithm itself and characterize the behavior of the algorithm on the test data.

2 Related Work

The closest point propagation technique of Breen [1] is based on the fast marching method of Sethian [10]. The main advantages of propagation algorithms are speed and simplicity which offsets the fact that they are usually applied to sampled volumes for which only a subset of the samples will be used. An exception to this is object morphing which requires operations on the entire volume. One disadvantage is that propagation methods rely on the user to set the appropriate sampling rate for a given object. In addition, the distance computation is sometimes sampled on a much finer lattice and then sub-sampled to the desired resolution of the distance approximation.

Perhaps closer in spirit to the approach of this paper is the surface reconstruction algorithm of Hoppe [5]. They define a signed distance function for scattered point data and reconstruct a plausible surface by taking the zero isosurface of the distance function. At the heart of the algorithm is a distance transform which relies on local best fit planes computed for each surface sample and its neighborhood. The reconstructed surface is extracted as the zero set of the approximate distance function. The approach in the present work requires normal vector information which is not usually present in data from range scanners and was not required by the method of [5].

3 The Curve Hierarchy

For this paper we have chosen to test the algorithm on isocontours of a regularly sampled scalar field [7]. In this section we describe the the curve hierarchy required as input to the distance transform. We begin by stating the properties of the input curve hierarchy for the general case, then describe in detail the isocontour hierarchy used in this paper.

3.1 Curve Hierarchy Requirements

Let C denote the set of points on the input curve(s) and D denote the domain in $\Re^2$ for which an approximate distance function is desired. The input curve must have the following properties:

1. **Continuity:** The curve(s) must be C^0 continuous everywhere and C^1 continuous everywhere except at a finite number of points.
2. **Partitions the domain:** The curve(s) must partition the domain D into an inside and outside labeled by negative and positive distances. This also implies that the curve end points may only occur on or outside the boundary of D.
3. **Bounding Boxes:** The curve must allow a hierarchy of bounding boxes to be specified. In the remainder of the paper we will index these bounding boxes by α. Each node in the hierarchy must define a bounding box B_α in $\Re^2$ which contains a subset of the curve.
4. **Normal Wedges:** Each curve hierarchy node α must define a bound on the directions of all unit normals of the curve contained in B_α. The bound is represented by a normal wedge (n_α, ψ_α) with central unit normal n_α and opening half angle $0 \leq \psi_\alpha \leq \pi$. A unit vector v is contained in a normal wedge for node α if the following condition is met: $v \cdot n_\alpha \geq \cos(\psi_\alpha)$.

Any number of data structures and curve definitions could be used to construct a curve hierarchy. We have chosen a hierarchy based on scalar field data generated by scientific simulation. The next subsection describes this isocontour hierarchy.

3.2 Constructing An Isocontour Hierarchy

Isocontours of a sampled scalar field with bilinear interpolation satisfy the first two requirements listed above as long as the domain D is the same as the scalar field domain or a subset of it. Two strategies for satisfying items 3 and 4 are possible: (1) Extract the isocontour and build a bounding box hierarchy from the geometry, or (2) Construct a spatial hierarchy on the scalar field and use it to generate bounding box hierarchies for all isocontours. The first strategy will result in the tightest bounding boxes because the geometry is exactly specified. We adopt the second strategy because spatial data structures such as min-max octrees are already commonly used to accelerate isocontouring operations and adding normal wedge information to them is straightforward.

We compute a min-max quadtree on the scalar field and compute bounds on the directions of the isocontour normals using the scalar field gradient. Figure 2 shows an example scalar field and three levels of the quadtree hierarchy. The normal wedges are defined by the scalar field gradients and are drawn inside each node with the opening angles denoted by dotted arcs.

The scalar field hierarchy has loose bounds on the normal direction because the direction bounds are valid for all isocontours within a particular node. This forces the

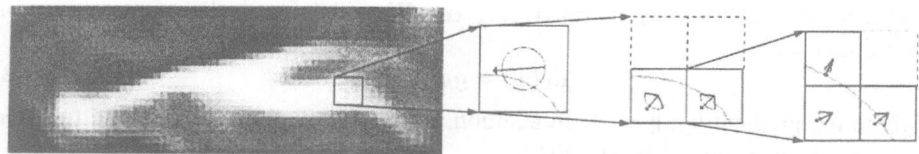

Fig. 2. (Left) A sampled scalar field rendered with one pixel/sample. A single quadtree node is highlighted. (Right) Three levels of the isocontour bounding box hierarchy.

distance transform algorithm to subdivide more often in areas where the original scalar field gradients are widely varying. This hierarchy is similar to the data structure used in [8] in that it involves minimal preprocessing and grants access to all of the isocontours in the scalar field.

4 The Distance Hierarchy

We begin this section by formally defining the signed distance function of a curve or set of curves C as follows:

$$\mathrm{d}(x) := \mathrm{sign}(x) * \min_{\forall y \in C}(\| x - y \|) \tag{1}$$

where C is the set of points on the input curve(s), x and y are points in $\Re^2$, $y \in C$, and $\mathrm{sign}(x)$ returns negative if x inside the curve, and positive if outside.

We will use T_k to refer to the triangle associated with a distance node k. All other quantities associated with a node shall be denoted by subscripting them with k.

4.1 Triangle Bintrees

The triangle bintree begins with a right isosceles triangle. Each level of the triangle bintree is created by edge bisection operations. A triangle is split by inserting a vertex at the midpoint of the hypotenuse and connecting it to the opposite vertex of the triangle. This operation is then applied recursively to the children. A new vertex is created when a triangle is split. This creates a crack problem because the triangle sharing the edge may not be split. The crack problem is solved by forcing splits until there are no vertices on any edges. In this paper we begin with a rectangular domain and relax the constraint that each triangle must be right isosceles.

4.2 Linear Approximation

A distance node maintains a list of curve bounding boxes which we will refer to as its *active list*. The bounding boxes in the active list contain all curve points which potentially contribute to the distance function in the node. In the remainder of the paper we will denote the active list of a distance node by A_k and define it as a list of indices of curve nodes as follows: $A_k := \alpha_0, \alpha_1, \ldots, \alpha_N$. We will define the linear approximation for a node k as:

$$\tilde{d}_k(x) = g_k \cdot x + c_k \qquad (2)$$

where the gradient $\|g_k\| = 1$. In addition, the error bound on the distance function within a distance node k must satisfy:

$$\| \max_{\forall x \in T_k}(\mathrm{d}(x) - \tilde{d}_k(x)) \| \leq \varepsilon_k \qquad (3)$$

In general, we want $\varepsilon_k < \mathrm{uerror}(k)$, where $\mathrm{uerror}(k)$ is a user defined error which depends on k and may depend on other parameters as well. When the distance function is too complicated to be linearly approximated, we fall back on a constant approximation ($g_k = 0$).

5   The Distance Transform

Given a distance node k, the transform algorithm recursively subdivides k until a user defined error criterion is met over the original triangle T_k. The subdivision may cause forced splittings of some triangles outside of T_k as mentioned in section 4.1. We assume that k contains an active list A_k of all curve nodes which may contribute to the distance function within T_k. The algorithm proceeds as follows:

1. **Establish Bounds:** For each curve node in A_k compute conservative lower and upper distance bounds. These bounds are computed with respect to the bounding regions B_α and the distance triangle T_k. These bounds are used to compute lower and upper bounds on the distance within T_k.
2. **Refine and Cull:** Place the curve nodes in A_k in a priority queue. Let the priority $p(\alpha) = 2r \sin(\psi_\alpha)$ be a bound on the error of a linear approximation of a node, where r the length of the node diagonal. This error bound stems from the fact that once a point on the curve is known the normal wedge constrains the possible directions of the curve. In priority order refine the nodes in A_k until $p(\alpha) < \mathrm{uerror}(k) : \forall \alpha \in A_k$. For each child node added to A_k compute conservative bounds as in step 1 and update the overall bounds on T_k. As the overall bounds tighten, *Bound Cull* any curve node in A_k that is unable to contribute to the distance function within T_k. When a curve node is bound culled label all adjacent curve nodes as gap nodes.
3. **Linear Approximation:** Compute a linear approximation of the distance function over T_k based on the refined active list nodes A_k. A point on the curve is sampled and the opening half angles of the curve normal wedges are used to compute a linear approximation with an error bound. A test is then made on A_k to insure that the bound is correct.
4. **Constant Approximation:** If no linear approximation with guaranteed error was computed, or the error of the linear approximation violates the user supplied error criterion, then compute a constant approximation as follows: Use the conservative bounds on the distance over T_k to compute a c_k and ε_k for equations 2 and 3.
5. **Recurse or End**: If the resulting approximation error does not satisfy the user defined error criterion, split distance node k and copy the active list A_k to both children. Repeat with step 1 for each child of k.

5.1 Bound Culling And Constant Approximations

In this section we describe how conservative guaranteed error bounds are established between a distance triangle T_k and the curve nodes in its active list A_k. We begin by defining when a point on the curve contributes to the distance function:

Definition 1. *A point $y \in C$ contributes to the distance function within the triangle T_k of a distance node k if there exists at least one point $x \in T_k$ for which y is the point of closest approach of the curve C.*

Furthermore, we say a curve hierarchy node α contributes to the distance function within T_k if at least one point $y \in C \cap B_\alpha$ contributes. We define unsigned conservative lower and upper bounds on the distance function with respect to a single curve hierarchy node as $lower_{k,\alpha} = \min(\| x - y \|)$ and $upper_{k,\alpha} = \max(\| x - y \|)$ for all $x \in T_k$ and all $y \in B_\alpha$. The lower and upper distance bounds are are conservative because they rely only on the bounding box B_α, and not on the actual curve contained within B_α. These bounds can be expensive to compute and become more expensive in three dimensions. We use bounding circles for the distance computations for simplicity.

The next step is to bound the distance function due to all curve nodes in the active list A_k. These bounds are given by $d_{lower} = \min(lower_{k,\alpha})$ and $d_{upper} = \min(upper_{k,\alpha})$ for each $\alpha \in A_k$. Thus, we have the following condition on the distance function within distance node k:

$$d_{lower} \leq \mathrm{d}(x) \leq d_{upper} \; ; \; \forall x \in T_k \tag{4}$$

A curve hierarchy node does not contribute to the distance function if the following condition holds: $lower_{k,\alpha} > d_{upper}$. Finally, the lower and upper bounds may be used to construct a piecewise discontinuous constant approximation as $\tilde{d}_k(x) = \frac{1}{2}\mathrm{sign}(x)(d_{lower} + d_{upper})$ and $\varepsilon_k = |d_{upper} - \tilde{d}_k(x)|$.

5.2 Computing Linear Approximations

The processes of computing a linear approximation and obtaining a guaranteed error bound are decoupled from one another. This is due to the fact that curves with complex foldings, disconnected components, and gaps created during bound culling produce distance fields which are difficult to bound correctly. In this section we describe how to compute linear approximations and error bounds. In the next section we describe how to guarantee that a bound is correct.

Equation (2) implies that the zero set of a linear approximation is the equation of a line with normal g_k. We set g_k to the normalized average of all central normals of the normal wedges in A_k and compute an opening half angle ψ_k with respect to the merged normal wedges in A_k. The constant c_k is obtained by sampling a point p on the curve contained in the active list A_k and solving for the zero set: $g_k \cdot p + c_k = 0$. The point p is computed by choosing an isocontour node in the active list and finding a contour intersection with one of its edges using linear interpolation. We bound the error of the approximation using the normal wedge information of the active list nodes. The extent r of A_k is computed as the maximum distance from p to the boundaries of the nodes in A_k. We compute the error bound as $\varepsilon_k = r \sin(\psi_k)$. In practice, this simple method works quite well near the curve. It is dependent on the variation of the normal directions of the curve hierarchy and tends to do less well for distance triangles farther from the isocontour.

5.3 Guaranteed Error Bounds

We assume a linear approximation has been computed and an error bound ε_k must be produced. We desire a tighter error bound than that presented in equation (4). We can achieve this by utilizing information about the shape of the curve contained in A_k. Figure 3 illustrates how an error bound ε_k constrains the location of the contributing curve points. A constant approximation has no estimate of the gradient and only constrains contributing curve points to the annulus centered at x denoted by the dashed circles. A linear approximation with a guaranteed error ε_k further restricts the possible locations of contributing points to the shaded area between the isocontours of the linear approximation at $\pm\varepsilon_k$.

Given a possible ε_k, the distribution of curve points in the active list must be analyzed to insure that at least one curve point falls in the shaded area of Fig. 3 for each point in T_k. Gaps or folds in the curve must be accounted for to insure that the bound on the distance is correct. The general procedure is as follows:

1. Compute a possible error bound ε_k.
2. Insure that the curve does not have folds or loops by requiring that the curve bend no more than 90 degrees from the estimated gradient g_k and is entirely front facing or entirely back facing with respect to g_k.
3. Insure that no gaps exist in the curve by requiring that no curve nodes in the active list A_k labeled as gaps exist in the shaded area of figure 3.
4. If both 2 and 3 are satisfied, then the ε_k computed in item 1 is a guaranteed bound.

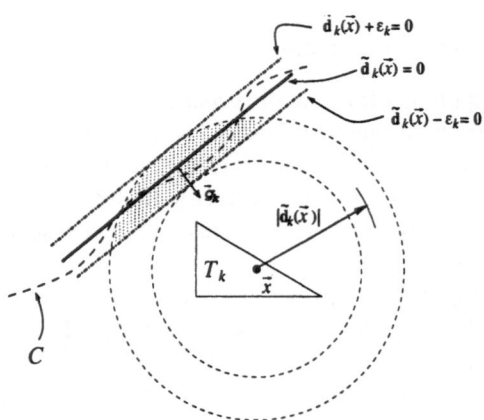

Fig. 3. Points of closest approach must fall inside the shaded area for a linear approximation with guaranteed error.

5.4 Producing a Continuous Approximation

The linear approximation within each triangle defines a range of distance values for each vertex of that triangle. The procedure for constructing a continuous distance function is the following. For each vertex the range of distance values is computed as the intersection of the distance ranges for each incident linear approximation. The midpoint of this range is chosen as the distance value of the vertex. This doubles the error bound on the approximation because the midpoint might occur at a lower or upper bound of one of the incident ranges. It is possible to use a more aggressive algorithm that uses the ranges of neighboring vertices to improve the accuracy of the continuous distance approximation.

6 Results

We tested our algorithm on scalar entropy data from a turbulent mixing simulation [7]. Figure 4 shows a 256×128 slice from one sub-domain of this computation with the isocontour at 50% of the maximum entropy. The error criterion we used in the examples was computed as follows: $\text{uerror}(k, \varepsilon_{min}, \lambda) := \max(\varepsilon_{min}, \lambda d_{lower})$. where d_{lower} is computed for each distance node k as defined in section 5.1. This error criterion allows less accurate approximations farther away from the curve and clamps approximations near the curve to a user defined minimum. We found that on average each distance triangle tracked $15 - 17$ curve nodes in its active list. This value was not dependent on the values of λ and ε_{min}. Our experiments indicate that the execution time and number of triangles varies approximately linearly with $1/\varepsilon_{min}$. That is, reducing the error by a factor of two increases the number of triangles and execution by approximately the same factor.

Figure 5 shows an adaptive distance function and the corresponding triangulation of the isocontour in Fig. 4. The transform took 37 seconds to compute and required approximately 60 megabytes of memory. The final distance triangulation contains 70361 triangles. The implementation was not optimized for memory usage and uses pointer based hierarchies. In figure 6 the same distance approximation is shown.

7 Conclusion

We have implemented a fully adaptive distance transform algorithm that produces piecewise discontinuous linear approximations in a top down fashion. Although the implementation was not coded for speed, it is clear that a three dimensional version of the transform would require more efficient data structures. The algorithm is tuned for approximations near the input curve and tends to rely on constant approximations farther away. This could be improved by using a different approximation strategy for boxes farther from the curve.

The algorithm does not handle discontinuities in the gradient of the distance function in an efficient manner. Gradient discontinuities violate the observation that the curve dependencies are reduced when a distance triangle is refined. An extreme example of this is a triangle which contains the center of a circle. The distance function within the triangle depends on the entire circle. At least one child triangle will contain the center regardless of the number of refinement operations.

8 Acknowledgments

This work was performed under the auspices of the U.S. Department of Energy by Lawrence Livermore National Laboratory under Contract W-7405-Eng-48. We wish to thank Valerio Pascucci for helpful discussions.

References

[1] David E. Breen, Sean Mauch, and Ross T. Whitaker. 3d scan conversion of csg models into distance volumes. In *Proc. 1998 Symposium on Volume Visualization*, pages 7–14, Oct 1998.

[2] Mark Duchaineau, Murray Wolinsky, David Sigeti, Mark C. Miller, Charles Aldrich, and Mark B. Mineev-Weinstein. Roaming terrain: real-time optimally adapting meshes. In *Proc. Visualization*, pages 81–88, 1997.

[3] Sarah F. Frisken, Ronald N. Perry, Alyn P. Rockwood, and Thouis R. Jones. Adaptive sampled distance fields: a general representation of shape for computer graphics. In *Proceedings of SIGGRAPH 2000*, pages 249–254, July 2000.

[4] Sarah F. Gibson. Using distance maps for accurate surface representation in sampled volumes. In *Proceedings of the 1999 Conference on Volume Visualization*, pages 23–30, Oct 1998.

[5] H. Hoppe, T. DeRose, T. Duchamp, J. McDonald, and W. Stuelzle. Surface reconstruction from unorganized points. In *Proceedings of SIGGRAPH 92*, pages 71–78, July 1992.

[6] Peter Lindstrom, David Koller, William Ribarsky, Larry F. Hodges, Nick Faust, and Gregory A. Turner. Real-time, continuous level of detail rendering of height fields. In *Proc. SIGGRAPH*, pages 109–118, 1996.

[7] A. A. Mirin, R. H. Cohen, B. C. Curtis, W. P. Dannevik, A. M. Dimits, M. A. Duchaineau, D. E. Eliason, D. R. Schikore, S. E. Anderson, D. H. Porter, P. R. Woodward, L. J. Shieh, and S. W. White. Very high resolution simulation of compressible turbulence on the IBM-SP system. In ACM, editor, *Super Computing 1999, Oregon*, pages ??–?? ACM Press and IEEE Computer Society Press, 1999.

[8] V. Pascucci and C. L. Bajaj. time-critical isosurface refinement and smoothing. In *Proceedings of Volume Visualization and Graphics Symposium 2000*. IEEE and ACM/SIGGRAPH.

[9] William J. Schroeder, William E. Lorensen, and Steve Linthicum. Implicit modelling of swept surfaces and volumes. In *Proc. Visualization*, pages 40–45, 1994.

[10] J. A. Sethian. A fast marching level set method for monotonically advancing fronts. *Proceedings of the National Academy of Science*, 93:1591–1595, 1996.

[11] Ross T. Whitaker and David E. Breen. Level-set models for the deformation of solid objects. In *Proc. of the 3rd International Workshop on Implicit Surfaces*, pages 19–35, June 1998.

[12] K. J. Zuiderveld, A. H. J. Konig, and M. A. Viergever. Acceleration of ray-casting using 3-d distance transforms. In *Proc. of Visualization in Biomedical Computing*, pages 324–335, 1992.

Editors' Note: see Appendix, p. 352 for colored figures of this paper

Progressive View-Dependent Isosurface Propagation

Zhiyan Liu, Adam Finkelstein, and Kai Li

Department of Computer Science, Princeton University

Abstract This paper proposes a new isosurface extraction algorithm that extracts portions of the isosurface in a view-dependent manner by ray casting and propagation. The algorithm casts rays through a volume to find visible active cells as seeds and then propagates their polygonal isosurface into the neighboring cells. Small pieces of the isosurface are generated by distance-limited propagation and joined together to form the final surface. We demonstrate that this progressive algorithm generates an approximate result quickly and refines it to the final correct image over time. In addition, the algorithm scales with the resolution of the display and supports adaptive-resolution visualization.

1. Introduction

Applications such as large-scale simulations typically generate scalar fields and store them as volumetric datasets. A 3D scalar field F can be represented by a volumetric dataset that has a set of data points and the corresponding scalar values sampled at each point in the set. To visualize the scalar field, a known method is to display isosurfaces where $F(x, y, z) = v$ for a given threshold v. To visualize the isosurfaces of massive datasets, the challenge is to develop an algorithm that extracts the isosurfaces efficiently, requires modest rendering power, and supports interactive, adaptive-resolution visualization on a high-resolution display system.

Much work has been done on extracting isosurfaces, but existing algorithms all have certain drawbacks. The *marching cubes* [11] algorithm visits all n cells in the dataset and triangulates the isosurface in each *active* cell, i.e. a cell that has values above and below the given threshold. Marching cubes is simple and straightforward, but examining all the cells in the dataset can be unnecessarily time consuming. Several subsequent algorithms reduce the time spent on finding the cells that intersect the isosurface. Wilhelm and Van Gelder used an *octree* [15] to leverage object-space coherence and discard sections of the dataset before examining them. Cignoni *et al.* proposed the *interval-tree* method [3] and Livnat *et al.* proposed to use *span space* [8]. In preprocessing, both of these algorithms sort all the cells according to minimum and maximum values and construct a search tree; then, for a given threshold, these methods search the tree to find all the active cells. Itoh and Koyamada used the *extrema graph* [5] approach to find seeds on the isosurface and propagate from these seeds. In the worst case the seed set could have size $O(n)$. Bajaj *et al.* described the *contour trees* method [1] for finding small seed sets for isosurface traversal.

Although these methods dramatically improve on the original marching cubes algorithm, they do not try to avoid generating occluded polygons, nor do they manage level of detail. As a result, they all generate the complete isosurface at the finest data resolution (one voxel). For very large datasets, generating and rendering the whole isosurface will prevent users from viewing the dataset at an interactive frame rate. Extraction can be slow, and the sheer number of polygons in the isosurface may overwhelm the hardware rendering capabilities.

Recently, two isosurface visualization algorithms are proposed to generate only the visible portions of the surface. Parker *et al.* proposed a *ray-casting* algorithm for isosurface extraction [13] that intersects viewing rays with the data volume and then computes the isosurface without generating an intermediate polygonal representation. For each ray intersecting the isosurface, a cubic equation is solved to find the normal at the intersection point. This approach is simple and requires no special rendering hardware. The authors have parallelized the algorithm to run on a 128-processor SGI Origin shared-memory multiprocessor to offer interactive frame rates for a 512×512 display. However, the running time of the algorithm is proportional to the number of pixels in the display, it therefore is not well suited for high-resolution displays.

Livnat and Hansen described WISE, a *view dependent* isosurface extraction algorithm that uses hierarchical tiles and shear-warp factorization for visibility testing, and then renders the polygons utilizing the graphics hardware [9]. They also used a 512×512 display. Traversing the dataset in a front-to-back order, (meta) cells are projected to the screen and tested against the current screen coverage map for visibility in software. Occluded (meta) cells are discarded. Visible meta-cells are examined recursively. All the triangles inside a visible cell are extracted and sent to the graphics hardware, and the current screen coverage map is updated accordingly. When the resolution of the display increases, both the coverage map calculation time and the space requirement for the hierarchical visibility mask will grow proportionally. Very recently, Livnat and Hansen have proposed SAGE [10], a view dependent algorithm that improves on the performance of WISE.

In this paper, we propose a new hybrid algorithm that shares several of the features of existing acceleration methods. The main idea is to use ray casting into an octree as a way to identify visible seed cells (rather than computing the complete isosurface as in the method of Parker *et al.* [13]) and then use propagation (as in [5]) to extend the isosurface from the seed cells. Unlike previous propagation methods that propagate to the whole isosurface, our method uses distance and viewing criteria to decide where to stop propagation, and thus generates only a small piece of the isosurface connected to each seed cell. These pieces are patched together to form a view-dependent region of the isosurface, which includes all the triangles that are visible as well as a small number of occluded triangles that are near the visible surface.

In addition to largely avoiding the rendering of occluded portions of the isosurface, our method supports two acceleration schemes suitable for interactive visualization. First, we show that we can quickly acquire a very good approximation of the visible isosurface from just a few initial seeds, and then progressively refine the surface as subsequent rays discover the remaining visible active cells. Second, for very high-resolution grids, we describe a form of adaptive-resolution rendering that relies on the octree organization of the data in order to extract and render triangles at a resolution chosen based on the scale of screen pixels relative to the voxel data.

2. The Algorithm

The proposed algorithm may be viewed as an extension to the propagation method [5]. Currently it works with structured rectilinear datasets. The main contribution is to make the propagation algorithm view dependent in a manner that is efficient and incremental while supporting adaptive-resolution visualization.

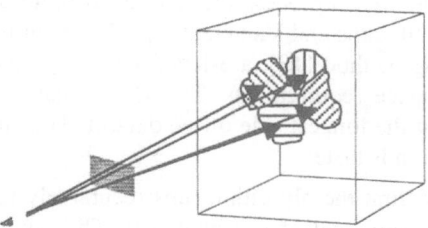

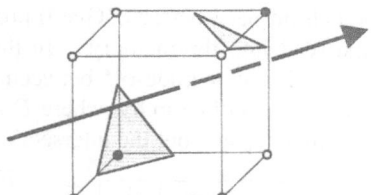

Figure 1: Illustration of the algorithm. Rays are cast into the dataset; a patch of surface is propagated from each seed cell found.

Figure 2: An example where the first active cell a ray intersects doesn't have an isosurface triangulation that intersects with the ray.

For the convenience of the description, let us first define the active cell as follows. Given a threshold v, we mark all the data points in the dataset with one of the two signs: "+" indicates that the scalar value at that point is above the threshold, while "- " indicates that the scalar value is below the threshold. We only consider the non-degenerated case where no one data point has exactly the value v. If a cell has vertices of different signs, then it's called an active cell, and the isosurface of threshold v will intersect this cell.

A key observation the propagation approach made was that if the vertices on a face of an active cell do not have the same sign, then the neighboring cell that shares the same face is also active. Therefore, the isosurface can be extended into the neighboring cell. This means that once an active cell is found as the seed, propagating the isosurface from that cell is efficient because one can avoid touching and examining inactive cells. However, neither the Extrema Graph nor the Contour Trees algorithm generates seeds that are guaranteed to be visible. An efficient propagation algorithm should traverse only the active cells that are visible.

Our algorithm executes in three stages, as Figure 1 shows. For each pixel in the screen space, first a ray is cast from the eye through the pixel into the dataset and the intersection is calculated. Next, the first active cell that contains a portion of the isosurface that intersects with the ray (if it exists) is used as the seed and propagated for a certain distance. Third, all the active cells that have been visited in this pass are examined, case numbers are generated and the parts of the isosurface in these cells are triangulated using Marching Cubes method.

2.1 Ray casting

Our method uses ray casting to identify active cells as seeds for propagation. The ray-casting step finds the first active cell in which the isosurface triangulation intersects with the ray. This cell is guaranteed to be visible. If it has not been visited, this active cell will be used as the seed for the propagation step. Note that this cell is not necessarily the first active cell a ray intersects, as Figure 2 shows. The first active cell a ray intersects may have an isosurface triangulation that doesn't intersect with the ray, which means the triangulation won't render to the corresponding pixel. By finding the first active cell that actually renders to the corresponding pixel, we guarantee that for each ray cast, the corresponding pixel has the correct color. After a ray has been cast for each pixel, the final image will be correct. That proves our algorithm is conservative.

In order to make the ray-casting step efficient, we preprocess the dataset to build a branch-on-need octree (BONO) proposed by Wilhelms and Van Gelder [15] when it is first read into the memory. In the ray-tracing method [13], a 3-level hierarchy was used. This is a trade-off between time and space requirements. The Octree has an $O(logD)$ level hierarchy, where D is the size of the longest side of the dataset. Thus it uses more space, but the intersection computation is faster.

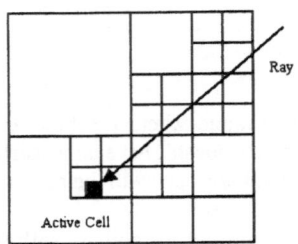

For each ray, first the algorithm runs recursively to find the first active cell that it intersects. The ray is first tested against the whole dataset. If it intersects the dataset and the threshold is between the overall minimum and maximum of the dataset, the sub-regions in the dataset that intersect the ray are examined in a front-to-back order. The algorithm performs intersection tests and value comparisons recursively until it finds an active cell that the ray intersects, as shown to the left. If the algorithm exits without finding a cell, then the ray does not intersect with any active cell in the dataset.

Once the active cell is found, we proceed to test whether the isosurface triangulation inside it actually intersects with the ray. If not, the next active cell the ray intersects is found and tested. This is done till an active cell whose isosurface triangulation intersects with the ray is found or the ray exits the dataset, which indicates that the isosurface doesn't cover the corresponding pixel.

We use a hash table to record which cells have already been visited. If the active cell found by the recursive algorithm has not been visited, we will mark it and use this cell as the seed for the propagation step. If the active cell has been visited, then no further action will be taken.

To accelerate the ray-casting step, we use integer coordinates for the data points. The eye and screen are mapped back to the object coordinate system using the current model view matrix. The coordinates of the data points are fixed and implicit: each data point is on a grid and has integer coordinates. The intersection is significantly faster than using the world coordinate system because every intersection test is with a cube with edges parallel to coordinate axes.

A typical way to visualize the data is to begin by casting sparse and evenly distributed rays in the screen space, then add more rays to increase the ray density gradually till a ray has been cast for each pixel or user interrupts the extraction.

For a given screen resolution, the order of rays to be cast can be predetermined: the first ray goes from the center of the screen, the next 4 rays are each from the center of one of the 4 quads, and so on. The granularity of ray casting determines the speed and the precision of isosurface extraction. Fine-grained ray casting takes time, but it yields precise isosurface representation. Our design is to let user control the density of ray. When the user changes the threshold or the viewpoint, all the calculations for the previous setup are immediately stopped and new ones begin. If the user doesn't interrupt, a ray will be cast for each pixel and the correct isosurface will be generated.

2.2 Propagation

Our algorithm uses a queue for propagation. Initially, the active cell found in the ray-casting step is the only one in the queue. For each cell in the queue, the algorithm dequeues it, sends it to the triangulation step, and checks all its active neighbors. If the active neighbor cells have not been visited and satisfy certain propagation criteria, they will be added to the queue.

The propagation criteria need to be chosen carefully. The further the propagation proceeds, the fewer inactive cells the algorithm has to examine. On the other hand, even though the seed is visible, the cells that it propagates to are not necessarily visible. More propagation may increase the chance of traversing occluded cells. Also, expanding out of the screen space or to the back-faced side of isosurface is not desirable.

Our algorithm sets a cut-off angle for a ray and names it the propagation distance. At each propagation step, the algorithm calculates the angle between the ray we cast and the vector from the eye to the current cell and stops adding it to the queue when the cut-off angle is reached.

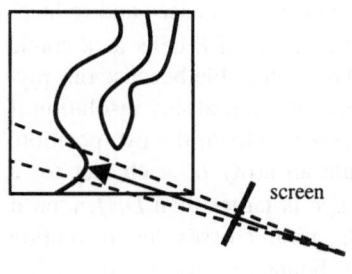

The figure to the left shows using a cut-off angle as the propagation distance. The propagation will not expand out of the region defined by the dashed lines. Suppose the first m rays have a combined propagation area that covers the whole screen, then after all the m rays are cast, the only possible reason for inaccuracy in the isosurface we computed is that there is an isolated part of the isosurface in front of the isosurface we expanded and it didn't intersect with any ray we've cast. To treat this situation we just cast more rays at different locations. If the user doesn't interrupt, at last one ray will be cast for each pixel and result is guaranteed to be correct. We may generate more triangles than that are actually visible, but it's interactive because the user gets approximate results that refine with time. Our results show that for several datasets, only a relatively small number of rays are needed.

In order to avoid generating occluded triangles, we calculate the angle between the current ray and the vector from the eye to the cell being propagated to detect whether the isosurface folds back. We name this angle the distance angle. If the isosurface folds back in a cell, that cell will not be added to the queue and the propagation from the cell will stop. The detection works as follows. If a cell C1 is examined and its neighbor C2 is added to the queue, then we say C1 is C2's predecessor in propagation, and C2 is C1's successor. Normally, the distance angle between the ray and the vector increases as the propagation proceeds. When the distance angle for a cell decreases comparing with its predecessor cell, then the isosurface is curving back and the cell should be discarded. However, to deal with bumpy surfaces, our algorithm has a small tolerance value. Only when the decrement exceeds the tolerance value do we stop.

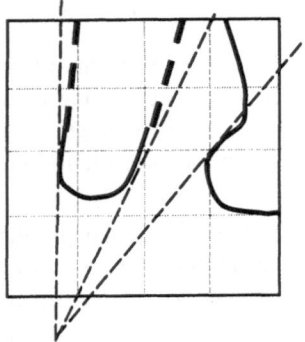

Propagation is an efficient way of extracting triangles; we are willing to pay the small price of rendering a few more triangles to keep the propagation going. The decrement is calculated from the largest distance angle among all the (indirect) predecessors of a cell, so even if each time the decrement is very small, they can accumulate and stop the propagation. This solves the problem where the isosurface curves back and propagates in a direction that's almost parallel to the ray, as shown on the left. The dashed part of isosurface won't be propagated to, whereas a small groove is tolerated and propagation goes on.

2.3 Adaptive-resolution isosurface

For a large dataset, it is possible that a far-away cell is of sub-pixel size when projected to the screen. If it is an active cell, then more than one triangle in the isosurface will render onto the same pixel. That is a waste of computing resources and does not increase the quality of image. Our algorithm detects such cases and reduces the data resolution to 2×2×2 (treating a meta-cell that consists of 8 cells as a single cell and ignoring all the inside values) or even lower. This is feasible because our ray-dataset intersection walks down an octree hierarchy. We can stop at any resolution if the (meta-) cell projects to less that one pixel on the screen. Given the eye position, screen position, and screen resolution, we can compute an array D, such that for a meta-cell of size $2^i \times 2^i \times 2^i$, if its distance from the eye is larger than $D[i]$, then it should be treat as one single cell. When the propagation crosses the resolution boundary defined by D, our algorithm stops at the boundary. At the resolution boundary, there will be cracks in the actual representation of the isosurface, i.e. the triangles from different resolutions may not connect to each other, but the cracks won't be visible because they are of sub-pixel size. Every pixel that the isosurface covers will be rendered to by the active cell that's found in the intersection phase using the ray that goes through the center of the triangle. This is similar to [9], where the set of triangles that are rendered is only a subset of all the visible triangles, and where a single point is used to represent a faraway meta-cell. The view-dependent methods generate results that user perceives as identical with the complete representation from his current viewpoint.

3. Results

We have implemented the algorithm above on a PC that runs Windows 2000, and conducted experiments with the implementation. The PC hardware includes a 933Mhz Pentium III CPU, an NVIDIA GeForce 256 graphics card, and 512 MB of main memory.

We applied our algorithm to the head section of the Visible Woman CT data, using a 512×512×209 dataset, which is at its original data resolution. A cut-off angle of 1.81 degrees was used for all the experiments. The visualization is done in full screen mode with a screen resolution of 1600x1200. In the absence of user interrupt, 1,920,000 rays will be cast, one from each pixel.

SKIN (v = 600.5, left column in Color Plate)

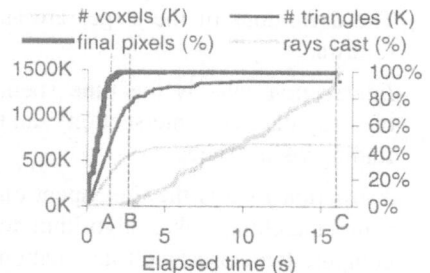

Graph 1: The front full view of the skin.

Bone (v=1224.5, right column in Color Plate)

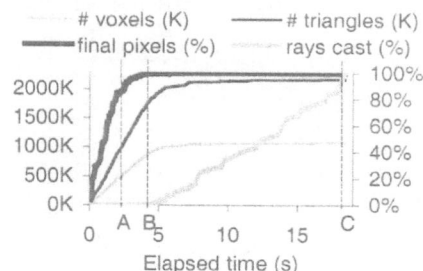

Graph 2: The side full view of the bone.

Point	Time (s)	Δs (K)	Pixels (%)	Rays (K)
A	1.1	514	85.3	0.2
B	2.4	1,068	99.5	15.6
C	16.2	1,380	100.0	1,920

Table 1: Different points in Graph 1.

Point	Time (s)	Δs (K)	Pixels (%)	Rays (K)
A	2.2	953	85.7	0.6
B	4.3	1,796	99.5	15.1
C	18.4	2,175	100	1,920

Table 2: Different points in Graph 2.

To quantitatively measure how close the an intermediate representation of the isosurface is to the correct and final representation, for each pixel that has been rendered to in the final representation, we check whether it has the same color in the intermediate image, and if so name it a final pixel. The percentage of the final pixels among all the rendered pixels indicates the correctness of the intermediate image. Graph 1 shows in an experiment of extracting the Visible Woman's skin, how the percentage of final pixels, the number of active cells visited, the number of triangles generated, and the percentage of rays cast change as the computation proceeds. The statistics for points A, B, and C are shown in Table 1. The corresponding screen images are shown in the color plates.

Graph 2 and Table 2 show the result from another experiment that extracts the bone structure from Visible Woman's head. The similarity between the graphs shows that our algorithm behaves consistently. Both cases show that the proposed algorithm works progressively and efficiently. After casting a few rays, our algorithm generates most of the isosurface.

In the skin extraction case, over 85% of the isosurface is extracted in 1.1 seconds with only 240 rays cast (point A), whereas 99.5% of the isosurface is extracted in about 2.4 seconds with about 0.8% rays cast (point B). To obtain 100% pixels, it took 16.2 seconds. The bone extraction case has similar curves but it took 2.2 seconds to obtain 85% of the isosurface and 4.3 seconds to extract 99.5% of the isosurface. The total extraction took 18.4 seconds. This is because the isosurface of the bone has about 60% more triangles to render than the skin.

The last 0.5% of pixels took much longer time to extract in both cases, but they make very little difference on the screen.

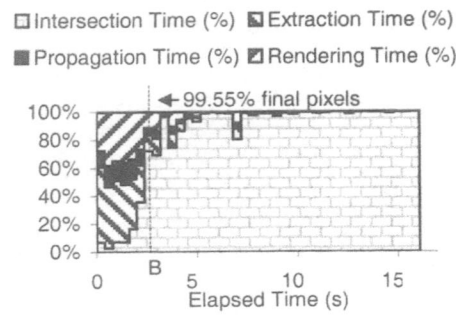

The graph on the left shows the breakdown of the computation time. The definitions of the stages are as follows:

Intersection time is the time spent on ray-dataset intersection and finding visible seeds.

Extraction time is the time spent on using marching cubes algorithm to compute the isosurface triangulation inside active cells.

Propagation time is the time spent on finding the active neighbors of a cell in the propagation queue and testing whether they satisfy the propagation criteria.

Rendering time is the time spent on rendering all the triangles generated.

Note that among these four time components, only intersection time is proportional to the number of rays cast. The other three are proportional to the number of triangles generated. When the full computation ends, the intersection calculation is the most time-consuming stage. However, most of the isosurface has been generated by the time the intersection stage becomes significant. The dashed line shows point B in Graph 1. Before point B, the time spent in intersection is negligible. At point B, the representation is 99.5% correct. After point B, the intersection calculation becomes dominant. In this experiment, because of the high resolution of the display (1600×1200), all the triangles are generated at the finest data resolution. This implies that our algorithm is suitable for large-scale displays. For a fixed dataset, when the resolution of the display increases, only the intersection time will increase, which has very little influence on the position of point B on the time axis.

4. Comparisons with previous algorithms

It is difficult to compare our approach quantitatively with other approaches, without implementing them all on the same hardware platform. However, we can make some qualitative comparisons. Here we focus on view-dependent work.

Our approach allows viewers to see the shapes of the isosurface after casting only a few rays, whereas in the naïve implementation of ray tracing the visual quality is linear in both the number of rays cast and the elapsed time. Also, our approach leverages the cost-effective rendering performance of PC graphics cards. Similar to the ray-tracing approach, our algorithm is image-space based and can be parallelized.

To perform a crude comparison with the WISE method of Livnat and Hansen [9], we ran our algorithm using the same size display (512×512 pixels) as they used in recent experiments [10] with the same Visible Woman dataset. The results of their experiments indicate that running on a SGI Onyx 2 they extract 344,628 triangles in 35.8 seconds, and then render this surface in 0.6 seconds. In contrast, our method, running on a Pentium III 933MHz PC with GeForce graphics card, extracts and renders 1,289,904 triangles in 4.3 seconds. Based on the relative clock rates on the platforms, we expect that our performance would be better on the SGI than that of the WISE algorithm. Very recent work by Livnat and Hansen introduces SAGE [10], a view dependent algorithm that improves on the performance of WISE (and given the relative hardware difference probably exceeds our performance as well), with a reported extraction time of 4.4 seconds and rendering time of 0.3 seconds for the same dataset. Because our visibility test is more conservative, our method extracts and renders many more triangles than the WISE and SAGE algorithms. However, our experiments indicate that triangle rendering is not a bottleneck, and inexact visibility allows us to quickly find large portions of the surface. Our algorithm very quickly provides a good approximation of the surface: 85% and 99.5% correctness were achieved at 0.9s and 2.2s respectively on the 512×512 display. Finally, comparing these numbers with those in Table 1 (1.1s, 2.4s), we show that while the number of the pixels increases by more than a factor of 7 (262,144 to 1,920,000), the points A and B were only delayed 20% and 10% respectively. This suggests that the progressive aspect of our algorithm scales well for very high-resolution displays.

5. Conclusions and Future Work

This paper describes a new isosurface extraction algorithm based on ray casting and propagation. We have shown that the new approach is progressive and efficient.

Our algorithm is suitable for high-resolution displays. The results reported in this paper were generated using a PC at full screen (1600×1200) resolution. We would like to adapt the algorithms presented here for use with tiled displays, as part of the Princeton Display Wall Project [7]. As an initial step, we ran the program on a large-scale (18-foot) display surface covered by 24 tiled projectors arranged on a 6×4 grid, yielding more than 20 million pixels. A server PC drives each projector, and each PC runs a copy of the isosurface extraction algorithm. When the isosurface is spread over several projectors, we find a corresponding performance improvement because each PC has a partial view of the surface and has fewer triangles to extract and render. However, when the isosurface falls entirely within one projector, the performance drops to that of a single PC. To address this problem, we are now working on a load-balanced parallel version of the algorithm.

Our algorithm is suitable for large datasets. Currently the entire dataset resides in memory, which limits the size of dataset we can visualize. Because surface propagation has strong data locality, we believe that it will be possible to adapt an out-of-core version of our algorithm.

Remote data visualization has become an important area of research because massive amounts of data are generated and distributed over the network. Since our algorithm aims to reduce the number of triangles generated as well as maintain a fast extraction speed, we believe it is suitable for remote data visualization. Moreover, surface propagation yields triangle patches that should perform well under geometry compression. Finally, we intend to exploit data-locality due to frame-to-frame coherence in interactive data exploration when adapting our algorithm for remote visualization.

Acknowledgements

This work is supported in part by Intel Corporation, the National Science Foundation (under grant CDA-9624099 and EIA-9975011) and Department of Energy (under grant ANI-9906704 and DE-FC02-99ER25387). The National Library of Medicine

232

provided the Visible Woman dataset. We are also grateful for advice from Tom Funkhouser as well as helpful discussions with Charles Hansen and Yarden Livnat.

References

[1] C. L. Bajaj, M. van Kreveld, R. van Oostrum, V. Pascucci, and D. R. Schikore. Contour Trees and Small Seed Sets for Isosurface Traversal. In *Proceedings of the 13th Annual ACM Symposium on Computational Geometry*, pages 212-219, ACM Press, Nice, France, 1997.

[2] Yi-Jen Chiang, Cláudio T. Silva and William J. Schroeder. Interactive Out-Of-Core Isosurface Extraction. In *Proceedings of IEEE 1998 Conference on Visualization*, 1998, Pages 167 – 174.

[3] P. Cignoni, P. Marino, C. Montani, E. Puppo, and R. Scopigno. Speeding Up Isosurface Extraction Using Interval Trees. *IEEE Transactions on Visualization and Computer Graphics*, 3(2): 158-170, 1997.

[4] Satyan Coorg and Seth Teller; Temporally Coherent Conservative Visibility (extended abstract). In *Proceedings Of The Twelfth Annual Symposium On Computational Geometry*, 1996, Pages 78 – 87.

[5] Takayuki Itoh and Koji Koyamada. Automatic Isosurface Propagation Using an Extrema Graph and Sorted Boundary Cell Lists. *IEEE Transactions On Visualization And Computer Graphics*, 1(4): 319 -327 December 1995.

[6] Takayuki Itoh, Yasushi Yamaguchi and Koji Koyamada. Volume thinning for automatic isosurface propagation. In *Proceedings Of IEEE 1996 Conference On Visualization*, 1996, Page 303-310.

[7] Kai Li, Han Chen, Yuqun Chen, Douglas W. Clark, Perry Cook, Stefanos Damianakis, Georg Essl, Adam Finkelstein, Thomas Funkhouser, Allison Klein, Zhiyan Liu, Emil Praun, Rudrajit Samanta, Ben Shedd, Jaswinder Pal Singh, George Tzanetakis and Jiannan Zheng. Early Experiences and Challenges in Building and Using A Scalable Display Wall System. *IEEE Computer Graphics and Applications, vol 20(4), pp 671-680, 2000*.

[8] Yarden Livnat, Han-Wei Shen, and Christopher R. Johnson. A Near Optimal Isosurface Extraction Algorithm Using the Span Space; *IEEE transactions on Visualization and Computer Graphics*, 2(1): 73-84, March 1996.

[9] Yarden Livnat and Charles Hansen. View Dependent Isosurface Extraction. In *Proceedings of IEEE 1998 Conference On Visualization*, 1998, Pages 175 – 180.

[10] Yarden Livnat and Charles Hansen. On View Dependent Isosurface Extraction for Large Scale Data. *Under submission*.

[11] William E. Lorensen and Harvey E. Cline. Marching cubes: A High-Resolution 3D Surface Construction Algorithm. In *Proceedings Of The 14th Annual Conference On Computer Graphics*, 1987, Pages 163 – 169.

[12] Michael Lounsbery, Tony DeRose and Joe Warren; Multiesolution Analysis For Surfaces Of Arbitrary Topological Type. *ACM Transactions on Graphics,* 6(1):34-73, January 1997.

[13] Steven Parker, Peter Shirley, Yarden Livnat, Charles Hansen and Peter-Pike Sloan. Interactive Ray Tracing For Isosurface Rendering. In *Proceedings Of IEEE 1998 Conference On Visualization*, 1998, Pages 233 – 238.

[14] Han-Wei Shen. Isosurface Extraction In Time-Varying Fields Using A Temporal Hierarchical Index Tree. In *Proceedings Of IEEE 1998 Conference On Visualization*, Pages 159 – 166.

[15] Jane Wilhelms and Allen Van Gelder. Octrees For Faster Isosurface Generation. *ACM Transactions on Graphics*. 11(3): 201-227, July 1992.

Editors' Note: see Appendix, p. 353 for colored figures of this paper

A Hardware-Assisted Visibility-Ordering Algorithm With Applications To Volume Rendering

Shankar Krishnan, Cláudio T. Silva, and Bin Wei

AT&T Labs-Research
180 Park Avenue
Florham Park, NJ 07932
{krishnas, csilva, bw}@research.att.com
http://www.research.att.com

Abstract. We propose a hardware-assisted visibility ordering algorithm. From a given viewpoint, a (back-to-front) visibility ordering of a set of objects is a partial order on the objects such that if object A obstructs object B, then B precedes A in the ordering. Such orderings are useful because they are the building blocks of other rendering algorithms such as direct volume rendering of unstructured grids. The traditional way to compute the visibility order is to build a set of visibility relations (*e.g.*, $B <_p A$), and then run a topological sort on the set of relations to actually get the partial ordering. Our technique instead works by assigning a *layer* number to each primitive, which directly determines the visibility ordering. Objects that have the same layer number are independent, and have no obstruction between each other. We use a simple technique which exploits a combination of the z- and *stencil* buffers to compute the layer number of each primitive. One application of our technique is to obtain a fast unstructured volume rendering algorithm. In this paper, we present our technique and its implementation in OpenGL. We also discuss its performance and some optimizations on some recent graphics hardware architectures.

1 Introduction

The original motivation for this work comes from volume rendering, but our work has other applications, which include image-based rendering acceleration, animations with selective display, efficient rendering with transparency [20]. The main contribution of this paper is a technique for computing a visibility ordering of a set of (acyclic) primitives by using features of the graphics hardware.

There are primarily two main approaches for exploring graphics hardware in volume rendering. One approach is to build new hardware, specialized for volume rendering. Quite possibly, the most visible example of this approach is VolumePro [15], which is based on the Cube-4 architecture of Pfister and Kaufman [16]. Another approach is to leverage existing graphics hardware, such as the texture-mapping based technique of Cabral *et al.* [1]. Although different, these two techniques shared the same volumetric data model, that is, each volumetric grid is basically a regularly spaced 3D matrix of voxels.

A technique that is able to leverage existing graphics hardware for volume rendering is the Projected Tetrahedra (PT) algorithm of Shirley and Tuchman [18], which uses the traditional 3D polygon-rendering pipeline. This technique renders a volumetric grid by breaking it into a collection of tetrahedra. Then, each tetrahedron is rendered by *projecting* its faces on the screen. This technique explores the graphics hardware for approximating the volume rendering lighting computations and generates high-quality images. Different from the previous approaches, the Shirley and Tuchman's method is not specific to a regular volumetric grid. In addition, it is quite efficient in terms of the number of triangles it needs to render per primitive. Wittenbrink [28] found experimentally that, on average, one needs 3.4 triangles per tetrahedron. On a fast graphics board, such as the recent Nvidia GeForce, one can potentially render several million tetrahedra per second.

In the domain of rendering of digital terrain models, the trends have been towards converting the data into some form of adaptive tessellations [12] instead of rendering a large collection of small triangles. Extending this notion to three dimensions, PT seems like the adaptive analog for volume rendering as opposed to approaches based on texture mapping hardware. In this sense, PT is conceivably a superior approach, even though it is currently being used to render only unstructured grids [21].

In order to apply PT, one needs to compute a visibility-ordering of the cells. Williams' Meshed Polyhedra Visibility Ordering (MPVO) algorithm [26] developed in the early 1990s provides a very fast visibility-ordering algorithm suitable for use in real-time rendering of unstructured grids. MPVO, which runs in linear time, works by exploiting the intrinsic connectivity of the unstructured grids and works well for well-behaved meshes (acyclic and **convex**). MPVO has recently been extended for general acyclic meshes by Silva *et al.'s* XMPVO [19], which lead to an $O(n + b^2)$ algorithm (where n is the total number of cells, and b is the number of cells in the boundary of the mesh). The work of Silva *et al.* relies on being able to compute a visibility-ordering of the boundary cells by first performing a sufficient set of ray shooting queries, then running a topological sort on the visibility relations found to infer the ordering. Comba *et al.* [7] further improved these results with BSP-XMPVO to $O(n + bp)$ (where p is the size of a small subset of the boundary cells), and leading to an order of magnitude improvement in sorting times over XMPVO. This technique requires a view-independent preprocessing which amounts to building a BSP tree of the boundary faces. Unfortunately, even BSP-XMPVO is not able to sort cells at millions of cells per second, which is necessary to drive high-end graphics boards at full speed. Another one of BSP-XMPVO's disadvantages is the fact that it is not possible to handle visibility ordering of dynamic meshes efficiently, which might arise from the extension to volumetric meshes of techniques such as the continuous level of detail algorithm of Lindstrom *et al.* [12] (these techniques usually require the geometry being rendered to change continuously as to match the user movement).

The fundamental computation which XMPVO and BSP-XMPVO are built on is the ability to obtain a visibility-order of the boundary cells. In this paper, we focus on how to find an ordering of the boundary cells using graphics hardware.

In this paper, we propose a new hardware-assisted visibility-ordering algorithm. At a high-level, our algorithm can be seen as a hardware implementation of the XMPVO algorithm, but there are some significant differences. XMPVO (and most traditional visibility ordering algorithms) first build a sufficient set[1] of pairwise visibility relations (*e.g.*, $B <_p A$), and then in a second phase, a topological sort is needed on the set of relations to actually get the ordering. Our technique instead works by assigning a *layer* number to each primitive, which directly determines the visibility ordering. To compute the layer number of each primitive, we make extensive use of the graphics hardware. In particular, we exploit a combination of the *z*- and *stencil* buffers.

In the rest of this paper, we first describe some related work in Section 2. In Section 3, we describe our new algorithm and some optimizations. In Section 4, we report some experimental results, including how our technique compares to XMPVO and BSP-XMPVO. We finish the paper in Section 5 with final remarks and our plans for future work.

2 Related Work

We let v denote the viewpoint and let ρ_u denote the ray from v through the point u. A *visibility ordering*, $<_v$, of a set of primitives $\mathcal{P} = \{p_1, p_2, \ldots, p_n\}$ from a given viewpoint, $v \in \Re^3$, is a linear order on $\mathcal{P}$ such that if $p \in \mathcal{P}$ visually obstructs $p' \in \mathcal{P}$, partially or completely, then p' precedes p in the ordering: $p' <_v p$. In general, $p' <_v p$, if there exists a ray ρ from the viewpoint v such that $\rho \cap p \neq \emptyset$, $\rho \cap p' \neq \emptyset$ and the intersection point of ρ with p is before the intersection point with p' along the ray.

Work on visibility ordering in computer graphics was pioneered by Schumacker *et al.* [22]. An earlier (complete) solution to computing a visibility-order was given by Newell, Newell, and Sancha (NNS) [13] which is the basis for several recent techniques [21]. The NNS algorithm starts with a rough ordering in z (depth) of the primitives, then for each primitive, it fine tunes the ordering by checking whether other primitives actually precede it in the ordering.

Building on [22], Fuchs, Kedem, and Naylor [9] developed the Binary Space Partitioning tree (*BSP-tree*), which is a data structure that represents a hierarchical convex decomposition of a given space (in our case, $\Re^3$) (see [8, 9, 17]). Each node ν of a BSP-tree $\mathcal{T}$ corresponds to a convex polyhedral region, $P(\nu) \subset \Re^3$; the root node corresponds to all of $\Re^3$. Each non-leaf node ν also corresponds to a plane, $h(\nu)$, which partitions $P(\nu)$ into two subregions, $P(\nu^+) = h^+(\nu) \cap P(\nu)$ and $P(\nu^-) = h^-(\nu) \cap P(\nu)$, corresponding to the two children, ν^+ and ν^- of ν. Here, $h^+(\nu)$ (resp., $h^-(\nu)$) is the halfspace of points above (resp., below) plane $h(\nu)$. Fuchs *et al.* [9] demonstrated that BSP-trees can be used for obtaining a visibility ordering of a set of objects (or, more precisely, an ordering of the fragments into which the objects are cut by the partitioning planes). The key observation is that the structure of the BSP-tree permits a simple recursive algorithm for "painting" the object fragments from back to front: If the viewpoint lies in, say, the positive halfspace $h^+(\nu)$, then we (recursively) paint first the fragments stored in the leaves of the subtree rooted at ν^-, then the object fragments $S(\nu) \subset h(\nu)$, and then (recursively) the fragments stored in the leaves of the subtree rooted at ν^+.

It is important to note that the BSP-tree does not actually generate a visibility order for the original primitives, but for *fragments* of them. Comba *et al.* [7] show how to

---

[1] *Sufficient* in the sense that it is possible to extend such pairwise relations into a valid partial order. In general, one has to formally show that this is the case. See [19].

recover the visibility order from the sorted fragments. There are a few issues in using BSP-trees for visibility-ordering. Building a BSP-tree is a computationally intensive process. Thus, handling dynamic geometry is a challenge. Using techniques from the field of "kinetic" data structures, Comba [6] developed an efficient extension of BSP-trees for handling moving primitives. At this time, his technique requires apriori (actually analytical) knowledge of the motion of the geometry to efficiently perform local changes on the BSP-tree as the primitives move.

Another technique for visibility order is described in Silva *et al.* [19]. In that paper, a well-chosen (small) set of ray shooting queries are performed, which compute for each primitive (at least) its successor and predecessor in the visibility ordering. By running a topological sort on these pairwise relations, it is possible to recover a visibility order. One of the shortcomings of this technique is that it might actually obtain a larger portion of the visibility graph than necessary to compute the ordering. Since the ray shooting queries are relatively expensive both in time and memory. This can be inefficient.

Another class of sorting techniques are based on power-sorting, see the work of Cignoni et al [2, 4, 3]. These techniques are quite fast, since they reduce the 3D sorting problem to a one dimensional sort, which can be done quite efficiently with quicksort. Unfortunately, these techniques make limiting assumptions about the shape of the actual grids (*e.g.*, a Delaunay triangulation, see [28]) and their use for general meshes would, in general, cause visibility-ordering problems. For highly tessellated unstructured grids, these errors in visibility-ordering are mostly imperceptible, but for adaptively sampled volumetric grids where big cells would be close to small cells, sorting errors might be large.

Snyder and Lengyel [20] present an incremental visibility sorting algorithm, similar in some respects to the NNS algorithm [13]. Their algorithm, despite having a worst-case running time of $O(n^4)$, is shown to be quite fast in practice. In order to cull the number of visibility relations they need to maintain, Snyder and Lengyel employ several optimizations, such as the use of kd-trees and the tracking of overlaps of the convex hulls of the geometric primitives. Their algorithm is able to explore temporal coherency, and in fact is optimized for dynamic geometry. They also propose a technique for correct rendering in the presence of cycles.

The VSbuffer technique of Westermann and Ertl [23] is related to our work. In their algorithm, they exploit the graphics hardware for performing depth-sorting of volumetric primitives by rendering the cells on a plane perpendicular to the scanline; then they use the imprinted cell ids and their geometric relationship to guide the volume integral calculation. Our volume rendering technique is quite different, since we do not use the hardware to sort all the volumetric primitives as they do, but only the boundary, and use MPVO relations for the interior of the volume. Because of this, we require adjacency information, which they do not. Although the two techniques are quite different, both of them share several of the same implementation issues, such as the use of pbuffers, the disabling of all lighting calculations, and the reading back of the OpenGL buffers to get primitive ids. Our experimental results show that our technique is considerably faster than the VSbuffer. Quite possibly, this is due to the fact that for typical datasets, we require a much smaller number of buffer reads.

Another related technique is presented by Mammen [14], where he uses a multi-pass rendering technique with a "moving" depth buffer to render transparent objects.

Building complicated data structures to solve the visibility-ordering problem is a fairly difficult task. Given that interactivity is of utmost importance in most applications, it would be prudent to try and solve this problem in hardware at some pre-specified resolution. As other researchers have found (see, for instance, Hoff *et al.* [10], Westermann and Ertl [24, 23]) exploiting the ever-faster graphics hardware available in workstations and PCs, can lead to simpler, and more efficient solutions to our rendering problems. Our work is motivated by this trend.

3 Our Algorithm

For the sake of discussion, we assume to obtain a front-to-back visibility order. The basic idea is to start with the complete collection of primitives, and extract the primitives in *layers*, that is, a maximally independent set of polygons which do not relate to each other in the visibility order. The algorithm works by extracting a single layer from the current set of primitives. We basically keep doing this until no more primitives can be removed. At this point, if the set of primitives without a layer number assigned is not empty, one of the following two conditions are true: (a) the remaining (un-classified) primitives are either orthogonal to the viewing direction, hence we can not really classify them with respect to each other or the rest of the polygons, or (b) they contain a cycle, and our algorithm does not handle cycles. (See Snyder and Lengyel [20] for a technique which can be used to handle the cycles.)

We now explain our algorithm in more detail. We assume that we have access to the z-, *stencil*, and *color* buffers. Also, for the sake of simplicity in presentation, we assume the input is composed of triangles, and all the transformation matrices have been handled by the code that is outside of this subroutine. We start with some basic notation. $\mathcal{T}$ is used to denote the set of triangles which have not been classified (notice that it changes over time); $\mathcal{F}$ is the current layer being extracted; $\mathcal{T}_i$, for a given i, is the set of triangles assigned to be in the ith layer. During our algorithm, the stencil buffer is sometimes disabled. But whenever it is enabled, it is set to increase the values on the stencil buffer any time a triangle would have been projected into those pixels. In OpenGL, the stencil buffer can be configured as such:

```
glStencilFunc(GL_ALWAYS, ~0, ~0);
glStencilOp(GL_KEEP, GL_INCR, GL_INCR);
```

In our algorithm, we make extensive use of the item buffer technique, where triangles are rendered with different colors, from which the original triangles can be identified by reading back the color buffer. We name this process as *reading and scanning* the buffer in the rest of our discussion. Reading buffers refers to performing the `glRead-Pixels` call, while scanning a buffer refers to the process of traversing the pixel arrays, to obtain the primitive ids and depth complexity. Here is our algorithm:

While $\mathcal{T} \neq \emptyset$, loop,
1. Clear the color buffer; disable the stencil buffer; configure z-test to GL_LESS, while clearing it to 1.0 (far).

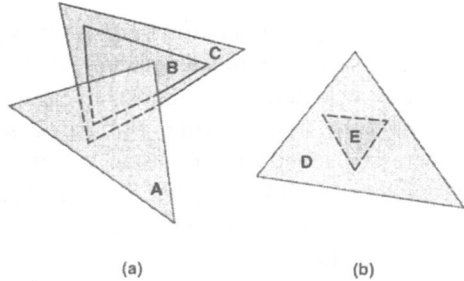

(a)   (b)

Fig. 1. (a) In this situation, triangle A occludes parts of triangles B and C, while B is completely occluded by C from the opposite direction. During the first scan, pixels (partially) covering B and C are present in the top layer, $\mathcal{F}$. Note that in step 4, we remove triangles from back to front. Since C completely occludes B, we have to go through step 4 multiple times to extract the correct layering. (b) Simple case where the depth complexity of $\mathcal{F}$ is always 1.

2. Render $\mathcal{T}$.
3. Read back the color buffer, and assign to $\mathcal{F}$ any triangle that belongs to the current color buffer. Note that these triangles are *potential* candidates to be in the current layer, since they might be obscured by some other triangle. (See Fig. 1.)

   A necessary and sufficient condition for $\mathcal{F}$ to be a layer is that the depth-complexity of $\mathcal{F}$ can be at most one. The idea in the next phase of our algorithm is to use the stencil buffer to test for this condition. In fact, by properly setting the z-buffer, it is possible to identify exactly the triangles which do not belong to the current layer by looking at pixels in the stencil buffer which have a depth-complexity larger than one.
4. Do
   - (4a) Clear and enable the stencil buffer; clear the color buffer; configure z-test to GL_GEQUAL, while clearing it to 0.0 (near).
   - (4b) Render $\mathcal{F}$.
   - (4c) Read back the color and stencil buffers. For each pixel in the stencil buffer which is larger than one, remove the corresponding triangle from $\mathcal{F}$, and re-insert it in $\mathcal{T}$. Since we rendered the scene from the back, we are necessarily removing a triangle that is covered by one or more other triangles. Note that if we never find a pixel which has depth-complexity higher than two, we can leave the loop at this point. Otherwise, we need to keep removing triangles from the back of $\mathcal{F}$, until the depth-complexity of each pixel is at most one.
   - (4d) Assign $\mathcal{T}_i = \mathcal{F}$ for the current layer number, and increment the layer number.

   While depth-complexity of $\mathcal{F} > 1$.
5. In case no triangles have been removed from $\mathcal{T}$ since step (1) of the algorithm (that is, the number of elements in $\mathcal{T}$ has not changed), we can stop the algorithm, and claim that the remaining triangles contain a cycle, or they are orthogonal to the view direction.

It is straightforward to turn the description of our algorithm above into working C++ code. If we have n triangles in a scene, the worst-case performance of our algorithm is $O(n^2)$, since all the triangles can be behind a single pixel. But this is rarely the case. Assuming the depth complexity of the scene is d, the complexity of the algorithm is much close to $O(nd)$. Each triangle is rendered multiple times, and can potentially be rendered $O(d)$ times. Often, rendering is not the bottleneck. As we show in Section 4, most of the time is spent in reading the color and stencil buffers, and scanning them (depending on image size, triangle count, and architecture limitations). Also, as layers are extracted, the actual footprint of a typical layer decreases quite rapidly (see Fig. 3). Thus, reading and scanning the whole buffers is a waste of time. We propose a simple modification of our algorithm which greatly improves the overall performance. It is based on the fact that once a pixel is not covered by a triangle after being rendered in step (2), it will never be covered again. Using this fact, it is advantageous to use a subdivision scheme of dividing the image into blocks, and keeping track of pixel coverage in every block, to avoid unnecessary reading and scanning. In most architectures, the larger the block size, the better the bandwidth in reading back the buffers, although this tends to max out usually somewhere around a 512×512 block. On the other hand, large blocks may not effectively reduce the unnecessary reading and scanning operations. Based on our experiments, a 64×64 blocking scheme works best on various hardware platforms.

4 Experimental Results

We use OpenGL to implement the depth sorting algorithm. We tested the performance on several workstations, including SGI Octanes and an HP PC. We are only presenting the data collected from the faster Octane and the HP PC. The SGI Octane we use has 300MHz MIPS R12000 CPU and 512MB main memory running IRIX 6.5 with an EMXI graphics board. The HP workstation has dual 450Mhz Pentium II Xeon processors and 384MB main memory running windows NT 4.0. The graphics subsystem is HP fx6. There are two versions of our algorithm. One is the naive implementation of the depth sorting algorithm and the other is the optimized version with the subdivision scheme (see Section 3) for better performance. We performed our experiments on two different window sizes: 256×256 and 512×512. For the optimized version of our algorithm, we also varied the block sizes. We used 32×32 and 64×64 for our experiments. There are five data sets in our experiments. We ran our program over a precomputed set of transformations. We collected the data over 30 frames. Table 1 lists some of the characteristics of these data sets. Generally speaking, the subdivision scheme reduces the total computation time. This is because the image layers after the top-layer extraction tend to be smaller and smaller in the frame buffer. With the subdivision scheme, we can read a fraction of the frame buffer as necessary and at the same time, the scanning area gets smaller. However, there are a few models like the *sphere* which are too symmetric for us to observe any performance improvement with our scheme. Figures 2 (a), (b), (c) and (d) list the percentage of the time spent on scanning layers and reading buffers for the two algorithms on the two machines. Scanning layers and reading buffers take most of the execution time. While the total percentage of time spent on scanning and read-

model	# of vertices	# of triangles	depth win256	depth win512
Bones	2156	4204	19.7	18
Mannequin	689	1355	10.8	12.4
Phoenix	8280	2760	9.6	11.1
Sphere	66	129	2.8	2.5
Spock	1779	3525	17.7	18.9

Table 1. Characteristics of the five models and their average depths for the window sizes of 256×256 and 512×512 over 30 frames.

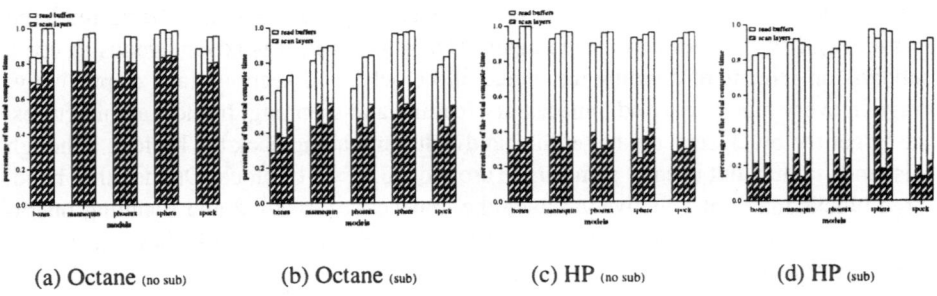

(a) Octane (no sub) (b) Octane (sub) (c) HP (no sub) (d) HP (sub)

Fig. 2. Percentage of the overall execution time spent on scanning layers and reading frame buffers of the algorithm with and without the subdivision scheme on Octane and HP.

ing the buffers is similar between the two architectures, we observe from the Figure 2 that the scanning time dominates in the SGI Octane, while in the HP, reading time is significantly higher. The most important reason for this discrepancy can be attributed to significant difference in the processor speeds. In most cases, the subdivision scheme speeds up the performance, sometimes over 4 times.

Unstructured Grid Volume Rendering. XMPVO [19] and BSP-XMPVO [7] are two volume rendering techniques based on extending MPVO [26] by sorting the boundary cells. The actual sorting techniques proposed in XMPVO and BSP-XMPVO are quite different, and lead to substantially different results. The XMPVO algorithm works by augmenting the visibility relation between cells by performing "ray shooting queries" between faces of the boundary cells. It is possible to replace the XMPVO sorting with our new approach, which essentially shoots *one ray per pixel*, and thus can lead to *inexact* sorting in some situations. The basic idea is to save the identity of a face that has been projected into a given pixel during the layering extraction. Then, while extracting higher layers later in the processing, add an arrow (ordering relation) to the face that projects in the same pixel, and has a higher layer. There are some choices on the actual accounting for the relations in a given implementation. One way would be to keep a number of relations equal to the number of projected pixels of each face. Again, note that we only need to care about boundary faces in this process, which in general is a very small number of faces compared to the total complexity of a given dataset (see [7] for details). Comparing with the results presented in [19] and accounting for the MPVO relations separately, our experiments indicate this discretized XMPVO is considerably faster than the one presented in [19], that is, about a factor of ten faster. It is not clear we are performing a fair comparison. BSP-XMPVO and XMPVO are truly "exact" techniques, while in our case, we could possibly miss generating ordering relations between cells that might need them. On the other hand, quite possibly the overall visibility ordering generated changes little, because the inner relations that MPVO generates are highly constraining. Even by classifying a subset of the cells in a correct layer is probably enough to avoid generating any sorting error. We believe this is one of the reasons that the MPVONC heuristic proposed by Williams [26] is so effective. See Cook [5] for an alternative sorting technique also based on a discretization of XMPVO.

5 Conclusion and Future Work

We have presented a hardware-assisted algorithm for visibility ordering. From a given viewpoint and view direction, we compute a partial ordering of the primitives which can then be rendered using the standard painter's algorithm. We have used a combination of the hardware z-, *stencil* and *color* buffers to compute this ordering. Our experiments on a variety of models have shown significant speedups in the ordering time compared to existing methods. The two main costs associated with our implementation are the cost of transferring the buffers to the host's main memory, and the time it takes the host CPU to scan them. It is possible to use the histogramming facility available in the ARB_imaging extension of OpenGL 1.2 to make the graphics hardware perform those computations (we refer the reader to Klosowski and Silva [11], and Westermann et al [25] for details). Unfortunately, those pixel paths are not optimized, and are often slower than our current implementation. If future hardware optimizes this functionality, it would be possible to further improve the performance of our technique.

Acknowledgements: The authors thank Nelson Max and Richard Cook for useful comments and discussions, and the referees for helpful suggestions.

References

1. B. Cabral, N. Cam, and J. Foran. Accelerated volume rendering and tomographic reconstruction using texture mapping hardware. *1994 Symposium on Volume Visualization*, pages 91–98. October 1994.

2. P. Cignoni and L. De Floriani. Power diagram depth sorting. In *10th Canadian Conference on Computational Geometry*, 1998.

3. P. Cignoni, C. Montani, D. Sarti, and R. Scopigno. On the optimization of projective volume rendering. In *Visualization in Scientific Computing '95*, pages 58–71. Springer Computer Science, 1995.

4. P. Cignoni, C. Montani, and R. Scopigno. Tetrahedra based volume visualization. *Mathematical Visualization – Algorithms, Applications, and Numerics*, pages 3–18. Springer Verlag, 1998.

5. R. Cook. Parallelizing and Implementing an Exact Topological Sort For The HIAC Volume Renderer. M.S. thesis, Department of Computer Science, University of California, Davis, 2001. (to appear)

6. J. Comba. *Kinetic Vertical Decomposition Trees*. PhD thesis, Department of Computer Science, Stanford University, 2000.

7. J. Comba, J. Klosowski, N. Max, J. Mitchell, C. Silva, and P. Williams. Fast polyhedral cell sorting for interactive rendering of unstructured grids. *Computer Graphics Forum*, 18(3):369–376, September 1999.

8. M. de Berg, M. van Kreveld, M. Overmars, and O. Schwarzkopf. *Computational Geometry: Algorithms and Applications*. Springer-Verlag, Berlin, 1997.

9. H. Fuchs, Z. M. Kedem, and B. Naylor. On visible surface generation by a priori tree structures. *Comput. Graph.*, 14(3):124–133, 1980. Proc. SIGGRAPH '80.

10. K. Hoff III, T. Culver, J. Keyser, M. Lin, and D. Manocha. Fast computation of generalized voronoi diagrams using graphics hardware. *Proceedings of SIGGRAPH 99*, pages 277–286, August 1999.

11. J. Klosowski and C. Silva. Efficient Conservative Visibility Culling Using The Prioritized-Layered Projection Algorithm. *Unpublished manuscript, 2000*.

12. P. Lindstrom, D. Koller, W. Ribarsky, L. Hughes, N. Faust, and G. Turner. Real-Time, continuous level of detail rendering of height fields. *SIGGRAPH 96*, pages 109–118, 1996.

13. M. E. Newell, R. G. Newell, and T. L. Sancha. A new approach to the shaded picture problem. In *Proc. ACM Nat. Conf.*, page 443. 1972.

14. A. Mammen. Transparency and Antialiasing Algorithms Implemented with the Virtual Pixel Maps Technique. *IEEE Computer Graphics & Applications*, pages 43–55, July 1989.

15. H. Pfister, J. Hardenbergh, J. Knittel, H. Lauer, and L. Seiler. The VolumePro real-time ray-casting system. *Proceedings of SIGGRAPH 99*, pages 251–260, August 1999.

16. H. Pfister and A. Kaufman. Cube-4 - A scalable architecture for real-time volume rendering. In *1996 Volume Visualization Symposium*, pages 47–54. October 1996.

17. H. Samet. *Spatial Data Structures: Quadtrees, Octrees, and Other Hierarchical Methods*. Addison-Wesley, Reading, MA, 1989.

18. P. Shirley and A. Tuchman. A polygonal approximation to direct scalar volume rendering. In *Computer Graphics (San Diego Workshop on Volume Visualization)*, volume 24, pages 63–70, November 1990.

19. C. Silva, J. Mitchell, and P. Williams. An interactive time visibility ordering algorithm for polyhedral cell complexes. In *Proc. ACM/IEEE Volume Visualization Symposium '98*, pages 87–94, November 1998.

20. J. Snyder and J. Lengyel. Visibility sorting and compositing without splitting for image layer decomposition. *Proceedings of SIGGRAPH 98*, pages 219–230, July 1998.

21. C. Stein, B. Becker, and N. Max. Sorting and hardware assisted rendering for volume visualization. *1994 Symposium on Volume Visualization*, pages 83–90. October 1994.

22. I. E. Sutherland, R. F. Sproull, and R. A. Schumacker. A characterization of ten hidden-surface algorithms. *Journal of the ACM*, March 1974.

23. R. Westermann and T. Ertl. The VSbuffer: Visibility ordering of unstructured volume primitives by polygon drawing. *IEEE Visualization '97*, pages 35–42, November 1997.

24. R. Westermann and T. Ertl. Efficiently using graphics hardware in volume rendering applications. *Proceedings of SIGGRAPH 98*, pages 169–178, July 1998.

25. R. Westermann, O. Sommer, and T. Ertl. Decoupling Polygon Rendering from Geometry using Rasterization Hardware. *Unpublished manuscript, 1999*.

26. P. L. Williams. Visibility-ordering meshed polyhedra. *ACM Transactions on Graphics*, 11(2):103–126, April 1992.

27. P. Williams, N. Max, and C. Stein. A high accuracy volume renderer for unstructured data. *IEEE Transactions on Visualization and Computer Graphics*, 4(1):37–54, January-March 1998.

28. C. Wittenbrink. Cellfast: Interactive unstructured volume rendering. In *Proc. Late Breaking Hot Topics, IEEE Visualization*, 1999.

Editors' Note: see Appendix, p. 354 for colored figure of this paper

Volume Rendering Data with Uncertainty Information

Suzana Djurcilov[1], Kwansik Kim[1], Pierre F. J. Lermusiaux[2], and Alex Pang[1]

[1]Computer Science Department, UCSC
[2]Division of Engineering and Applied Sciences, Harvard University

Abstract. This paper explores two general methods for incorporating volumetric uncertainty information in direct volume rendering. The goal is to produce volume rendered images that depict regions of high (or low) uncertainty in the data. The first method involves incorporating the uncertainty information directly into the volume rendering equation. The second method involves post-processing information of volume rendered images to composite uncertainty information. We present some initial findings on what mappings provide qualitatively satisfactory results and what mappings do not. Results are considered satisfactory if the user can identify regions of high or low uncertainty in the rendered image. We also discuss the advantages and disadvantages of both approaches.

1 INTRODUCTION

Visualization is used for gaining an understanding of large amounts of data in a short period of time. Scientific datasets often have associated with them a measure of quality, reliability or uncertainty which also needs to be made a part of the visual output.

Uncertainty can be caused by many factors in the data collection and processing: from unreliable instrumentation and problems in transportation, to errors caused by the interpolation and modeling algorithms. While these errors can sometimes be ignored, it is important to alert the users to the trustworthiness of the image upon which they need to make a decision.

While the uncertainty is an essential part of the data, it has often been ignored while processing or displaying. This can be misleading to the user unaware that parts of the dataset contain unreliable information. For accurate interpretation it is important to display the original data together with its uncertainty. Uncertainty visualization techniques present data in such a manner that users are made aware of the locations and degree of uncertainties in their data so as to make more informed analyses.

In this paper we concern ourselves with uncertainty visualization using one particular rendering method, namely direct volume rendering. We present two general options: one which is calculated at the rendering time and presented as part of the volume rendering of the primary value, which we call inline processing; and one which combines the volume renderings of the primary value and of the uncertainty value as a post-processing method.

2 BACKGROUND

Visualizing uncertainty is a recognized challenge in the visualization community, and recently, more visualization research have focused on this area. For example, Cedilnik and Rheingans [1] looked at different ways of imparting uncertainty over 2D fields using procedural methods to distort overlaid grid lines, Interrante [3] discussed how one might use natural textures over a map to show uncertainty, Djurcilov and Pang [2] looked at different ways of incorporating uncertainty information in contour lines and isosurfaces of sparse data sets, Wittenbrink et al. [11] included uncertainty in direction and uncertainty in magnitude into glyph designs, and Pang et al. [7] described some general methods for incorporating uncertainty into visual displays.

The approaches above involve some modification of how the data is represented, and through this modification, impart the uncertainty information. The modifications are typically applied to geometric primitives and attributes such as grid lines, contour lines, glyphs, and textures. Unfortunately, volume rendering does not produce any intermediate geometric primitives that could be modified in order to represent uncertainty. Therefore, this paper explores different alternative techniques for including uncertainty information directly in volume renderings.

3   DATA WITH UNCERTAINTY

3.1   Ocean data and dynamical model

During July and August of 1996, ocean data were collected in the Middle Atlantic Bight (MAB) south of New England, as part of the "ONR Shelfbreak PRIMER Experiment" [6]. The dominant dynamical feature in the MAB consists of a temperature and salinity front, separating the shelf and slope water masses. This front is often located above the shelfbreak, near the 100 m isobath (see Figure 13). It is usually tilted, in the opposite direction of the bottom slope. The main objective was to study the influence of oceanographic variability on the propagation of sound between the shelf and slope regions. Intensive cruise surveys were carried out daily in a 45 km by 30 km domain between the 85 m and 500 m isobaths.

The physical variables or fields are the temperature, salinity, velocity and pressure. They are dynamically evolved by the numerical ocean model of the Harvard Ocean Prediction System [8]. Atmospheric fluxes based on buoy data are imposed in surface.

3.2   Uncertainty forecasts

To dynamically evolve the physical uncertainty, an Error Subspace Statistical Estimation (ESSE) scheme [5] is employed. This scheme is based on a reduction of the evolving error statistics to their dominant components or subspace. Presently, statistics are measured based on a variance or least-squares criterion [9]: a subspace is then characterized by the dominant eigen decomposition of a covariance matrix. The objective is then to dynamically forecast the principal component decomposition of the uncertainty of the physical fields.

In the present MAB case, these error principal components are initialized combining data and dynamics. To account for nonlinearities, they are forecast by an ensemble of Monte-Carlo forecasts.

In the visualizations presented here, only temperature and salinity uncertainty forecasts are used. However, since physical fields are coupled, the effects of velocity errors are included in these forecasts and accurate estimates of temperature and salinity errors can thus be obtained.

As a first endeavor, we utilize the variances of the Monte-Carlo ensemble as a scalar representation for uncertainty at each point.

4 INLINE APPROACH

The classic volume rendering equation is:

$$C(a, b) = \int_a^b E(s) e^{-\int_a^s \delta(x) dx} ds \qquad (1)$$

where $C(a, b)$ is the color intensity contributions through a line from position a to b. E is the color emission function and δ the differential opacity function. Equation (1) calculates an integrated color for each pixel by summing up opacity weighted emittance values. One form or another of this equation is used to generate volume rendered images of 3D scalar fields. Because the data set generally consists of a single scalar field, the same scalar field is used to determine both the opacity and the material emittance values. This is typically achieved by transfer functions that map the scalar data value to both opacity and color.

Because our uncertainty is also represented as a 3D scalar field, we have the opportunity to map field values to color and uncertainty values to opacity, and experiment with different transfer functions. We refer to this approach as "inline" in the sense that the uncertainty information is directly incorporated into the rendering process. In this section, we describe two inline experiments.

4.1 1D Transfer Functions

In this experiment, we mapped salinity values to E and uncertainty values to δ. We then use a 1D transfer function to separately map the salinity and uncertainty values. We experimented with a transfer function that maps increasing uncertainty to increasing opacity, and composited the resulting images to a black background with white grid lines.

As a point of reference, Figure 1 is a traditional volume rendering of the mean salinity field. Figures 2 and 3 both show a volume rendering of the uncertainty in the salinity field. Figure 2 maps uncertainty values above 0.2 to high opacity values, while Figure 3 maps uncertainty values above 0.5 to high opacity values. High uncertainty regions show up as a bluish cloud. Dark regions have lower uncertainty. We note that most of the uncertainty lies along the salinity front on top of the shelfbreak.

One can map uncertainty to opacity in a number of ways. In this experiment, we mapped higher uncertainty values to higher opacity values. Field values such as salinity and temperature are mapped to color. A black background with white grid lines is used to accentuate the fact that more transparent regions have lower uncertainty. Note that

246

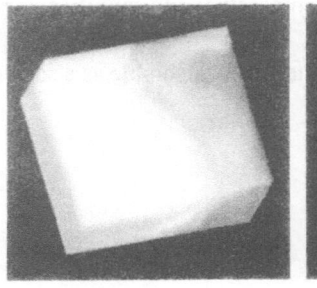 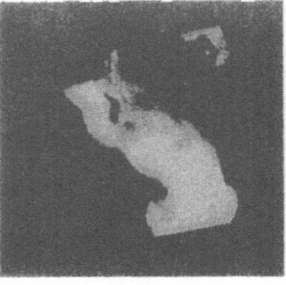 

Fig. 1. Mean salinity. **Fig. 2.** Uncertainty ≥ 0.2. **Fig. 3.** Uncertainty ≥ 0.5.

regions with low uncertainty do not automatically produce more transparent regions because of the potential occlusion with accumulated opacities from different viewing angles.

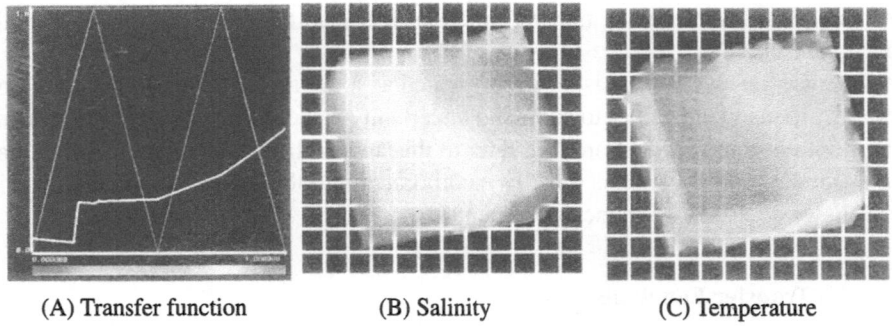

(A) Transfer function (B) Salinity (C) Temperature

Fig. 4. (A) Shows the transfer function for both the field values and uncertainty values. All values have been normalized to lie between 0 and 1. The increasing white curve maps higher uncertainty to higher opacity. (B) Volume rendering of the mean salinity field mapped to color and uncertainty in salinity mapped to opacity. (C) Volume rendering of the mean temperature field mapped to color and uncertainty in temperature mapped to opacity.

One can also experiment with an increasing uncertainty to decreasing opacity mapping so that the regions of uncertainty show up as transparent regions rather than opaque regions. The choice of mapping increasing/decreasing opacity seems to depend on the volume data to be studied. Looking at the uncertainty of the temperature field alone confirms that, indeed, the regions of high uncertainty in the right columns of Figures 4 and 10 (color plate) are in the greenish opaque regions. On the other hand, the fine structural details in the uncertainty field are washed out and lost in the resulting rendering. In addition, there is some ambiguity in interpreting the image. The ambiguities can be attributed to a number of factors including varying thickness of the volume from a given viewpoint, the depth within a volume of a region of high uncertainty, interaction of the color and opacity compositing. The image in Figure 10 is similar to Figure

4 except that we used a transfer function which produces more contrast between high and low uncertainty regions. We also removed the white grid lines to see if it is better without them or not.

4.2 2D Transfer Functions

In this experiment, we use 2D transfer functions similar to those used by Kindlmann and Durkin [4]. However, instead of looking at the first and second derivatives of the data, we look at data versus uncertainty values. Figure 5 is a 2D scatter plot showing the distribution of mean salinity versus uncertainty in salinity. We use this 2D scatter plot as the basis for our transfer function, mapping different regions of the scatter plot to different color values. Figures 11 and 12 show different 2D transfer functions and the corresponding volume rendered images of the combined salinity and uncertainty fields.

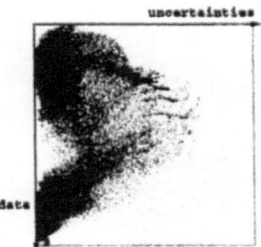

Fig. 5. Scatter plot of mean salinity (Y-axis) versus uncertainty (X-axis). Mean salinity values increase towards the bottom, while uncertainty values increases towards the right.

Unlike 1D transfer functions where we mapped uncertainty to opacity, 2D transfer functions primarily use color to show regions with varying uncertainty. For example, the middle images of Figures 11 and 12 (color plate) use a constant opacity regardless of uncertainty. However, opacity can be used to also emphasize or de-emphasize uncertainty. For example, the right images of Figures 11 and 12 use a step function that maps low uncertainty data to an almost transparent value, and high uncertainty data (greater than 0.2, as in Figure 2) to high opacity. The result is a volume rendering of the salinity data, but with obvious structural features showing the location of the high uncertainty regions. In Figure 12 blue and cyan regions have higher uncertainty. Middle and right images use the same uncertainty to opacity mapping as the corresponding images in Figure 11.

5 POST-PROCESS APPROACH

Due to the use of transparency, images produced by volume rendering algorithms have a soft and smooth quality to them. This aspect lends itself into exploring the use of discontinuity as a means of representing uncertainty. We use discontinuity in several ways by introducing speckles, noise and texture as options used in post-processing of an image to highlight areas where data is uncertain.

248

5.1 Inserting Speckles/Holes

This task is accomplished in several steps:

1. Produce a standard volume rendering of the field values (see Figure 14).
2. Produce a gray scale volume rendering of uncertainty values from the same view-point (see Figure 6). Note that converting a color volume rendering of the uncertainty field to gray scale will not produce the same desired effect.
3. Dither the gray scale rendering into a black and white bitmap image with inverted values (see Figure 7). The purpose of this step is to create a rendering in which each black dot will be a representation of uncertainty in that neighborhood. The dithering itself makes sure that the dots are evenly distributed and visually pleasing.
4. Generate a composite image by multiplying the color volume rendering with the bitmap image pixel by pixel (see Figure 15).

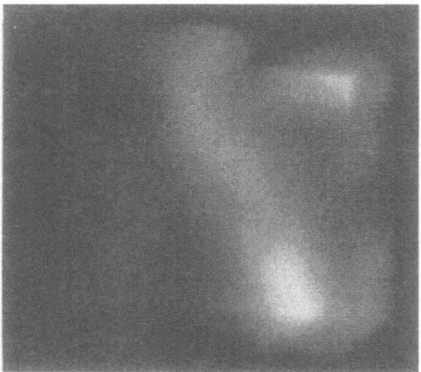

Fig. 6. Gray-scale rendering of uncertainty.   **Fig. 7.** Inverted bitmap rendering.

Figure 15 shows the outcome of the operations 1 to 4 - an image in which the volume rendering of the primary data value is modified to show pixel-sized holes in areas of high uncertainty. The user is still able to grasp the overall structure of the primary value throughout the dataset, and yet has an understanding of where the data is unreliable.

One possible pitfall of this method is that at a distance the small holes may blend into the image and cause the volume rendering to appear darker in regions with high error. This may be undesirable and can be improved by increasing the size of the holes, thus making it more apparent that the disturbance is not a coloring artifact, but indeed an intended feature of the image. We show one such example in Figure 16 where the holes are increased four-fold in order to emphasize the uncertainty. This image was produced by first sub-sampling the volume rendering of the uncertainty values (output of step 2) by a factor of four, then proceeding with the dithering, after which the image is brought back to its original size and multiplied as in step 4. The end-result is an image where the holes are four pixels large.

We would like to point out that in these examples we have used black to color the holes and match the background color. It would be up to the user to decide the choice of color for the speckles, but we recommend black as an intuitive choice for representing holes.

5.2 Adding Noise

Noise seems a natural option for conveying uncertainty - our minds easily accept the idea that a picture containing noise is less reliable than a clear one. Noise also has the convenient property that it can be introduced into an image without worry for side-effects, as its random nature eliminates any possibility for artifacts appearing as regular patterns.

We apply this idea to the volume rendered images by selectively disturbing the images in the area if high uncertainty. The output (see Figure 17) shows how randomized color dots can be added to an area, thus causing it to appear uneven and fuzzy.

The pseudo code for the algorithm is quite simple.

For each color pixel of the original volume rendering:

1. Find the matching pixel in the uncertainty rendering.
2. Rescale the uncertainty gray shade value to between $0 .. P$. P is the probability that the original color will be changed.
3. Replace the original color pixel with a random color with probability P.

This algorithm ensures that the areas with high uncertainty (lighter gray shades) on the uncertainty image are translated into regions with higher numbers of disturbed pixels in the original rendering. The use of probability allows a portion of the pixels to retain their original color even in areas of high uncertainty, so that the overall color context is not lost. In our experiments, we found that setting P to 20 produced a desirable effect. This will preserve at least 80% of the original color pixel values, and yet introduce enough noise in high uncertainty areas. An alternative scheme is to use the uncertainty value as an amount (rather than as a probability) to change the original color value in color space.

5.3 Adding Texture

Similarly to the previous option, we explore the use of textures in the post-processing context. We use 2D grainy, gray scale textures with varying intensity or contrast levels to represent different levels of uncertainty. Low contrast represents low uncertainty, while high contrast represents high uncertainty. We then use the texture brightness (value in HSV space) to alter the brightness of the original color image (value in HSV space). Naturally, in areas of very low or no uncertainty we do not apply any modifications.

The algorithm for adding textures to represent uncertainty in a volume rendered image is also carried out on a per-pixel basis. The difference from the previous method is that the different levels of texture contrast have to be created first. Each texture is tiled so that they are at least as large as the volume rendered image.

For each color pixel of the original volume rendering:

1. Find the matching pixel in the uncertainty rendering.
2. Bin the uncertainty value to one of 5 contrast levels, i.
3. Find the corresponding pixel from texture map i.
4. Adjust the brightness of the original pixel to that obtained from the texture map.

The example in Figure 18 uses a sandstone texture to alter the original volume rendering. Figures 8 and 9 show the sandstone texture at 2 of the 5 different uncertainty levels. In our experiments, we found that 5 levels of contrast to represent different uncertainty levels was sufficient. Beyond 5 levels, it was difficult to distinguish additional levels of uncertainty.

Fig. 8. Low uncertainty texture.

Fig. 9. High uncertainty texture.

6 DISCUSSION AND CONCLUSION

We have described some experiments on how one might include volumetric uncertainty information in a volume rendering. They can be classified as either inline or post-process. Of course, one can also use a pre-process approach where the two volumes are first combined to produce a single scalar volume. Different strategies may be employed to combine the two volumes. For example, one can perform a point wise multiplication of the two fields and volume render the result. In this case, low values would indicate either low data value, low uncertainty value, or both. Converse is true for high values. We did not experiment with this approach because it would be difficult to distinguish between data and uncertainty values in the resulting images.

One can argue which is the better approach: inline or post-process ? The inline method has the advantage that the uncertainty information is integrated into the volume rendering calculation, taking into account their 3D positions within the volume, and hence the results are more faithful. On the other hand, more research is needed to design transfer functions that will unambiguously show the uncertainty information together with the data values. The post-process approach has the advantage of producing images that intuitively show the locations and extent of uncertainty in the volume renderings. However, it is not as faithful to the data in the sense that the uncertainty presentations are really just image embellishments on the volume rendering of the data. For example, if there is a region of high uncertainty embedded within the volume, the post-process approach does not accurately capture the interaction of this region of uncertainty with the corresponding embedded data values.

In this paper, we applied different ideas of incorporating uncertainty into volume rendering using the data set from ocean modeling. Of course, the techniques are applicable to data sets from other domains as well. Some of the questions seeking further research include: How many levels of uncertainty are necessary and can one perceive? What transfer function best combines data and uncertainty, and perhaps their derivatives? And if one has a probability distribution function at each voxel, such as the Monte-Carlo ensemble, how does one go about visualizing such a data set? Finally, while volume rendering does not produce any geometry to be rendered, it does produce derived data in its rendering pipeline. These derived data, when combined with the uncertainty information, can also be used to depict uncertainty information [10]. This approach should also be investigated further.

7 ACKNOWLEDGEMENTS

We would like to thank Craig Wittenbrink for discussions on transfer functions for volume rendering, as well as Dr. Pat Haley and the members of the HOPS group led by Prof. A. R. Robinson for guidance on the HOPS software. We would also like to thank the members of the Advanced Visualization and Interactive Systems laboratory at Santa Cruz for their feedback. We are grateful to the Office of Naval Research for their support during the Uncertainty Pilot Working Group initiative, under grant N00014-00-1-0764 and N00014-00-1-0771. This project is supported in part by NASA grant NCC2-5281, LLNL Agreement No. B347879 under DOE Contract No. W-7405-ENG-48, NSF ACI-9908881, and DARPA grant N66001-97-8900.

References

1. Andrej Cedilnik and Penny Rheingans. Procedural annotation of uncertain information. In *Proceedings of Visualization 00*, pages 77–84. IEEE Computer Society Press, 2000.
2. Suzana Djurcilov and Alex Pang. Visualizing sparse gridded datasets. *IEEE Computer Graphics and Applications*, 20(5):52–57, September 2000.
3. Victoria Interrante. Harnessing natural textures for multivariate visualization. *IEEE Computer Graphics and Applications*, 20(6):6–11, November/December 2000.
4. G. Kindlmann and J.W. Durkin. Semi-automatic generation of transfer functions for direct volume rendering. In *IEEE Symposium on Volume Visualization*, pages 79–86, 170. IEEE, 1998.
5. P.F.J. Lermusiaux. Data assimilation via error subspace statistical estimation, Part ii: Middle Atlantic Bight shelfbreak front simulations and ESSE validation. *Monthly Weather Review*, 127(7):1408–1432, 1999.
6. E. Levy, G. Gawarkiewicz, and F. Bahr. The ONR shelfbreak PRIMER experiment: shelfbreak frontal dynamics in the Middle Atlantic Bight. URL: http://matisse.whoi.edu/primer_cd, 1999.

7. A. Pang, C.M. Wittenbrink, and S. K. Lodha. Approaches to uncertainty visualization. *The Visual Computer*, 13(8):370–390, 1997.
8. A.R. Robinson. Physical processes, field estimation and an approach to interdisciplinary ocean modeling. *Earth-Science Review*, 40:3–54, 1996.
9. A. Tarantola. *Inverse Problem Theory. Methods for Data Fitting and Model Parameter Estimation*. Elsevier Science Publishers, 1987.
10. Craig M. Wittenbrink. IFS fractal interpolation for 2D and 3D visualization. In *IEEE Visualization '95*, pages 77–84, Atlanta, GA, November 1995. IEEE.
11. Craig M. Wittenbrink, Alex T. Pang, and Suresh K. Lodha. Glyphs for visualizing uncertainty in vector fields. *IEEE Transactions on Visualization and Computer Graphics*, 2(3):266–279, September 1996. Short version in SPIE Proceeding on Visual Data Exploration and Analysis, pages 87-100, 1995.

Editors' Note: see Appendix, p. 355f. for colored figures of this paper

Adaptive Volume Rendering using Fuzzy Logic Control

Xinyue Li and Han-Wei Shen

Department of Computer and Information Science
The Ohio State University
Columbus, Ohio 43210
USA
E-mail: xli@cis.ohio-state.edu and hwshen@cis.ohio-state.edu

Abstract. This paper presents an automatic error tolerance specification system to control the performance of hierarchical volume rendering. Rather than requiring the user to provide an explicit error tolerance numerically, we let the user to specify only the target rendering speed. Our system can then calculate an appropriate error tolerance adaptively to satisfy the user's performance goal. The system is realized using fuzzy logic control, which enables run-time adaptation based on iterative feedback control and knowledge acquired from past experience. We describe the process of constructing the fuzzy logic control system, and show that the system can successfully steer the performance of volume rendering.

1 Introduction

Direct volume rendering is an effective technique for analyzing three dimensional scalar data, as it allows the user to visualize the underlying field's global structures without the need to generate intermediate geometry. While effective, volume rendering is computationally intensive, which makes interactive manipulation and display of large-scale volume data difficult. In the past, researchers have proposed various software and hardware solutions to accelerate volume rendering. Among these methods, the use of hierarchical data structures and rendering algorithms has proven to be effective as it permits run-time tradeoffs of image quality and rendering speed. In general, this tradeoff is controlled by an error tolerance, which is used to select an appropriate level of details from a multi-resolution volume hierarchy. To generate a rendering result, the volume hierarchy is traversed. If low resolution subvolumes at particular levels of the hierarchy have error measurements smaller than the error tolerance, the subvolumes are used for rendering. Otherwise, the volume hierarchy traversal is continued until all the subvolumes with lower errors are identified.

In general, the error tolerance is specified by the user at run time based on the desired rendering speed and visualization quality. However, as the error measurements for the volume hierarchy are both data and error metric dependent, the rendering speed and image quality corresponding to a particular error tolerance are difficult to predict without extensive knowledge of the underlying data. In addition, the dependency of the volume rendering performance on the image size and transfer function further complicates the process of error tolerance specification. Moreover, in an interactive visualization session where the user's requirements in image quality and computation speed change frequently, it is impractical to assume the user can constantly keep track of an appropriate error tolerance in a time-critical manner.

This paper presents an automatic error tolerance specification system for volume rendering. Instead of specifying a numerical error tolerance which requires a good understanding of the error metric and the underlying data, the user only needs to provide the desired performance in terms of frame rate. Our system is able to locate an appropriate error tolerance automatically to satisfy the desired performance goal. In addition, our system can adjust the error tolerance on the fly when the user varies the viewing parameters. We develop a *Fuzzy Logic Control* algorithm, which allows the system to adapt based on iterative feedback control and the knowledge acquired from past experience. In the following, we first overview related work on hierarchical volume rendering. We then provide a brief overview of fuzzy logic control fundamentals. Details about our adaptive rendering system are provided in section 4, and experimental results are discussed in section 5.

2 Related Work

Researchers have proposed various hierarchical methods to accelerate volume rendering. Levoy [1] used a pyramid data structure to record the volume's local occupancy information based on a binary classification of the voxels, which allows space-leaping and adaptive termination of ray tracing. Laur and Hanrahan [10] proposed a *hierarchical splatting* method, where a pyramid data structure is used and each node of the pyramid stores the corresponding volume block's mean and standard deviation. At run time, a user-supplied error tolerance is provided and an octree is fit to the pyramid. Based on the local error measurements, different regions of the volume can be rendered in different resolutions. Danskin and Hanrahan [2] proposed an *importance sampling* method, where the sampling rate along a ray is changed according to the local and accumulated opacities. A special type of pyramid, called Range$_{27}$ pyramid, is used to facilitate the adaptive sampling. Wilhelms and Van Gelder proposed a selective octree traversal method for hierarchical volume rendering and compression with an extension of storing voxel and cell trilinear functions at the octree nodes [3].

More recently, researchers have proposed to use hierarchical data representations to assist hardware-based volume rendering via 3D texture mapping. The main motivation of the research is to overcome the limitation of texture memory capacity imposed by the underlying hardware. LaMar *et al.*[4] proposed to use a pyramid data structure to construct a texture hierarchy. The leaf nodes of the hierarchy constitute the original volume data and internal nodes store down-sampled volume bricks. The run-time selections of volume bricks are guided by the distances of the bricks to the view point and the center of focus, and their projected angles. Weiler *et al.*[5] proposed a level-of-detail rendering method using a similar data structure with a guarantee of consistent interpolation between different resolution levels. Finally, Ellsworth *et al.*use a Time-Space Partitioning Tree [6, 7] with OpenGL texture object optimization and fast color metric approximations to accelerate time-varying volume rendering.

While all these techniques allow user control to trade image quality for speed, the relations between the control parameters and the rendering speed are not known. As a result, the user often needs to carry out the adaptation on a trial-and-error basis. In this paper, we propose to use a fuzzy logic control method to achieve automatic error tolerance specification for adaptive volume rendering.

3 Fuzzy Logic Control

Unlike the conventional control methods, whose effectiveness strongly relies on the accuracy of the analytic control model, fuzzy logic control is used when developing such a model is difficult or impossible. Like other control mechanisms, fuzzy logic control is essentially a feedback control system as presented in Figure 1. The object to be controlled is called the *system*, denoted as S, which is the volume rendering algorithm in our case. The controller, denoted as C, is to guarantee a desired response of the output y, i.e., keeping the output y close to the reference point w (keeping e small). In our application, y and w are the current and desired frame rates. The output u of the controller C, is the control action, which is the error tolerance in our application. In essence, fuzzy logic control relies on a set of IF...THEN inference rules [8], which have the general form:

if x is A then y is B

where x is the input variable, and y is the output variable. The values A and B are expressed linguistically rather than in numerical forms. Examples of linguistical values are *very low*, *low*, *medium*, *high*, and *very high*. Using the concept of fuzzy sets proposed by L.A. Zadeh in 1965 [9], these linguistic values can be translated into numerical values to perform calculations. The concept of fuzzy set is considered as a generalization of the classical crisp set, which allows objects to take

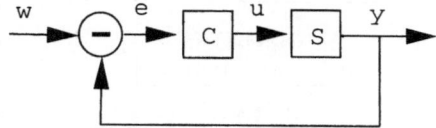

Fig. 1. A feedback control system

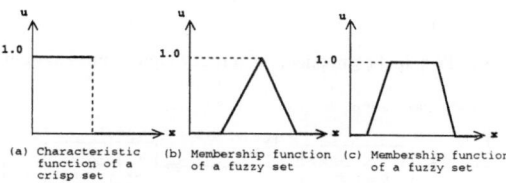

Fig. 2. Crisp and fuzzy sets

partial membership in vague concepts (ie. fuzzy sets). For a crisp set, the degree of an object belongs to the set is either 0 or 1, this relationship can be denoted using a characteristic function as shown in figure 2 (a). For a fuzzy set, however, the degree that an object belongs to the fuzzy set is often a real number between 0 and 1, which is called the membership value in the set. This is denoted using membership functions, as shown in figure 2 (b) and (c).

To build a fuzzy logic control system, a *knowledge base* needs to be first constructed. The knowledge base is composed of two parts, data base and rule base. Data base includes all the membership functions of inputs and outputs of the fuzzy logic control system, which are used to denote the relationship between an accurate value and the linguistic sets. The rule base is used to store all the inference rules of the fuzzy control system, with each rule has the general IF ... THEN ... form. Given the knowledge base which contains both the data base and rule base, fuzzy logic control system at run time will continuously monitor the current system state (the input) and perform the following three steps:

(1) fuzzification: This process is to convert the accurate value of input into fuzzy value, i.e. membership values of the fuzzy sets.
(2) fuzzy reasoning: This is to use the fuzzy inputs and the knowledge base to get the fuzzy output of the control system. This is the kernel of fuzzy control.
(3) defuzzification: The output of a fuzzy logic control system is a fuzzy value which cannot be used directly. This fuzzy output value is converted into an accurate control value by the process of defuzzification.

Figure 3 shows the principal components of a fuzzy logic control system.

4   Automatic Error Tolerance Specification for Adaptive Volume Rendering

In octree-based volume rendering, the trade off between image quality and frame rate is controlled by the error tolerance. A high error tolerance can result in a high frame rate and vice versa. Normally, the user is asked to specify the error tolerance at run time. However, when the user does not have sufficient knowledge about the data set, it can become difficult to predict the relationship between an error and the corresponding frame rate. In this case, the user often needs to go through several trials by specifying arbitrary numerical error tolerance values until a desired frame rate, as well as the image quality, are received. In the following, we present our fuzzy logic control method to realize adaptive frame rate control. In our system, the user only needs to specify the desired rendering speed (frame rate) rather than numerical error tolerance values. Based on the

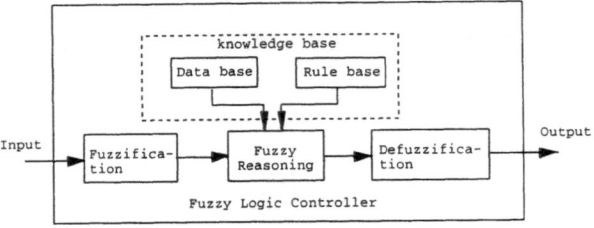

Fig. 3. Principal components of a fuzzy logic control system

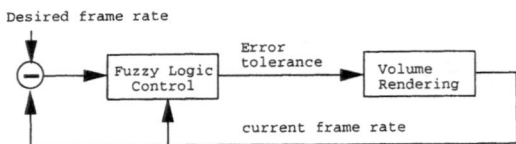

Fig. 4. Adaptive volume rendering system overview

desired frame rate, our system can automatically generate an appropriate error tolerance and the resulting rendering time is approximately at the rate specified by the user. We first briefly overview the hierarchical volume rendering algorithm that we use. We then describe our error specification algorithm in detail.

4.1 Hierarchical volume rendering

The volume hierarchy that we use is similar to the one used by Laur and Hanrahan [10], which is a standard complete octree that recursively subdivides the volume spatially until all subvolumes reach a predefined minimum size. The leaf nodes of the tree point to the raw volume data, and the internal nodes of the tree store the mean values of the corresponding subvolumes. The error metric associated with each internal node is the voxels' standard deviation. At run time, the user specifies an error tolerance, and this error tolerance is compared with the error estimates recorded in the tree nodes. If an internal node's error estimate is smaller than the error tolerance, the mean value is used to represent the corresponding subvolume. Otherwise, the traversal of the octree continues. If a leaf node is reached, the original data in the subvolume are used. To perform the rendering, we use a simple 3D texture hardware acceleration technique proposed by Ellsworth, Chiang, and Shen [7]. This method uses a polygon slicing and 3D texture mapping technique and breaks the volume into small blocks as the result of octree traversal. If an internal node is selected, flat shaded polygons are drawn. Otherwise, slicing polygons within the volume blocks are rendered and mapped with 3D textures. We note that our error specification algorithm described in the following does not use any specific properties of this volume rendering method. Therefore, the algorithm can be applied to other hierarchical volume rendering methods as well.

4.2 Error Specification System Outline

Figure 4 shows an overview of our run-time adaptive volume rendering system. It's a feed back system and the main part of the system is the control unit. The input of the control unit is the current frame rate, and the difference between the user desired frame rate and the current frame rate. The output of the control system is an error tolerance used for volume rendering. In essence, the system works as follows:

(1) After the user specifies a frame rate, the control system randomly chooses an initial error which the rendering system uses to generate an image. The frame rate based on this initial error is then fed back to the input of the control system.

(2) Based on the current frame rate and the difference between the current frame rate and the desired frame rate, the control system adjusts the error tolerance. The basic idea is: "When the difference is positive, which means current frame rate is higher than the desired frame rate, we need to decrease the error tolerance". The higher the difference, the larger the error tolerance will drop and vice versa.

(3) The new error tolerance will be used by the volume rendering system again and will result in a new frame rate. If the difference between this new frame rate and the desired frame rate is less than a minimum range, the control system will stop adjusting the error tolerance. Otherwise, this new frame rate is fed back to the input of the control system and repeat step 2 and 3.

Fuzzy Logic Control (FLC) is chosen to implement the above control system. We choose fuzzy logic control for two reasons. First, the input of the control system is a fuzzy variable. There is no explicit threshold to decide that the difference of frame rates is large or small. For instance, if we decide that the value that distinguishes small difference and large difference is 0.8, traditional crisp logic will classify a difference of 0.79 as small, and a difference of 0.81 as large, which is not a desirable decision. Fuzzy logic can solve the problem better as it uses multivalued logic to model problems. For example, instead of saying that a frame rate difference of 0.8 is "small", fuzzy logic would state that " A frame rate difference of 0.8 is 40% belonging to small and 60% belonging to large". The second reason for using fuzzy logic control is that it is difficult to find a mathematical model to describe the relationship between the frame rate and the error tolerance due to the combination effect of 3D projection, transfer function, visibility, and so on in volume rendering. Fuzzy logic has proven to be quite effective to model such a complex and perhaps nonlinear system.

4.3 The Fuzzy Logic Control System

In this section, our fuzzy logic control system is described in detail. We first present the construction of its knowledge base, and then describe the three main steps, *fuzzification*, *fuzzy reasoning*, and *defuzzification*, for making run time control decisions.

(1) Construction of Knowledge Base As mentioned before, the knowledge base in a fuzzy logic control system consists of a data base and a rule base. The data base includes the membership functions of inputs and outputs, and the rule base contains the inference rules. Our fuzzy control system has two inputs and one output. The input variables are the current frame rate and the frame rate difference (current frame rate minus desired frame rate). The output variable is an error tolerance modulator, which is used to adjust the error tolerance. We need to create one set of membership functions for each input/output variable.

(1.a) Membership functions of input variables: To define the membership function of the frame rate difference, we divide the range of the frame rate difference into five linguistic sets, Negative Big (NB), Negative Small (NS), Zero(ZO), Positive Small (PS), and Positive Big (PB), and the membership functions are shown in figure 5. In the figure, the horizontal axis indicates the difference between the current and the user specified frame rates. The scale of the horizontal axis (frame rate difference) can be scaled differently based on the rendering method and the input data. The vertical axis denotes the membership value of a given frame rate difference. This value is used to indicate the degree to which a difference belongs to a linguistic set. For example, from the figure we can see that the degree of frame rate difference of 0.8 to PS is 0.6, and to ZO is 0.4. In this example, a precise frame difference value is mapped to two linguistic sets.

To generate a new error tolerance, the current frame rate also needs to be taken into account. This is because given the same frame rate difference, the amount of error modulation needed will be different in different frame rate ranges. In our system, current frame rate is divided into three categories, "High (H)", "Medium (M)", and "Low (L)", and the membership functions are shown in figure 6.

(1.b) Membership functions of output variables: We adjust the error tolerance by using an error tolerance modulator M, which is the output of the control system. If the current error tolerance

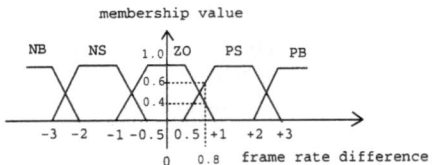

Fig. 5. Membership functions for the input: frame rate difference

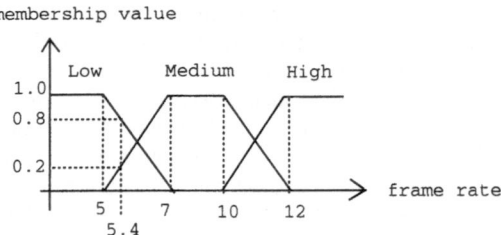

Fig. 6. Membership functions for the input: current frame rate

is E, and the control system output is M, then the new error tolerance becomes $E + E \times M$. To define the membership function of the output variable, the values of the modulator is categorized into fifteen linguistic sets, LNB, LNS, LZO, LPS, LPB, MNB, MNS, MZO, MPS, MPB, HNB, HNS, HZO, HPS, and HPB. We use fifteen output linguistic sets because there are three levels of current frame rate (L, M, H), and five levels of frame rate difference (NB, NS, ZO, PS, PB), which results in fifteen different combinations. We use a singleton membership function to represent each of the linguistic set, as it is easy to perform defuzzification[11]. Figure 7 shows five output singleton membership functions.

To determine the singleton values for the output linguistic sets, we need to perform profiling runs. In essence, the purpose of profiling is to obtain knowledge about the approximate error tolerance modulation needed in order to achieve the desired frame rate. Table 1 shows sample information received from experiments using a sample dataset when the desired frame rate drop is one. Column 1 shows that the current frame rate is categorized into three levels (H, M, L), column 2 shows the initial and target frame rates, column 3 shows the error tolerances used to generate the initial and target frame rates in our profiling with a fixed view, and column 4 shows the corresponding error modulation values. In this example, the frame rate difference (-1) is considered to be "Negative Small (NS)". To obtain the singleton values for the output linguistic sets HNS, MNS, and LNS, we average the error tolerance modulation values in column 4 for each of the three frame rate categories (High, Medium, and Low). Other singleton values are obtained in the same way. We note that although we use a sample view to get the error tolerance-frame rate information in the profiling runs, the information is sufficient for us to construct a sound fuzzy control system.

(1.c) Creation of rule base: In this step, we construct the fuzzy reasoning rules that govern the relations between the input and output variables. As our present system has two inputs and one output, the form of each rule is: " IF current frame rate is A and frame rate difference is B, then the error modulation is C", where A is chosen from "L", "M", "H", B is chosen from "NB", "NS", "ZO", "PS", "PB", C is a singleton output. A sample rule can be written as:

If the current frame rate is L, the frame rate difference is PB, then the error tolerance modulator is LPB. (decrease the error tolerance to a high degree)

259

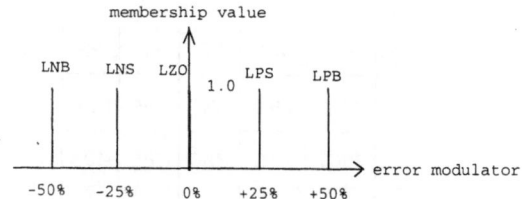

Fig. 7. Output singleton membership functions for error modulator

frame rate	frame rate change	error tolerance change	modulation
	11 to 10	0.06 to 0.04	-33%
High	10 to 9	0.04 to 0.02	-50%
	9 to 8	0.02 to 0.008	-60%
	8 to 7	0.008 to 0.004	-50%
Medium	7 to 6	0.006 to 0.0025	-58%
	7 to 6	0.006 to 0.0025	-58%
	6 to 5	0.0028 to 0.0015	-40%
	5 to 4	0.0015 to 0.001	-35%
Low	4 to 3	0.0012 to 0.0005	-58%

Table 1. Sample profiling results

Table 2 shows the complete rule base. Each entry in the matrix is the linguistic output of the error modulator for the corresponding current frame rate (row) and the frame rate difference (column).

(2) Fuzzification At run time, the accurate values of current frame rate and frame rate difference are converted into fuzzy values based on their membership functions which have been defined in the data base. For example, from the membership functions of frame rate difference shown in figure 5, we can see that a frame rate difference of 0.8 belongs to two fuzzy subsets, PS and ZO, and the membership values are 0.6 and 0.4 respectively, which denote the degrees to which the frame rate difference belongs to the linguistic sets. By this way, an accurate difference value can be mapped to some linguistic subsets and this completes the fuzzification of frame rate difference. The fuzzification of current frame rate can be performed in a similar way.

(3) Fuzzy Reasoning Based on the input frame rate and frame rate difference, fuzzy reasoning will take place to compute an new error tolerance modulator. In essence, there are three steps in fuzzy reasoning, In the following, we use frame rate difference of 0.8 and current frame rate of 5.4 as an example to explain the three steps:

(3.a) Find the firing level of each rule: The firing level of a rule is determined by the satisfaction of each of the component in the antecedent of the rule. From the membership functions shown in figure 5 and figure 6, we can see that a frame rate difference of 0.8 is related to two linguistic sets, PS and ZO, and the membership values are 0.6 and 0.4 respectively; the current frame rate of 5.4 is related to two linguistic sets, L and M, and the membership values are 0.8 and 0.2 respectively. As a result, four rules, rules LPS, LZO, MPS, MZO, shown in table 2 will be activated. The firing level of a rule is calculated by combining the satisfaction of each of the antecedent component. Normally we use Min aggregation for combination. For instance, in the case that the degree of "current frame rate" in L is 0.8 and "frame rate difference" in PS is 0.6, then the firing level for the rule LPS is Min(0.8, 0.6) = 0.6.

(3.b) Find the output of each rule: Looking at table 2, we can know that the output for rule LPS is "the error tolerance modulator is LPS", for rule LZO the output is "the error tolerance modulator is LZO", and so on.

Output ↘ FRD CFR	NB	NS	ZO	PS	PB
L	LNB	LNS	LZO	LPS	LPB
M	MNB	MNS	MZO	MPS	MPB
H	HNB	HNS	HZO	HPS	HPB

FRD: Frame Rate Difference
Output: Error Modulator Output
CFR: Current Frame Rate

Table 2. Fuzzy reasoning rule base

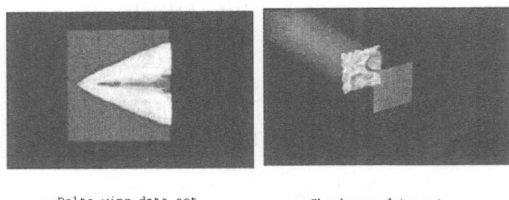

Delta wing data set   Shockwave data set

Fig. 8. Images for the delta wing and shockwave data sets

(3.c) Aggregate the individual rule: Based on the activated rules and the firing level of each rule, the overall output of the control system is an aggregation of the output from step 1 and 2. In our example, the degree of output error tolerance modulator belongs to LPS is 0.6, to LZO is 0.4, to MPS is 0.2, to MZO is 0.2. Note that at this stage, the system output is still a fuzzy value, we need to have a defuzzification step as described in the next step to get a usable numerical output for the error tolerance modulator.

(4) Defuzzification The defuzzification process is needed for calculating a numerical output from the aggregation of individual fuzzy output. In our case, it is to determine the actual value of the error tolerance modulator. We use center of area (COA)method [12] to perform defuzzification. That is, given the firing level F_i for the rule(i), and the rule's corresponding singleton output S_i, the defuzzified value is:

$$M = \sum (F_i \times S_i) / \sum F_i$$

For instance, if the singleton outputs for LPS, LZO, MPS, and MZO are -25%, 0%, -10%, and 0%, and the firing levels for the rules are 0.6, 0.4, 0.2, 0.2, then the final error tolerance modulator is:

$$M = (0.6 * -25\% + 0.4 * 0\% + 0.2 * -10\% + 0.2 * 0\%)/(0.6 + 0.4 + 0.2 + 0.2)$$

As a result, the new error tolerance is calculated as: new error tolerance = current error tolerance $\times (1 + M)$.

5 Results and Discussion

We have implemented our automatic error specification algorithm with a hardware based volume rendering method mentioned before. All the experiments shown in this section were performed

delta wing			shockwave		
desired	result frame rate	iterations	desired	result frame rate	iterations
4	4.27	5	4	3.78	6
5	4.52	4	5	5.27	7
6	6.24	4	6	5.97	7
7	7.44	4	7	7.19	7
8	8.03	4	8	7.84	8
9	9.12	6	9	9.03	6
10	10.42	4	10	10.46	5

Table 3. The desired frame rate, result frame rates, and iterations

on an SGI Octane with one MIPS R12000 processor, 512 Mbytes memory, and 4Mbytes texture memory. Two regular Cartesian gridded datasets were used in our tests. One is a delta wing data set with resolution $111 \times 126 \times 51$, and the other is a shockwave data set with resolution $512 \times 64 \times 64$. Figure 8 shows the sample images for both data sets. We tested the effectiveness of our fuzzy logic control system under two scenarios. One is to let the user specify an initial frame rate requirement when the system just starts up, and the other is to allow the user to incrementally change the frame rate requirement on the fly. In the first scenario, our error specification system initially chose a random error tolerance and then performed the feed back control to adjust the output. Our goal is to have the system converge to the desired frame rate with a minimum number of iterations. Table 3 shows the results for both the delta wing and shockwave data sets.

From the results, we can see that even with a randomly picked error tolerance in the beginning, our automatic error specification system can effectively identify appropriate error tolerances after a small number of iterations. Table 4 shows the results when the user incrementally changed the desired frame rate dynamically for both the delta wing and shockwave data sets. Our fuzzy control system can adjust the error tolerance efficiently.

delta wing				shockwave			
initial	target	result frame rate	iterations	initial	target	result frame rate	iterations
12	10	10.20	2	8	6	5.6	3
10	8	8.13	1	7	5	4.74	1
8	6	6,.28	1	6	4	3.6	1
6	4	4.39	1	5	3	3.42	1
4	6	5.87	1	4	2	2.27	1
6	8	8.30	1	2	4	4.22	1
8	10	9.75	2	4	6	6.39	3
10	12	11.82	2	6	8	7.70	1

Table 4. The initial frame rate, target frame rate, result frame rate, and iterations when the user incrementally changed the desired frame rate

Our fuzzy logic control system constantly monitors the rendering performance at run time and adjusts the error tolerance when necessary. For instance, when the user scales the volume data to different sizes, the rendering time will change since the speed of most of the volume rendering algorithms are sensitive to the projection area of the volume. Table 5 shows the results from the delta wing and shockwave data sets when the user dynamically changed the scale of the object and the desired rendering speed was ten frames per second. Five different scales resulted in five different error tolerances, and therefore five different levels of details in the rendering results. Our

	delta wing			shockwave		
scale	result frame rate	error tolerance	iterations	result frame rate	error tolerance	iterations
scale 1	10.42	0.022	1	9.58	0.821	1
scale 2	10.04	0.0098	1	9.94	0.469	1
scale 3	9.61	0.0067	1	9.52	0.445	1
scale 4	9.68	0.0056	1	9.7	0.359	1
scale 5	9.62	0.0031	1	10.11	0.397	1

Table 5. The actual frame rate, the error tolerance. and the number of iterations when the user interactively change the scale of the volumes. The user desired frame rate is 10.

automatic error specification algorithm was able to maintain the frame rate close to the user's expectation.

6 Conclusions and Future Work

We have presented an automatic error tolerance specification system for hierarchical volume rendering using fuzzy logic control. Our system can dynamically track the performance of the rendering program and calculate appropriate error tolerances to satisfy the user's performance goal. Our system allows the user to control the performance tradeoff more intuitively without the need to have extensive knowledge about the data. We have showed that fuzzy logic control has a great potential to assist in constructing adaptive volume rendering algorithms.

Future work includes integrating the control system with a variety of volume rendering methods and applications. We will also explore the use of fuzzy logic control to precisely maintain the image quality. Furthermore, we intend to integrate the control system with operating system support to create a resource aware adaptive visualization framework.

References

1. M. Levoy. Efficient ray tracing of volume data. *ACM Transactions on Graphics*, 9(3):245–261, 1990.
2. J. Danskin and P. Hanrahan. Fast algorithms for volume ray tracing. In *Proceedings of 1992 Workshop on Volume Visualization*, pages 91–99. ACM SIGGRAPH, 1992.
3. J. Wilhelms and A. Van Gelder. Multi-dimensional trees for controlled volume rendering and compression. In *Proceedings of 1994 Symposium on Volume Visualization*, pages 27–34. ACM SIGGRAPH, 1994.
4. E. LaMar, B. Hamann, and K. Joy. Multiresolution techniques for interactive texture-based volume visualization. In *Proceedings of Visualization '99*, pages 355–361. IEEE Computer Society Press, Los Alamitos, CA, 1999.
5. M. Weiler, R. Westermann, C. Hansen, K. Zimmerman, and T. Ertl. Level-of-detail volume rendering via 3d textures. In *Proceedings of 2000 Symposium on Volume Visualization*, pages 7–13. ACM SIGGRAPH, 2000.
6. H.-W. Shen, L.J. Chiang, and K.L. Ma. A fast volume rendering algorithm for time-varying field using a time-space partitioning (tsp) tree. In *Proceedings of Visualization '99*. IEEE Computer Society Press, Los Alamitos, CA, 1999.
7. D. Ellsworth, L. Chiang, and H.-W. Shen. Accelerating time-varying hardware volume rendering using tsp trees and color-based error metrics. In *Proceedings of 2000 Symposium on Volume Visualization*. ACM SIGGRAPH, 2000.
8. L.A. Zadeh. The calculus of fuzzy if/then rules. *AI Expert*, 7:23–27, March 1992.
9. L.A. Zadeh. Fuzzy sets. *Information and Control*, 8:338–353, 1965.
10. D. Laur and P. Hanrahan. Hierarchical splatting: A progressive refinement algorithm for volume rendering. In *Proceedings of SIGGRAPH 91*, pages 285–287. ACM SIGGRAPH, 1991.
11. K. Tanaka. *An introduction to fuzzy logic control for practical applications*. Springer New York, 1997.
12. R. Yager and D. Filev. *Essentials of fuzzy modeling and control*. J. Wiley New York, 1994.

Editors' Note: see Appendix, p. 357 for colored figure of this paper

I/O-Conscious Volume Rendering

Chuan-Kai Yang and Tzi-cker Chiueh

Department of Computer Science, State University of New York at Stony Brook, Stony Brook,
NY 11794-4400, USA
emails: {ckyang, chiueh}@cs.sunysb.edu

Abstract. Most existing volume rendering algorithms assume that data sets are memory-resident and thus ignore the performance overhead of disk I/O. While this assumption may be true for high-performance graphics machines, it does not hold for most desktop personal workstations. To minimize the end-to-end volume rendering time, this work re-examines implementation strategies of the ray casting algorithm, taking into account both computation and I/O overheads. Specifically, we developed a data-driven execution model for ray casting that achieves the maximum overlap between rendering computation and disk I/O. Together with other performance optimizations, on a 300-MHz Pentium-II machine, without directional shading, our implementation is able to render a 128x128 grey-scale image from a 128x128x128 data set with an average end-to-end delay of 1 second, which is very close to the memory-resident rendering time. With a little modification, this work can also be extended to do out-of-core visualization as well.

1 Introduction

Despite the fact that volumetric data sets are inherently huge, most previous ray casting algorithms research reported performance numbers, assuming that data sets are entire memory-resident. This assumption is not valid when individual data sets are too large to fit into main memory (*out-of-core rendering*), or when users need to browse or explore a large number of data sets. Such assumptions tend not to hold especially on personal workstations, where volume visualization technology is gradually gaining grounds.

The motivation of this work is to develop a high-performance volume rendering system on commodity PCs without special hardware support, with a focus on reducing the *end-to-end* rendering delay, including the disk overhead of bringing the data sets in and out of the host memory. The key technique to minimize the performance impacts of disk I/O is to overlap disk operations with rendering computation so that the disk I/O time is masked as much as possible. To achieve this goal, a volumetric data set is decomposed into blocks, which are stored on disks and accessed as indivisible units. As data blocks are retrieved from disks, rendering computation on those blocks that are brought in earlier proceeds simultaneously. In this execution model, the *minimum* total rendering time for a disk-resident data set is the sum of the rendering time when the data set is entirely memory-resident, and the time required to fetch the first data block.

Surprisingly, the above overlapping execution model is difficult to get right in practice. This paper presents one such optimal execution model: *data-driven block-based volume rendering*, which hides most of the disk I/O delay while at the same time ensures that a data block is completed exercised once it is brought into memory from the disk. The bottom-line result is that on a 300-MHz Pentium-II machine, without directional shading, this implementation strategy is able to complete the task of rendering a 128x128x128 data set into a 128x128 image in 1 second on the average, including the disk I/O time.

The rest of this paper is organized as follows. Section 2 reviews previous volume rendering work that paid attention to disk I/O issues. Section 3 describes the design dimensions of I/O-conscious volume rendering algorithms, and their associated performance tradeoffs. Section 4 proposes a simple extension of this work to do out-of-core visualization as well. Section 5 shows the results of a detailed performance evaluation of the prototype implementation, which is built on top of a Pentium-II machine running Linux. Section 6 concludes this paper with a summary of the major research results. Due to space limitation, we have omitted some details. Please refer to full paper at http://www.ecsl.cs.sunysb.edu/tr/TR89.pdf.

2 Related Work

The main focus of this work is to reduce the disk I/O performance overhead in volume rendering computation, particularly ray casting algorithms. *Out-of-core rendering* refers to the case where the rendering machine's physical memory can not hold the entire data set and thus need to perform disk I/O *during* the rendering process. Cox [4], [3] studied this problem by examining the performance impacts of the operating system interfaces on the disk I/O cost, as well as related file cache management issues. In contrast, our work attempts to use algorithm-specific prefetching to ensure that the data blocks could be brought in before they are needed. The proposed prefetching mechanism is closely tied with the rendering computation, and is completely algorithm-specific. This tightly integrated approach also sets itself apart from other more general-purpose disk prefetching research, as done in [6], [8], [9, 1] and [7]. Another way to reduce the performance overhead due to disk I/O is to use compression to cut down the I/O traffic volume, as done in [10], [5], [2] and [11]. Our work assumes that the ray casting algorithm is more computation-intensive than I/O-intensive, and therefore spending additional decompression computation or restricting the data viewing scope to lower disk traffic is not considered a desirable tradeoff. Rather, we focus on how to *mask* the disk I/O delay.

3 I/O-Conscious Ray Casting Algorithm

3.1 Optimization for Memory-Resident Ray Casting Algorithm

To reduce the end-to-end volume rendering time, the performance of the ray casting algorithm when the data set is completely memory-resident should be optimized to the extent possible. We have added the following performance optimizations to arrive at a high-quality and high-performance ray caster, as the baseline case.

The first optimization replaces floating-point computation with integer arithmetic, specifically in tri-linear interpolations. By replacing the floating-point numbers in tri-linear interpolation, which are between 0.0 and 1.0, with 8-bit integers, we improve the overall performance by almost an order of magnitude in certain cases on a Pentium-II machine, because our ray caster uses only integer arithmetic, and Intel processor's floating-point hardware traditionally lags significantly behind its integer counterpart. This optimization, however, does not affect the rendering quality. For example, Figure 1 and figure 2 show two images rendered through floating-point arithmetic and integer arithmetic without much perceptible differences. The second performance optimization

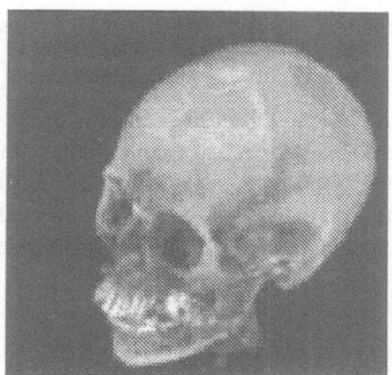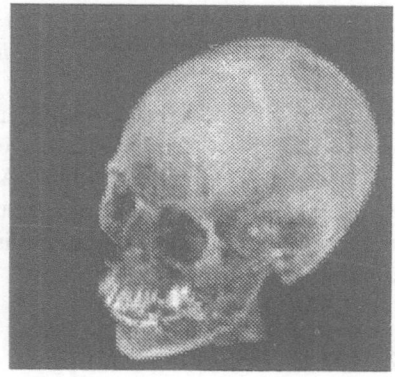

Fig. 1. Floating point computation.   **Fig. 2.** Integer computation.

attempts to exploit the instruction-level parallelism using the MMX instruction set extensions available on the Pentium-II processor. MMX is capable of executing multiple low-resolution fixed-point operations in parallel on a high-resolution data-path, e.g., 4 16-bit multiplications on a 64-bit multiplier. By using integer arithmetic and four kinds of MMX instructions: PMULHW, PMULLW, PMADDWD and PSUBW, we create a new version of tri-linear interpolation which takes only 37 instructions. Unfortunately the performance of this code on Pentium-II does not improve much over the non-MMX version, and in some cases actually worsens. Please refer to the full paper for a detailed explanation.

When volumetric data sets are represented as 3D arrays, the address generation logic for the samples used in tri-linear interpolation is susceptible for optimization. Specifically, the eight samples used in tri-linear interpolation have a fixed and simple offset relationship among themselves. By exploiting these relationships to generate the memory addresses of the eight samples involved in tri-linear interpolation, we are able to improve the rendering performance by up to 15%.

The last optimization avenue that we explored is related to caching. We discovered that the ray casting performances for different viewing directions could differ by as much as 30%, although they require the same amount of computation. To improve the cache performance, we have tried to cast a group of rays concurrently rather than one ray at a time, so that each time a cache block is brought in, it can be utilized as much as possible. However, for reasons as explained in the full paper, the ray group approach does not improve the overall performance. Table 1 shows the performance improvement

Optimization	Performance Improvement
Replace Floating-Point with Integer	4 to 6 times faster
Using MMX	0% faster on (Pentium-II)
	60-80% faster on (Pentium)
Hand-Code Address Generation	up to 15%
Caching	No obvious overall improvement

Table 1. *Performance improvements from various optimizations to a generic ray caster implementation on a 300-MHz Pentium-II machine.*

from each of the performance optimizations. For a $128 \times 128 \times 128$ data set with 1-byte voxel and a 128×128 rendered image, the measured ray casting time is 0.68-1.0 sec on a 300-MHz Pentium-II machine. At the same time, the time to retrieve the same data set from the disk is 0.33 sec, assuming that the data set is laid out sequentially. Therefore, it is essential to minimize disk I/O's visible performance overhead to reduce the end-to-end rendering time.

3.2 I/O-Conscious Ray Casting

The general strategy to mask disk I/O delay is to overlap disk I/O with rendering computation. Each volume data set is decomposed into 3D sub-cubes or *macro-voxels*, which are stored contiguously on the disk. However, when a macro-voxel is brought into memory, the voxels are *scattered* into their corresponding positions in the 3D array. In the ideal case, when a macro-voxel is being fetched from the disk, the CPU performs rendering computation on the macro-voxel that is brought in previously, and thus hides all the disk I/O delay. Therefore, the minimum end-to-end rendering time when the input data set is disk-resident is the time to fetch the first macro-voxel plus the time to render the data set when it is completely memory-resident. However, achieving such an ideal overlap between disk I/O and rendering computation remains elusive in practice.

The fundamental mechanism to mask the disk I/O delay is to prefetch the macro-voxels in advance before they are actually needed for ray casting computation. To ensure that the rendering computation should never be stalled due to unavailability of required voxels, the sequence of macro-voxels that are prefetched should be identical to the traversal pattern of rendering computation. In other words, the prefetch stream should traverse the volume data set in exactly the same way as the rays cast. To achieve this effect, the prefetching module should execute the same traversal code as used in the ray caster. Given a macro-voxel size, $B \times B \times B$, it can be shown that as long as the origins of the rays that are cast for prefetching purpose are at most B pixels apart on the image plane, and the sampling distance along these rays remain at 1, then these rays can cover all macro-voxels in the input data set. During prefetching-induced traversal, the algorithm checks whether each sample on each ray steps into a new macro-voxel. If so, the algorithm brings in the new macro-voxel from the disk; otherwise it continues sampling along the ray.

In summary, the I/O-conscious ray casting algorithm consists of two modules, one for casting rays and the other for prefetching macro-voxels according to the way rays are cast into the input volume data sets. There are three dimensions along which one can implement these two modules. The Cartesian product of the alternatives along each dimension constitutes the entire design space.

Software Structure Because the ray casting module is data-dependent on the prefetching module, careful scheduling between these two modules is essential to mask the disk I/O delay. The current implementation uses the two-thread approach because switching between these two threads incurs a fixed but small thread-level context switch overhead, compared to the two processes approach.

Volume Traversal Strategy The ray casting module can either shoot one ray at a time or a group of rays concurrently. As more rays are cast simultaneously, more states are required to maintain the progress of each ray, and the accumulated color and/or opacity values. On the other hand, the ray group approach enables more processing parallelism in that as the number of concurrently cast rays increases, the CPU is less likely to be idle for the lack of useful work to do. Unlike the CPU cache case, the overhead of state maintenance is well worth the benefits it brings. Therefore, the ray group approach is chosen in the current implementation.

Control Flow There are two ways to pass control between the prefetch and ray casting modules. The traditional approach is *program-driven*, which views the ray casting module as the dominating entity that assumes control most of the time, and occasionally passes control to the prefetch module to bring in the next macro-voxel. This approach requires the system to check each ray in the ray group to see whether the macro-voxel it needs to proceed is available, and if so, advances the ray as far as it can, and then repeats the cycle. When the entire ray group stops, the ray casting module yields the CPU through busy-waiting, until the next macro-voxel is brought into memory. The other approach for passing control is the *data-driven* approach, which advances each ray exactly the same way as the previous approach, but attaches the ray to the macro-voxel that it is waiting for when it stops. Every time a macro-voxel arrives, the system continues the processing for the set of rays that are previously attached to this macro-voxel. The main performance advantage of the *data-driven* approach is that it allows the use of larger ray groups, which improve the processing parallelism, without incurring excessive synchronization checks, which will be the case for the *program-driven* approach. Our current implementation thus chooses the *data-driven* approach for control flow transfer.

Given these design decisions, the I/O-conscious ray casting algorithm works as follows. The prefetch and ray casting modules are implemented as separate threads. The prefetch thread traverses the volume data sets in exactly the same way as the ray casting thread, except that the adjacent rays it shoots are B pixels apart, where B is the dimension of the macro-voxel. The ray group size is the same as the size of the image plane. That is, the ray casting thread starts with as many rays as there are pixels on the image plane. Each ray is initially attached to the first macro-voxel that it encounters while traversing through the volume data set. As the prefetch thread traverses the input data set, it fetches from the disk macro-voxels that have not been brought into memory previously. Every time a macro-voxel arrives, the ray casting module continues the rays that are currently attached to the macro-voxel. Each such ray will advance as far as possible, until it runs into another macro-voxel that is not resident in memory, at which point the ray is attached to the missing macro-voxel, or it runs to completion.

Figure 3 illustrates this process assuming a 2D data set and a 1D image plane. The prefetch thread shoots only rays in circles whereas the ray casting thread shoots every ray. When the 1-st ray, initiated by the ray casting thread, reaches the 1-st macro-voxel, it checks whether the macro-voxel is already brought into memory. If yes, it steps through the 1-st macro-voxel along the 1-st ray. Otherwise, the ray casting thread enqueues the state of the 1-st ray to the work queue of the 1-st macro-voxel. Figure 3 shows the content of each macro-voxel's work queue when each ray first touches the volume data set boundary. In this case, when the 2-nd macro-voxel is loaded into memory, Ray 3, 4, 5 and 6 will be dequeued in that order and proceed as far as possible until they reach another macro-voxel that is not memory-resident.

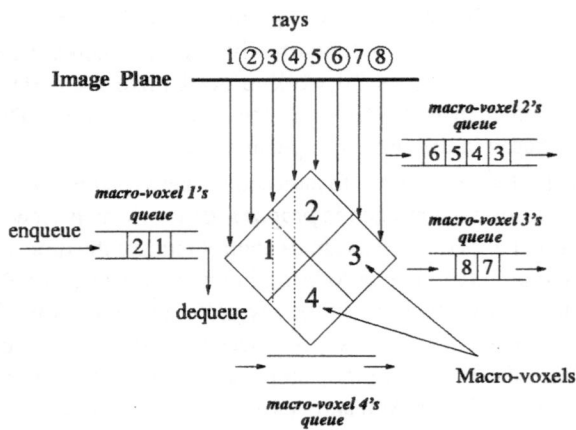

Fig. 3. *A data-driven rendering.*

4 Extension to Out-of-Core Rendering

Because the ray group size is the entire image plane, this means that whenever a macro-voxel is brought in, *all* the rays that need this macro-voxel to advance will be processed before the next macro-voxel arrives. This ray processing pattern leads to two important advantages. First, it exposes the maximum amount of parallelism by identifying all possible rays that are ready to continue. Second, it makes it possible to use a simple FIFO replacement policy for macro-voxels in the case of out-of-core rendering, because once a macro-voxel is "touched," it is no longer needed in future ray processing. For the macro-voxel access pattern to be truly FIFO-like, macro-voxels need to be overlapped with each other by 1 voxel to ensure that each macro-voxel is self-contained during tri-linear interpolations even for rays that pass through the boundaries. That is, a $K \times K \times K$ logical macro-voxel actually contains $(K + 2) \times (K + 2) \times (K + 2)$ voxels physically. However, in general, the access pattern to macro-voxels is not always FIFO-like, because some macro-voxels that are brought in earlier may be partially blocked by others that are scheduled to be fetched in later. Consider ray 4 in Figure 3. If the first macro-voxel brought in is macro-voxel 1, then because macro-voxel 2 that ray 4 needs is still not in the memory, macro-voxel 1 will still be needed for ray 4 after its traversal of macro-voxel 2, thus making the macro-voxel access pattern not FIFO-like. For the macro-voxel access pattern to be truly FIFO-like, the prefetch thread should bring in the macro-voxels according to their distances to the image plane. That is, the closer a macro-voxel is, the earlier it should be brought into memory.

Instead of sorting all the macro-voxels based on their distances to the image plane, we use the same idea as used in *Splatting* where voxel (in our case, macro-voxel) projection order can be pre-determined and there are only a fixed number of orders possible with respect to all viewing directions. Macro-voxels are then dealt with in that order and attached rays are processed/advanced accordingly.

5   Performance Evaluation

We have implemented a prototype ray caster that incorporates various I/O-conscious performance optimizations described in the previous section. All the following performance measurements are collected from a 300-MHz Pentium-II machine, except those for application-specific file prefetching. The shading model we used is post-shading model, i.e., only density values are interpolated during ray traversal, and then mapped to color and opacity values. We applied linear color and opacity transfer functions and mapped the density value range [0,max] to opacity value range [0,1], where max is the maximal density value. Only grey-scale images are generated and no directional shading is performed.

To overlap disk I/O with rendering computation, volume data sets should be brought into memory incrementally in smaller units, i.e., macro-voxels. Every time one macro-voxel of the input data is available, rendering computation based on this macro-voxel can proceed immediately, presumably in parallel with the disk access for the next macro-voxel. Although smaller disk access granularity facilitates the exploitation of parallelism between CPU and I/O, it has an undesirable effect: the disk access efficiency may suffer because a single sequential disk read of an input data set is now decomposed into a sequence of disk reads, one for each macro-voxel. On the other hand, when CPU processing and disk I/O are fully overlapped, larger macro-voxel increases the start-up overhead, or the time to bring in the first voxel. In the extreme case, the macro-voxel is of the same size of the entire data set, which degenerates into conventional "load and render" approach.

Table 2 shows the loading time measurements for a $128 \times 128 \times 128$ data set under different view angles. We found that $64 \times 64 \times 64$ appears to be the best choice considering both the total I/O time and the start-up overhead. In all the following experiments, we assume $64 \times 64 \times 64$ macro-voxels. Smaller macro-voxels do not perform well because their associated disk access patterns tend to cause excessive random disk head movements.

Macro Voxel Size	Orthographic		Non-orthographic	
	0 0 1	1 0 0	1 1 1	0.3 -0.8 0.4
128 × 128 × 128	0.33(0.33)	0.33(0.33)	0.33(0.33)	0.33(0.33)
64 × 64 × 64	0.30(0.070)	0.39(0.071)	0.40(0.070)	0.36(0.070)
32 × 32 × 32	0.30(0.020)	0.37(0.020)	0.60(0.030)	0.79(0.044)
16 × 16 × 16	0.34(0.039)	0.48(0.042)	3.25(0.037)	3.40(0.039)
8 × 8 × 8	0.25(0.038)	0.51(0.038)	3.25(0.037)	3.50(0.035)
4 × 4 × 4	0.28(0.018)	0.93(0.016)	4.20(0.025)	4.90(0.040)

Table 2. *Total time (sec) to load a 2MB data set (128 × 128 × 128) into memory with different macro-voxel sizes. Numbers in parentheses are the start-up overhead.*

To evaluate the performance of the proposed I/O-conscious ray casting algorithm on an end-to-end basis, we measured the rendering times for three data sets using the conventional approach, which loads the entire data set and performs rendering, and using the data-driven ray casting approach. Then we calculate the optimal bound for the data-driven approach, which is the time to load the first macro-voxel and the maximum of the two: the time to render a volume data set assuming it is entirely memory-resident, and the time to load the remaining macro-voxels. The results are shown in Table 3. As the size of the data set increases, the performance difference between the data-driven ray casting algorithm and the conventional ray casting algorithm widens, because the disk I/O cost is playing an increasingly important role.

Table 3 also demonstrates that the current implementation of the data-driven ray casting algorithm is close to the theoretical optimal bound. The performance difference between the current implementation and the optimal bound also decreases as the data set size increases. This discrepancy comes from the prefetch thread's computation, and additional macro-voxel boundary checks and state maintenance overhead during ray traversal.

To understand the performance gain of the proposed I/O-conscious ray casting algorithm as processors get faster, we render only every other pixel on the image plane, to simulate a factor of 4 improvement in rendering computation. The end-to-end delay measurements for three data sets, *CThead*, *Lobster* and *Brain* and for different view angles are shown on the last two rows in Table 3. For large data sets, the performance gain of the proposed approach, compared to the conventional approach, increases because the disk I/O cost becomes more dominant and therefore the ability to mask it is more important to minimize the end-to-end delay.

Table 4 shows the performance comparisons between the data-driven and program-driven approaches for three different data sets, *CThead*, *Lobster* and *Brain*, and for different view angles. In general, the performance difference between the two approaches increases as the viewing direction moves away from the major axes, because the traversal pattern of the prefetching thread tends to differ more from that of the rendering thread. As a result, the program-driven approach is more likely to be delayed because the prefetch thread is less likely to bring in all the macro-voxels in time for the rendering thread.

Table 5 shows how the ray group size affects the total rendering time under different viewing directions. The results show that the rendering performance improves with the

	CThead (2MB) 64 × 64 image		Lobster (4MB) 128 × 128 image		Brain (8MB) 128 × 128 image	
Viewing direction	Conven. /Bound	Data-driven	Conven. /Bound	Data-driven	Conven. /Bound	Data-driven
0 0 1	1.33/1.10	1.10	2.97/2.43	2.60	5.63/4.36	4.78
1 1 1	1.01/0.75	0.91	2.49/1.90	2.07	4.86/3.59	3.88
0 0 1	0.61/0.33	0.46	1.3/0.79	0.92	2.43/1.33	1.60
1 1 1	0.56/0.33	0.58	1.3/0.80	1.17	3.37/1.33	2.10

Table 3. *Comparison of rendering time (sec) on PII 300MHz between the I/O-conscious data-driven ray casting algorithm, its optimal bound, and the conventional load-and-render ray casting algorithm, for different data sets under different viewing directions.*

	CThead (2MB) 128 × 128 × 128		Lobster (4MB) 256 × 256 × 64		Brain (8MB) 256 × 256 × 128	
Viewing direction	Data-driven	Prog.-driven	Data-driven	Prog.-driven	Data-driven	Prog.-driven
0 0 1	1.10	1.25	2.33	2.34	4.78	4.80
1 1 1	0.91	1.40	2.07	2.74	3.88	4.98

Table 4. *Rendering time comparison (sec) between the program-driven and data-driven approaches for three data sets under different viewing directions.*

increase in the ray group size. That is, the performance gain from the ability to exploit more parallelism always out-weighs the additional state maintenance overheads as the ray group size increases.

Table 6 shows the rendering times for a 256 × 256 × 256 using the out-of-core rendering algorithm under different viewing directions and different memory capacity. That fact that the rendering times are within 8% of each other demonstrates this algorithm's insensitivity to the main memory size.

Ray group size	0 0 1	1 1 1
128 × 128	1.10	0.99
64 × 64	1.31	1.15
32 × 32	1.42	1.23
16 × 16	1.46	1.23

Memory capacity	0 0 1	1 1 1
1 MB	8.74	8.02
2 MB	8.80	8.09
4 MB	8.90	8.22
8 MB	9.10	8.67
16 MB	8.80	8.64

Table 5. *Rendering time for a 128 × 128 × 128 data set with different viewing directions and different ray group sizes.*

Table 6. *Rendering times for a 256 × 256 × 256 data set with different viewing directions and different amounts of memories.*

6 Conclusion

In this paper, we studied the problem of hiding disk I/O delay associated with large-scale volume data set rendering. We attacked this problem by considering in two steps:

make the rendering as fast as possible assuming the data set is already memory resident; mask the I/O latency as much as possible by taking data loading overhead into account. We tackle the former part of the problem by (1) approximating floating-point computation with integer arithmetic without causing perceptible loss of quality on the generated images; (2) speeding up the address generation for the eight voxels used in tri-linear interpolation by exploiting the fixed relationships among them; and (3) employing M-MX instructions to execute multiple instructions simultaneously. To effectively mask the I/O delay, one has to overlap the disk accesses with rendering computation. Data sets are divided into "sub-blocks" or "macro-voxels" to allow separate rendering and I/O threads to work on different macro-voxels. To hide the disk I/O delay, the prefetch thread should preceed the rendering thread for each macro-voxel accessed. We have developed an innovative data-driven approach to exploit as much parallelism as possible while at the same time reducing unnecessary synchronizations checks to the minimum. By incorporating all these optimizations, given a $128 \times 128 \times 128 \times 1$(bytes) data set, our system is able to render a 128×128 grey-scale image in one second on the average using a Pentium II 300MHz machine. For larger data sets, the rendering time scales proportionally. Moreover, we found our system not only can mask the I/O overheads effectively, but also can perform out-or-core rendering effectively without much modification.

References

1. P. Cao, E. W. Felten, A. Karlin, and K. Li. A study of integrated prefetching and caching strategies. *ACM SIGMETRICS Conference on Measurement and Modeling of Computer Systems*, May 1995.
2. Tzi-Cker Chiueh, Chuan-Kai Yang, Taosong He, H. Pfister, and A. Kaufman. Integrated volume compression and visualization. *Visualization '97*, pages 329–336, October 1997.
3. M. Cox. Managing big data for scientific visualization. *ACM SIGGRAPH '98 Course*, August 1997.
4. M. Cox and D. Ellsworth. Application-controlled demand paging for out-of-core visualization. *Visualization '97*, pages 235–244, October 1997.
5. J. Fowler and R. Yagel. Lossless compression of volume data. In *Proceedings of Visualization '94*, pages 43–50, October 1994.
6. D. Kotz and Carla Schlattr Ellis. Practical prefetching techniques for parallel file systems. *First International Conference on Parallel and Distributed Information Systems*, December 1991.
7. Tulika Mitra, Chuan-Kai Yang, and Tzi-Cker Chiueh. Application-specific file prefetching for multimedia programs. In *IEEE Multimedia 2000*, July 2000.
8. Todd C. Mowry, Monica S. Lam, and Anoop Gupta. Design and evaluation of a compiler algorithm for prefetching. *The Fifth International Conference on Architectural Support for Programming Languages and Operating Systems*, pages 62–73, October 1992.
9. R. H. Patterson, G. Gibson, E. Ginting, D. Stodolsky, and J. Zelenka. Informed prefetching and caching. *15th ACM Symposium on Operating System Principle*, December 1995.
10. A. Trott, R. Moorhead, and J. McGinley. Wavelets applied to lossless compression and progressive transmission of floating point data in 3-d curvilinear grids. *Visualization '96*, pages 385–388, October 1996.
11. S. K. Ueng, K. Siborski, and K. L. Ma. Out-of-core streamline visualization on large unstructured meshes. *ICASE Report*, April 1997.

Interacting with Stock Market Data in a Virtual Environment

Keith Nesbitt

Department of Computer Science and Software Engineering,
University of Sydney, Sydney, NSW. 2006. Australia
knesbitt@cs.newcastle.edu.au

Abstract. Virtual Environment technology enables new styles of user interfaces that provide multi-sensory interactions. For example, interfaces can be designed which immerse the user in a 3D space and provide multi-sensory feedback. Many information spaces are multivariate, large and abstract in nature. It has been a goal of Virtual Environments to widen the human to computer bandwidth and so assist in the interpretation of these spaces by providing models that allow the user to interact 'naturally'. One goal for this interaction may be to uncover useful patterns within the data. This paper describes a Virtual Environment system called the "Workbench" and explains three models of stock market data that have been developed for this environment. The aim of this work is to provide models that allow analysts to explore for new trading patterns in the stock market data. Some early results of this work are discussed.

1 Introduction

Virtual Environment technology provides a new style of human-computer interface, the primary goal of which is to significantly increase the communication bandwidth between human and computer.

Virtual Environments attempt to create a natural way of interacting with computers using the human body and all its senses. In Virtual Environments users do not operate computer applications via an interface, rather people participate, perform tasks and experience activities within a computer generated world. The idea is to immerse a person in an environment that allows natural interaction and participation in order to perform tasks.

Many different types of Virtual Environment systems have been built and the technology has been applied in a wide range of fields [1,2]. This paper begins by describing a "Workbench" environment developed at the German National Research Center for Information Technology (GMD) [3, 4].

Like many modern industries, stock market analysis is characterized by an increase in the size of the data sets available. This data is large and multivariate. Analysts and traders attempt to make profitable trades by determining relationships within the data. The domain of 'Technical Analysis' focuses on market activity to determine the balance of supply and demand for a financial instrument. This assists traders in making assessments on probabilities and risks about likely market directions. Estimating the direction and size of price movements from patterns in the data is useful for trading over various time frames. While many traditional techniques have developed to trade patterns within this data, finding new rules or patterns in stock market data may lead to new and more profitable trading systems. Section 3 of this paper provides an overview of the field of "Technical Analysis".

Virtual Environments like the "Workbench" offer new ways of presenting and exploring abstract data. With appropriate models it is possible to efficiently utilize our human capability for pattern recognition. Section 4 describes three new models that have been developed using traditional stock-market data. These models have been developed and demonstrated on the "Workbench" Virtual Environment. The paper concludes with a discussion of the early results, associated work and future directions.

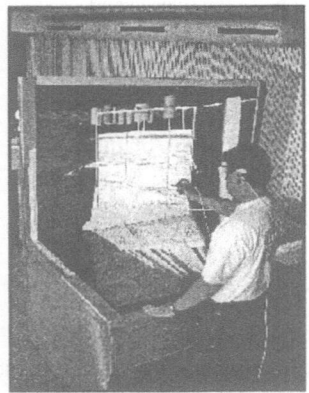

Fig. 1. The Workbench at the German National Research Center for Information Technology. (Photo courtesy of GMD)

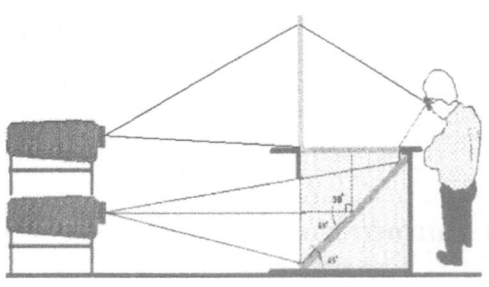

Fig. 2. The layout of an L-shaped Workbench showing how two projected images are combined to create a single displayed model.

2 The Virtual Workbench

The concept of the "Workbench" originated from research by Wolfgang Krüger at the German National Research Center for Information Technology (GMD) [3]. The Workbench enables a user to perceive and interact with a 3-D image that appears to float just above a table (Fig. 1). The computer generated image is projected onto a mirror beneath the workbench, where it is reflected upwards to an horizontal rear-projection screen which forms the table top surface (Fig. 2). The original design used only a single flat projection surface. However, this arrangement allows only models that have minimal height to be displayed. This design was augmented with a second rear-projection screen at the back of the bench. This creates an L-shaped surface and allows models to be displayed in a more vertical orientation. The extra projector provides a further advantage as it also increases the brightness and resolution of the display.

The image (typically 1024 x 768 pixels) for both the horizontal and vertical surface are generated by a high-end Silicon Graphics Onyx workstation, equipped with Infinite Reality Graphics hardware. The workstation also receives information from an electromagnetic tracker unit, which provides position and orientation of the user's head and hands in the Workbench's virtual workspace. To perceive a 3-D stereo image, the user must wear liquid crystal shutter glasses synchronized with the workstation's graphic output. The workstation generates a separate image for each eye, and alternates display of each image in synchronization with the liquid crystal shutters.

The distinguishing features of a workbench versus a traditional workstation based application are that:

1. The user perceives models in three dimensions.
2. The view of the data is controlled by the user's head position. As the user moves his/her viewpoint, the data is displayed as if seen from this position.
3. The user can interact directly with the virtual objects displayed above the tabletop. Models can be selected, rotated, translated and zoomed using virtual tools. These virtual tools are associated with a physical prop such as a pen with a selection button.

3 Technical Analysis

'Technical analysis' is defined as "the study of behavior of market participants, as reflected in price, volume and open interest for a financial market, in order to identify stages in the development of price trends" [5]. Users of Technical Analysis seek to make profitable trades by studying market activity to determine the balance of supply and demand of a financial instrument. This field originated with Dow Theory and has developed to the extent that a number of different techniques now exist to assist with trading across different time periods.

Technical Analysis is sometimes called 'charting' and, as the name suggests, often involves inspection of charts. The charts typically show price on the vertical axis and time on the horizontal. Price for a single time period is shown as a vertical line, or bar, which is drawn from the minimum price to the maximum price for the period. The period bar is augmented with ticks showing opening and closing price (Fig. 3). A time period represented may be a very short period, of he order of minutes, or longer periods such as a day, a week, months or years. The chosen period length reflects the trading strategy, longer periods if the emphasis is on long term trading or shorter periods for trading short-term market trends.

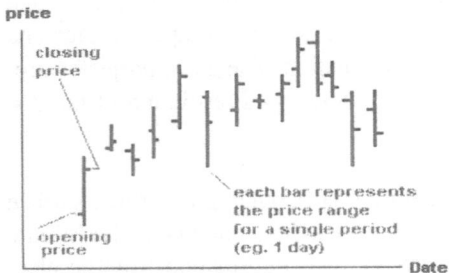

Fig. 3. A traditional daily bar chart.

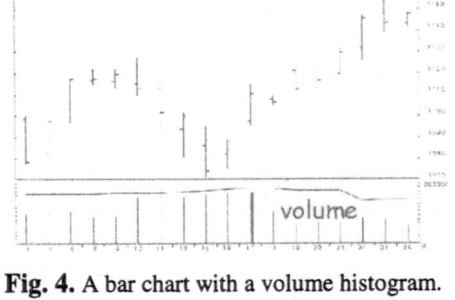

Fig. 4. A bar chart with a volume histogram.

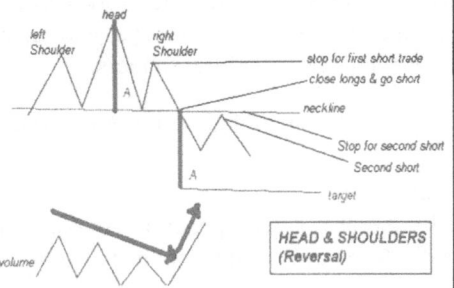

Fig. 5a. Trading strategies for the head and shoulder pattern.

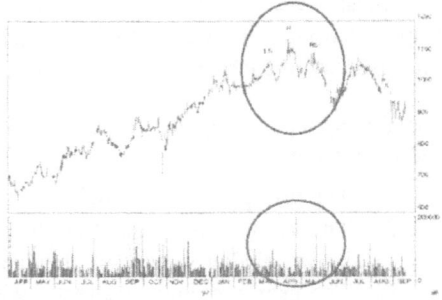

Fig. 5b. A head and shoulders pattern shown in the bar chart of price and volume data.

Charting techniques rely on the well-understood concept of a two-dimensional abstract space to present relationships. Price is used for the vertical axis and time is represented along the horizontal. Choosing time as one axis creates a time series of the data and each data bar can be compared to another in some temporal ordering. Other simple rules and relationships within this space are also understood. The height of the bar represents the variation of price for the period. The price from one period to the next can be compared by the placement of the bar along the vertical axis. Larger structural patterns can also be found such as the upward progression of consecutive bars which represents an up trend in prices. Many other patterns have been identified which characterize turning points of such trends and are useful for the trader to identify [12]. An example of a trend reversal pattern is shown in Fig. 5a and Fig. 5b.

These charts may also be augmented by a volume histogram that shows the volume of trades for each period at the base of the chart (Fig. 4). A number of derived indicators are also used to assist in analysis. These include a curve showing the moving average of closing price for consecutive periods. Moving averages help filter out short term fluctuations in price and provide information about longer-term trends (Fig. 6). Taking the difference in closing price between periods is the basis of another set of indicators. They provide an indication of the relative movement of price and are known as momentum indicators (Fig. 7). Like moving averages they can be derived for a range of different time steps.

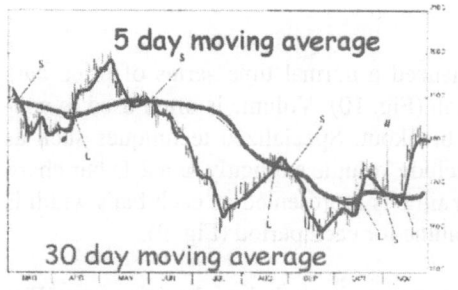

Fig. 6. Daily bar chart with two moving averages.

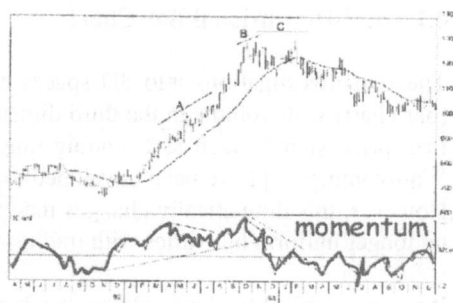

Fig. 7. Daily bar chart with momentum indicator

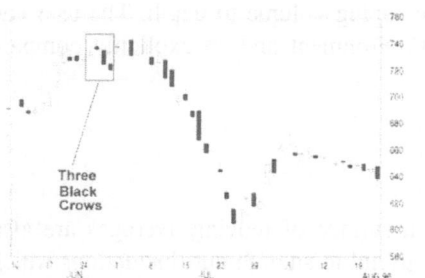

Fig. 8. Candlestick charting techniques use metaphors to describe patterns. Shown here is the "3 black crows".

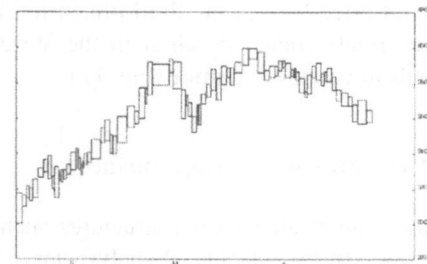

Fig. 9. Equivolume charting where trading volume is represented by width of the price box.

While bar charts are the most frequently used charting techniques, there are also a number of specialised visualizations that have been successfully applied. Candlestick charts [6] are similar to bar charts but were developed independently in Japan. The bars or candles have a 'body' which is defined to be between they open and close price. The candle 'wicks' extend beyond the body to show maximum and minimum price. Black candles indicate price has fallen from open and close of the market. White candles occur when price has risen during the trading period. Metaphors are characteristically used to describe useful patterns such as "dark cloud cover" or "three black crows" (Fig. 8).

4 Interactive Stock Market Models

Three new stock market models have been developed and they are now described. The first is a simple extension of traditional bar charts into three dimensions called the '3-D Bar Chart'. The second model, the 'Moving Average Surface' uses a series of moving averages created from price data to create a surface. The final model allows for real-time monitoring of 'bids' and 'asks' during market trading as is called the 'Depth of Market Landscape'.

4.1 The 3-Dimensional Bar Chart

The first investigations into 3D spaces enhanced a normal time series of price bars (bar chart) with volume in the third dimension (Fig. 10). Volume is often used to confirm price signals such as a trading range breakout. Specialized techniques such as 'Equivolume' [5] have been developed to include volume explicitly in a 2-D bar chart. However, this dramatically changes the way time is represented as each bar's width is no longer uniform but varies with trading volume for each period (Fig. 9).

It is more typical to chart volume as a histogram separately below the price bars (Fig. 4, Fig. 5b). While this allows price to be observed in relation to volume it requires moving the eyes back and forth between volume and price when trying to distinguish a correlation between the two variables of price and volume. It is a simple matter to extend price bars in the third dimension by mapping volume to depth. The user can then simply rotate the chart in the Virtual Environment and so explicitly compare trends in volume and price (Fig. 11).

4.2 The Moving Average Surface

More complicated spatial structures such as a surface of moving averages are also possible. In some technical analysis tasks it is useful to smooth out fluctuations which occur in price at each time step. 'Moving averages' [5] can be used to do this. Closing price is simply averaged over some number of time periods. Some trading systems rely on the intersection of different moving averages to signal the beginning and end of trends. For example a one, fifteen and thirty day moving average may be calculated and plotted. Signals are generated by where these moving average lines cross. The choice of how many time periods to include in the averages can vary, creating somewhat arbitrary signals. It is desirable to analyze a wide range of moving averages to choose appropriate signals for trading a particular instrument. A two-dimensional display of multiple moving averages soon becomes crowded if more than a few curves are plotted. Occlusion makes it hard to distinguish between curves (Fig. 12).

To overcome this problem a surface of moving averages was constructed with traditional price bars positioned at the center of the surface. This surface is constructed by joining together a number of strips. Each strip represents a different moving average curve. These moving average strips are joined to create a continuous surface. The surface is a reflected about the central axis so that each edge represents a moving average of 30 days. As the surface moves towards the central bar chart the number of days in the moving average is reduced, from 30 to 29, to 28 and so on until a 1 day moving average is placed adjacent to the central bar chart (Fig. 13, 14, 15). By definition the closing price on these central price bars corresponds to a one-day moving average.

Signals generated from this view of the data still need to be clarified. However, trading signals are often only treated as indicators for action rather than absolute rules.

What this model provides is another way of looking at an indicator like moving average, allowing an analyst to consider a range of possible values for the parameters involved in it's calculation. After viewing this model it was suggested that the area above or below curve might be useful to consider as a 'new' type of trading indicator.

To assist in the analysis of this 'new' indicator a variation of this moving average landscape was created. Using the same series of moving averages from 1 to 30 days. Price bars are extended in one axis to create boxes that cover the width of the moving average surface (Fig. 16, 17, 18). Looking directly at one edge of the model shows a typical bar chart time series with the edge of the moving average surface seen as a line plotted through the closing price of each bar. By rotating the model the user can see price compared against the edge of the plane which represents the 30-day moving average.

While exploring this model it was found that in an up trend the price bars are predominantly above the plane and in down trends they are predominantly below. This has suggested a new way to evaluate trends by considering the extent that price bars lie above or below the moving average plane.

4.3 The Depth-of-Market Landscape

Traders of financial instruments often have very different trading time frames. Long term traders for example will be have a strategy of entering the market as it begins a primary up trend. This minimizes transaction costs as few trades are made and profits can result from both dividends and from the general upward trend of prices over time. The time frame may be months or years. Short-term traders on the other hand attempt to trade much shorter fluctuations in prices for profit. Here the time frame is of week or days. Both the 3D Bar Chart and the Moving Average Surface extend traditional pattern analysis techniques of charted stock market data. They can be useful for examining trends over different time frames by altering the period of the price bar. Some traders are interested in very short term trading opportunities that may result from fluctuations in market prices over five or 10 minutes.

The 'Depth of Market landscape' was developed to explore the potential of trading opportunities that occur in very short time frames. In particular, the depth-of-market landscape allows the user to explore for new patterns in depth of market data that the short-term trader could exploit for profit.

The *"Depth of Market"* refers to the number of buyers and sellers currently trying to trade a particular financial instrument. A financial instrument may be something like a company share or future contract. The current selling price for an instrument can be considered as a balance between the price buyers will pay and the price sellers will accept. A buyer makes a *"bid"* to purchase a specified volume of an instrument. At the same time sellers try to sell a certain volume of an instrument for which they *"ask"* a particular price. The balance of buyers *"bids"* and sellers *"asks"* determine the state of the current market. Often there may be a difference in the buying and selling price and this difference is known as the *"spread"*.

Current displays of depth of market data usually present the *"bids"* and *"asks"* in a simple table format that orders the list of bids and asks by price. The last selling price is also displayed indicating at what price the last trade was made. The landscape consists of a set of strips at each time step. Each strip has three components that represent a volume and price histogram for *"bids"*, *"asks"* and also *"trades"*. As time changes a new strip is added to represent the depth of market at that time frame (Fig. 19). This makes the depth-of-market landscape a 3-D model of a surface that evolves over time. It is expected that patterns may occur both in the static spatial structure and the evolution of the surface over time.

The landscape has the natural analogy of hills and valleys in the real world. The landscape evolves with time as the balance of buyers and sellers changes. These changes create waves that move on the surface of the landscape and can indicate changing trends in the short-term market. This is better understood if we consider some simple scenarios. Where there is a high volume of both buyers and sellers which is symmetric about the last sale price we expect price to remain fairly static as buyers and seller exchange trades. This may represent a price point about which distribution or accumulation is occurring. If, however, there were a valley between a peak of buyers and a peak of sellers this would indicate a market spread. Over time we could see this evolve into different situations. There could be no change in the market in which case we would expect few trades. If the peak of sellers moves towards the buyers we may expect prices to be driven down. Or alternatively the buyers may move their bids towards the available sellers and this may drive prices upward.

Explorations with this model are still continuing and early results are encouraging, though it requires a very dynamic market - that is with many frequent trades so the landscape can evolve at an 'interesting' rate. Another useful way to exploit this model may be for real-time monitoring of a market.

5 Discussion

Like many new emerging computer technologies much hype and speculation has surrounded the value and application of Virtual Environments. Realistically, everyday use of these environments for applications such as technical analysis is not likely in the short term. High cost, many useability issues and the lack of commercial software make it infeasible for rapid adoption of these environments. A shorter-term possibility is the use of such environments to investigate the discovery of new relationships in abstract data, such as that provided by the stock market. In such cases the potential reward may offset the risks against success.

Metaphors that providing totally new ways of exploring financial data may help reveal patterns that have not previously been understood. This may in turn create new and unique trading opportunities. Many systems have been developed which use algorithmic or heuristic rules for trading the market based on price trends. Once new patterns or rules are discovered the opportunity then exists to incorporate them into such automatic trading systems.

Further work needs to be done in developing and testing these models. It has been shown that a number of new opportunities for interpreting financial data can be provided by Virtual Environments. Early feedback from users confirms that these models are intuitive and easy to understand. A 'new' pattern - the area above and below the moving average surface is indicated for further investigation. The Depth of Market landscape requires further work to improve the look of the model and determine its usefulness. However, early indications from user feedback are encouraging.

6 Conclusion

New opportunities for developing multi-sensory human-based tools have been made possible with new user-interface technologies. Virtual Environments immerse a person in a computer interface that allows natural interaction and participation within that environment to perform tasks. Interaction and perception in the real world is based on the use of multiple senses. We use our eyes and ears, the sense of touch and smell to perform activities. Virtual Environments attempt to mimic interactions in the real world and has seen the development of interfaces that support interaction for many of the human senses. While most activity has centered on three-dimensional visual models, there are also a growing number of applications where auditory and force displays are being used to help in data interpretation.

With multi-sensory interfaces we can potentially perceive and assimilate multivariate information more effectively. The hope is that mapping different attributes of the data to different senses, such as the visual, auditory and haptic (touch) domains will allow large data sets to be better understood. However, multi-sensory interpretation is a very complex field and involves understanding the physiological capabilities of each sense and the perceptual issues of individual and combined sensory interactions. Associated work is looking at extending these stock market models described here to provide multi-sensory feedback [7].

282

To assist in designing more intuitive multi-sensory interactions a classification of natural metaphors has been developed [8]. Associated with this classification are guidelines for integrating these metaphors in a way that best supports the human perceptual capability [9]. The harder question still remains, that is, to experimentally prove that this approach results in 'better' models for human-based data-mining.

7 Acknowledgements

The ideas in this work have resulted from close collaboration with Bernard Orenstein of Agents Incorporated in Sydney. Bernard has provided both support and invaluable expertise in the field of technical analysis. The integration of these models into the 'Workbench' was made possible with the assistance and support of Martin Göbel and Bernd Fröhlich from the Virtual Environment group. This group is part of the Institute for Media Communication (IMK) at the German National Research Center for Information Technology (GMD) [4].

References

1. Durlach, N.I., Mavor, A.S. Virtual Reality. Scientific and Technological Challenges. National Academy Press, Washington, DC. 1996.
2. Stuart, R. The Design of Virtual Environments. McGraw-Hill, New York. 1996. ISBN 0-07-063299-5
3. Krüger, W., Bohn, C., Fröhlich, B., et al. The Responsive Workbench: A Virtual Environment. IEEE Computer, pp.42-48, July, 1995.
4. Internet web site: Institute for Media Communication, GMD, German National Research Center for Information Technology. http://viswiz.gmd.de
5. Technical Analysis : Course Notes from Securities Institute of Australia course in Technical Analysis. (E114), 1999. http://www.securities.edu.au
6. Specialised Techniques in Technical Analysis : Course Notes from Securities Institute of Australia course in technical Analysis (E171), 1999. http://www.securities.edu.au
7. Nesbitt, K. V. and Orenstein B.J. Multisensory Metaphors and Virtual Environments applied to Technical Analysis of Financial Markets. Proceedings of the Advanced Investment Technology, 1999. pp 195-205. ISBN: 0733100171.
8. Nesbitt, K. V. A Classification of Multi-sensory Metaphors for Understanding Abstract Data in a Virtual Environment. Proceedings of IV 2000, London. 2000.
9. Nesbitt, K. V. Designing Multi-sensory Models for Finding Patterns in Stock Market Data. Proceedings of International Conference on Multimodal Interfaces, Beijing. 2000.

Editors' Note: see Appendix, p. 358 for colored figures of this paper

Case Study: Visualization and Information Retrieval Techniques for Network Intrusion Detection

Travis Atkison, Kathleen Pensy, Charles Nicholas,
David Ebert, Rebekah Atkison, Chris Morris

Computer Science and Electrical Engineering Department
University of Maryland, Baltimore County
1000 Hilltop Circle, Baltimore, MD 21250
{atkison, kpensy1, nicholas, ebert, ratkis1, cmorris}@umbc.edu

Abstract. We describe our efforts to analyze network intrusion detection data using information retrieval and visualization tools. By regarding Telnet sessions as documents, which may or may not include attacks, a session that contains a certain type of attack can be used as a query, allowing us to search the data for other instances of that same type of attack. The use of information visualization techniques allows us to quickly and clearly find the attacks and also find similar, potentially new types of attacks.

1   Introduction

The proliferation of the Internet over the last few years has brought many new and improved services to the populace, but with the good, there must be the bad. There has been a new type of crime to hit the information superhighway, *Network Intrusion*. Network intrusion occurs when an unauthorized entity gains access to one or more components of a network.

The motivation behind these experiments was to develop more effective network intrusion detection tools through the combination of information retrieval and information visualization techniques. The goal of our work is to use multi-dimensional visualization to detect attempts, successful or not, at network intrusion.

Our system combines the Telltale information retrieval system and the Stereoscopic Field Analyzer (SFA) information visualization system to create an effective intrusion detection solution. Telltale is a dynamic hypertext environment that provides full-text information retrieval from a text corpus [MILL99, PEAR97]. Telltale computes the similarity between a given document and a query based on the frequencies of n-grams (n character sequences of text). SFA uses glyph-based volume rendering to visualize multi-variant, multidimensional data, enabling more complex data relationships and information attributes to be visualized than in traditional 2D and surface-based visualization systems.

In the next section, we describe our test data, and provide additional details on Telltale and SFA. In Section 3, we describe the two phases of our experiments. In Section 4 we discuss our results. Finally, in Section 5 we present our conclusions and plans for future work.

2 Background

We have explored and evaluated the effectiveness of combining information retrieval (IR) techniques with information visualization techniques as a solution to the Network Intrusion Detection problem. Below, we describe the sample network data set that we have used and the details of the IR and information visualization tools we chose for our experiments.

2.1 Data

The data that was used in our experiments came from the 1998 off-line intrusion detection evaluation (IDEVAL), which was conducted by MIT Lincoln Laboratory under DARPA sponsorship. An intrusion detection evaluation test bed was developed under this program which generated normal traffic similar to that of a U.S. government site containing hundreds of users on thousands of hosts.

The contents of network traffic such as SMTP, HTTP, and FTP file transfers were either statistically similar to live traffic, or sampled from public-domain sources. Telnet sessions were generated from statistical profiles of user types that were used to generate interactive sessions. These statistical profiles indicated the frequency of occurrence of different UNIX commands (e.g. mail, lynx, ls, cd, vi, cc, and man), typical login times and telnet session durations, typical source and destination machines, and other information [LIPP00].

More than 300 instances of 38 different automated attacks were launched against victim UNIX hosts during a simulated nine-week exercise. Attack scenarios were developed for different attackers. For example, one attacker collected information and left a back door; another was a novice hacker who broke in and then left, and a third was a disgruntled employee [CUNN99].

The following attack families were included in the evaluation: *user to root, remote to local, denial of service,* and *probe/surveillance*. A *user to root* attacks occurs when a local user on a machine tries to obtain privileges normally reserved for the UNIX root or super user. In *remote to local* attacks, an attacker who does not have an account on a victim machine sends packets to that machine in order to gain local access. *Denial of service* attacks are designed to disrupt a host or network service. *Probe/surveillance* attacks occur when an unauthorized user scans a network of computers to gather information or find known vulnerabilities [LIPP00], perhaps in order to then launch one of the other attacks. For a more detailed explanation and definition of these families of network attacks, see Kendall's thesis [KENN99].

2.2 SFA

The SFA visualization system is a tool for visualization of multidimensional and volumetric data [EBER96]. SFA combines glyph-based volume rendering with a minimally-immerse interaction metaphor to provide interactive visualization, manipulation, and exploration of multi-variant, volumetric data. SFA uses a glyph's location, 3D size, color, shape and opacity to encode up to nine attributes of scalar data per glyph [EBER97]. Attribute mappings can be changed in real-time, data can be filtered, and subsets can be created, allowing the user flexibility in the display of the data set.

By using glyph-based volume rendering, SFA does not suffer the initial costs of isosurface rendering or voxel-based volume rendering, while still offering the capability of viewing the entire volume. Glyph rendering also allows the simultaneous display of multiple data values per volume location. SFA allows the three-dimensional volumetric visualization, interactive manipulation, navigation, and analysis of multi-variant, time-varying volumetric data, increasing the quantity and clarity of the information conveyed from the visualization system [EBER96]. SFA has been successfully used for both scientific and information visualization tasks. We have previously applied SFA to the information visualization tasks for visualizing document similarities [EBER97] and visualizing document authorship with very successful results.

2.3 Telltale

Telltale is an IR system that provides full-text search in text corpora that may be garbled by OCR or transmission errors, or may contain text written in languages other than English. Unlike most IR systems, Telltale uses n-grams, rather than keywords or phrases. An n-gram is defined as a sequence of n consecutive characters, typically including whitespace, punctuation, and so forth. Two documents (or a document and a query) are considered similar if a sufficiently large number of the same n-grams (more than would be expected due to chance) appear in both documents. There is no notion of stemming, or stopword processing, as in word-based IR systems. As a result, n-gram based IR systems are, in general, less language-specific than other IR systems [PEAR97]. (Typical IR systems reduce the number of terms to be indexed by excluding so-called "stopwords" which appear in virtually every document and therefore have little or no discriminating power. However, the set of stopwords varies from one language to another, and is therefore a source of language dependence.)

The data being analyzed in these experiments is not ordinary natural language text. In fact, the data is drawn from tcpdump output, so there are timestamps, IP addresses, and acronyms in much greater quantity than in ordinary text. One of our main objectives was to see how well an n-gram based IR system would handle such data.

3 Experiments

The IDEVAL data set that was used in all our experiments was initially pared down to a subset that included only Telnet packets, i.e. packets that involved port 23 as either the source port or destination port. The IDEVAL data set consists of seven weeks of TCP traffic for training, and another two weeks of TCP traffic for testing. We used five weeks' worth of the seven weeks of training data, resulting in about three million Telnet packets. (Limits in our database software prevented us from using the remaining weeks of data.)

3.1 Phase 1

Initial experiments on the reduced data set involved the writing of several Perl scripts and analyzing initial processing results. These scripts created histograms on various combinations of attributes of the data. Here we defined attributes of the data to be analogous to columns in a database, e.g. timestamp, protocol, and so forth. Histograms gave us a general feel of the distribution of the data set. The most insight was gained when we used the scripts to create histograms on the combination of source IP address, source port, destination IP address and destination port. From this particular combination we were able to detect a number of *denial-of-service* attacks. This combination proved powerful in that this particular type of attack could be discovered reliably in a wide variety of situations. However, there are denial of service attacks (such as UDP floods, for example) that cannot be detected using simple histograms of tcpdump data, so from a network intrusion detection standpoint the scripts were limited. The numerous other families of network attacks still remained hidden within the corpus.

3.2 Phase 2

The experiments described above gave us a foundation for developing a robust and powerful methodology for detecting network intrusions. The chief insight was that

Telnet sessions could be regarded as documents. As a result, a corpus of tcpdump traffic possibly containing "attack" sessions can be regarded as a corpus of documents, and a session that includes an attack can be regarded as a query.

From the Telnet packets we extracted from the IDEVAL data set, we developed a procedure for reconstructing the Telnet sessions in their entirety. This conversion from packets to sessions involved creating a database to hold our network data and then developing the scripts that would extract and convert the individual sessions. The database consisted of two relations, one for the connections, and the other for the packets. A one-to-many mapping existed between the connection and packet relations. Once the database schemas had been developed, the database was created on a MySQL [MYSQL] database server loaded onto a four-node Beowulf cluster. Our Telnet packet data was then loaded onto the database server. Several Perl scripts were written to extract the sessions from the database and then convert them from their native hexadecimal format to ASCII. This extraction and conversion allowed us to analyze the data using our information retrieval tools.

Using Telltale, we calculated similarity scores based on how similar or dissimilar the sessions (documents) were to the attacks (queries). The IDEVAL data includes a list of the known network intrusion attacks (e.g. ffbconfig, dictionary, portsweep, etc.), times that the particular attacks occurred, as well as the source and destination machines on which the attacks occurred. This known set, or truth set, of network attacks allowed us to create a set of queries with known answers, i.e. we knew which attacks occurred and which sessions were involved. We created five session corpora, where each corpus contained approximately fifty sessions. In each, perhaps five or ten sessions were attacks. We assigned sessions to corpora based on size of session, and timestamp. The size of session was an important attribute because we needed to have sessions within the corpora that were comparable in size to the various attacks. If we had placed sessions distinctly different from the attack in our corpora, our results would have been skewed. Timestamps also played an important role because those sessions shortly before an attack might hold peripheral information that could be useful in detecting the impending attack(s).

We loaded each of the five session corpora into Telltale, one at a time, and used the known attacks as queries. The output from each query was a list of similarity scores, i.e. the similarity between the "attack" query and the various sessions. If, for example, there were three ffbconfig attacks, then we received three lists of scores. If the attack session itself occurred in the corpus, the normalized similarity score for that session was very high.

To visualize the relationship, if any, between attacks, we loaded the similarity score lists into SFA, using each list as its own dimension. The scores for the three ffbconfig attacks, for example, were assigned arbitrarily to the x, y, and z dimensions in SFA. Traditionally, the first three dimensions of a data set are mapped in this way, within the three-dimensional environment. With SFA, other dimensions can be mapped to such parameters as color, size, transparency, shape, and vector components. Had we had more ffbconfig attacks to use as queries, we could have assigned them to any of these six remaining dimensions. Through the SFA system interface we were able to, in real time, map our three ffbconfig attacks to different dimensions. With this system flexibility we gained a better understanding of our session corpora from multiple views of the same data. Changing the data mappings and interactively exploring the visualized data provided easier analysis of the data and enabled pre-attentive visual similarity processing and fast visual clustering.

4 Results

As expected, we found that if an attack occurred in the corpus, then we had no difficulty finding the attack session using that same attack as a query. For example, using a given ffbconfig attack as a query, we were able to find that same attack in the corpus if it was present. The most useful result was that we were also able to spot other ffbconfig attacks.

We also discovered that we were able to spot attacks that were within the same family as the query. For example, we were able to discover an eject attack when we used a ffbconfig attack as the query. These two attacks are variants of a user-to-root network attack, and in fact both are buffer overflow attacks, so the system is useful in detecting "families" of attack types. Our system should also be effective for detecting new attacks based on variants of known attacks.

Figure 1 (see Appendix) shows the results of visualizing three ffbconfig attacks as the queries against one of the session corpora. Notice the cluster of glyphs, each of which corresponds to a session in the corpus, grouped around the origin. These sessions have normalized similarity scores near zero when compared with the attack query. In the far right, front and top corners of Figure 1 we see other glyphs, which are the three ffbconfig attacks themselves. However, the attacks are not in the extreme corners of the display, as one would expect. Glyphs for each attack are attracted to the other axes, pulling each glyph slightly away from its corner.

An attack query of a given type can find itself, and other instances of that same type of attack. Furthermore, we were able to spot attacks of different types within that same family of attacks. For example, a ffbconfig attack can be used to find itself, other instances of ffbconfig attacks, *and* other buffer overflow attacks, such as eject, as shown in Figure 2 (see Appendix). Figure 2 shows the results of using three ffbconfig attacks as the queries against a session corpus. All of the sessions without an attack within this family are clustered at the origin. The remaining session glyphs contain attacks within this family. There are three ffbconfig attacks in the session corpus visible near the extreme right, lower left and upper corners. Somewhat closer to the origin, but still distinct from it, are sessions containing other buffer overflow attacks, such as eject and fdformat. The spatial location of these attack sessions also shows the similarity of each of the different ffbconfig attacks and may help determine from which known ffbconfig attack a new attack was derived. Some of these similarities can be attributed to the possibility that core pieces of programming code used by network attackers to construct these network intrusions are similar. We suspect that the similarity does not stop here; therefore, further experiments along these lines are needed.

5 Conclusion and Future Work

These results support the claim that there are in fact underlying patterns associated with different network attack families, and that these patterns can be detected and visualized. Our experience with visualizing Telnet sessions indicates that displaying attacks in a higher-dimensional space leads to insights that would be harder to come by in a two-dimensional visualization.

Our next step will be to investigate system scalability with respect to attacks that don't take place in Telnet sessions. We also plan to investigate detection of attacks in closer to real time. To do this, we will experiment with methods that add the network sessions to the corpus just after their completion. This improvement will allow system administrators to identify possible attacks by simply looking at the visual output. It may also be possible to show a system administrator results of the form: "with n% probability the following unfinished 'session' is an attack of type y". The use of similarity isosurfaces within the SFA display could be used as a visual cue to show probability of attack sessions. Even if we can't identify a new attack by type, it would be desirable to identify the attack by probable family.

Other experiments that will be preformed are with different policies for aging of sessions from the session corpus. Such a policy is necessary since otherwise we end up with an infinitely large session corpus. If models of certain attacks can be developed over time, it may be that detailed sessions of those types of attacks are no longer needed.

We will also explore the addition of more session details as metadata within the visualization display, and explore the effectiveness of different glyph attributes for conveying important intrusion detection session attributes. We have only explored a small portion of the potential benefit of information visualization for discovering network intrusion attacks and we will continue to refine the visualization process to more effectively highlight intrusions.

290

References

[CUNN99]   R. K. Cunningham, R. P. Lippmann, D. J. Fried, S. L. Garfinkel, I. Graf,
           K. R. Kendall, S. E. Webster, D. Wyschogrod, M. A. Zissman,
           "Evaluating Intrusion Detection Systems without Attacking your
           Friends: The 1998 DARPA Intrusion Detection Evaluation," SANS,
           1999.

[EBER96]   Ebert, D., Shaw, C., Zwa, A., and Starr, C. "Two-handed Interactive
           Stereoscopic Visualization," IEEE Visualization '96 1996.

[EBER97]   Ebert, D, Kukla, J., Shaw, C., Zwa, A., Soboroff, I., and Roberts, DA.,
           "Automatic Shape Interpolation for Glyph-based Information
           Visualization," IEEE Visualization 97 Late Breaking Hot Topics,
           October 1997, Phoenix, AZ.

[KEND99]   K.Kendall, "A Database of Computer Attacks for the Evaluation of
           Intrusion Detection Systems", S. M. Thesis, MIT Department of
           Electrical Engineering and Computer Science, June 1999.

[LIPP00]   Richard P. Lippmann, David J. Fried, Isaac Graf, Joshua W. Haines,
           Kristopher R. Kendall, David McClung, Dan Weber, Seth E. Webster,
           Dan Wyschogrod, Robert K. Cunningham, and Marc A. Zissman,
           "Evaluating Intrusion Detection Systems: The 1998 DARPA Off-Line
           Intrusion Detection Evaluation," in Proceedings of the 2000 DARPA
           Information Survivability Conference and Exposition, 2000, Vol 2.

[MILL99]   Ethan L. Miller, Dan Shen, Junli Liu, Charles Nicholas, and Ting Chen,
           "Techniques for Gigabyte-Scale N-gram Based Information Retrieval on
           Personal Computers," Proceedings of the 1999 International Conference
           on Parallel and Distributed Processing Techniques and Applications
           (PDPTA '99), Las Vegas, NV.

[MYSQL]   "MySQL Reference Manual",
           http://www.mysql.com/documentation/index.html

[PEAR97]   Claudia Pearce and Ethan Miller, "The TELLTALE Dynamic Hypertext
           Environment: Approaches to Scalability," in Advances in Intelligent
           Hypertext, J. Mayfield and C. Nicholas, eds. Lecture Notes in Computer
           Science 1326, Springer-Verlag.

Editors' Note: see Appendix, p. 359 for colored figures of this paper

DDDiver:
3D Interactive Visualization of Entity Relationships

Marc Coomans and Harry Timmermans

Eindhoven University of Technology, Faculty of Architecture, Building and Planning,
Mail station 20, P.O. box 513, 5600 MB Eindhoven, Netherlands
{M.K.D.Coomans, H.J.P.Timmermans}@tue.nl

Abstract. In this paper we present DDDiver, a tool for the interactive visualization and editing of Object-Oriented databases. It was developed to visualize and manipulate large loosely-structured data sets with multiple relation types. This makes the tool especially useful in application areas that involve product data models, design information systems, and semantic networks. DDDiver can visualize such relational data sets in a 3d graph. The layout mechanism used for the graph is not based on a deterministic mathematical algorithm, but on the distinction between a number of relation kinds, and on user interaction. The intuitiveness and quickness of the visualization tool was further improved by adding animated visual feedback effects.

1 Introduction

Information visualization is one of the relatively new areas of research and development in computer science. The visualization of large, abstract data structures is often regarded as one of the crucial tasks in bringing computers closer to the general public [1]. Unfortunately, many commercial software applications that deal with complex relational data sets do not offer the appropriate visualizations, with which a user can easily observe the true structure of the data at hand. These poor visualizations are the result of both a lack of graph drawing knowledge in the field, and of the lack of graph drawing support in software development tools.

Some specialized systems have come to the fore in the last years which can visualize large complex data networks very well. Some examples of such systems are the H3Viewer[2], NicheWorks[3], and the LaTour system[4]. These are very useful tools for the visualization of web site structures, organizational diagrams, etc. Unfortunately they are much less useful for the visualization of product data models and semantic networks, since they focus on data sets in which all relations are of a single kind.

In the domains of product data modeling and knowledge modeling, relational data sets are constructed with several kinds of relations. These data sets usually also lack any rigid internal structure (e.g. no tree structure), and they can be subject to cyclic relationships. The systems with which these complex databases are developed, typically offer only limited visualization facilities like tables and treeview visualizations (showing only a single relation type at a time). It is expected that the user interfaces of

292

these applications would considerably be improved if a more suitable visualization technique was implemented.

The DDDiver system is an experimental visualization system for the manipulation of Object-Oriented databases that are non-homogeneous in terms of both objects and relations. The occurrence of multiple relation types was the starting-point for its visualization method. In the next section, we first give a brief overview of the existing visualization techniques. In section 3, we discuss the visualization method offered by DDDiver, and consecutively in section 4, we present a practical application of DDDiver in a CAD system.

2 Existing Techniques

In software applications, relational data sets are conventionally displayed either in table format or in 2D graphical layouts. Tables can display lists of object-relation-object sets in a very efficient way and are very readable at the same time (figure 1a). Manipulation of data in relation tables is typically done by editing names and tags in the table fields; either manually or by selecting possible values from dropdown menus. The disadvantage of tables is that they do not very well support the discovery

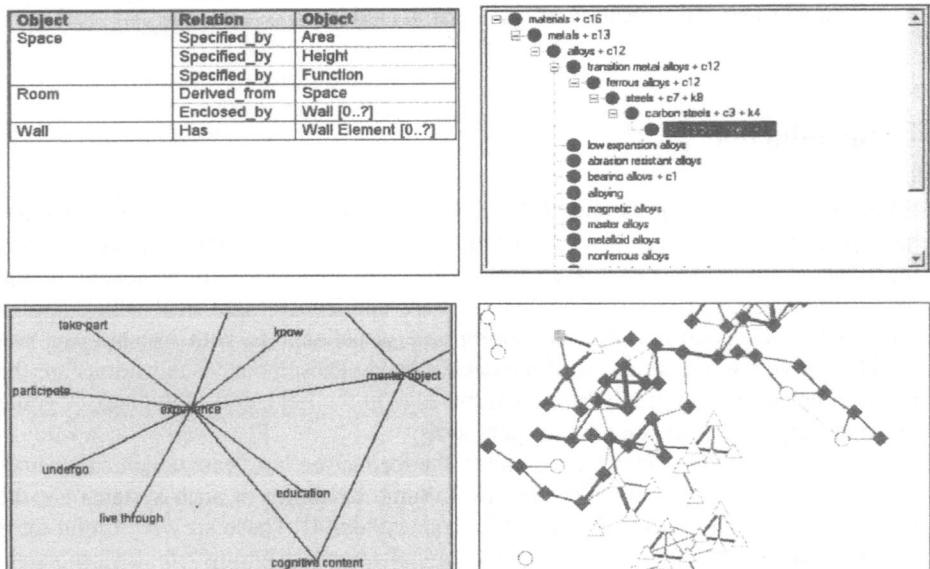

Fig. 1. Existing visualization methods for large relational data sets: a. relation table (top-left), b. treeview (top right), c. graph layouts with tags require lots of screen space (bottom left), d. large graphs are hard to visualize in an aesthetically appealing way (bottom right)

of implicit data structuring. Graphical layouts do provide a much better insight in the structure of the data sets. Object clustering and relation sequences can much easier be denoted.

The most used graphical layout in applications is the "treeview" (figure 1b). The treeview is well suited for hierarchical data sets. Its screen efficiency is close to that of tables. Non-hierarchical (cross-linked) data sets can best be visualized in graph layouts. One disadvantages of graphs layouts is that they are often less efficient in screen area usage, especially when nodes and edges have tags (figure 1c). Another is that cross-linked graphs tend to become messy and difficult to read when the number of objects and relations gets large (figure 1d).

The construction of aesthetically appealing graphs is addressed by a specific discipline of the mathematics and computer science community, known as "Graph Drawing" [5]. In spite of all the results achieved in this discipline, practical applications of large graph visualizations remain difficult. Most of the graph drawing techniques assume that the complete set of nodes and relations can reasonably be represented in a readable and understandable manner on the display medium [6]. Real-life applications deal with databases that are much too big to be displayed at once.

The visualization of real-life data sets requires interactive visualization techniques. At each moment only a small part of the total data set is displayed. The user is provided with navigation tools with which, in subsequent steps, details can be explored and unnecessary parts can be hidden.

When the classical graph drawing algorithms are used in practical applications, they are confronted with large data sets. In [4], the authors pointed out the 3 problems that often become problematic in these practical applications: (1) low speed due to the computational complexity, (2) lack of lay-out predictability for the user, (3) no navigation method supported.

Only a small number of visualization techniques have been presented that do not suffer from these problems. Graham [7] and Eades [6] presented two of these rare exceptions. Both have proposed a 2D drawing technique in which child nodes are located around parent nodes on respectively circles and circular wedges. In both cases, the spatial distribution of the child nodes around their parent depends on the context (number of the child nodes' children, ...), and on the browsing history (which "uncle" and "nephew" nodes that have been looked at before).

Another restriction of the classical graph layout algorithms is that they are designed for networks that are homogeneous in both nodes and relations. Edge labeling is typically not taken into account either. Unfortunately, Graham's and Eades' layout mechanisms suffer from these restrictions as well.

3   The DDDiver Drawing Technique

The DDDiver system was developed as a visualization tool for product data models, design information databases, and semantic networks. In product data modeling, a multitude of product characteristics is collected in a single product database. The assembled information typically relates to multiple views on the product (technological, functional, ...) and/or to multiple life-time stages of the product. The multitude of characteristics and information types is usually structured by the object orientation paradigm [8]. Besides the generalization/specialization relationship, product models use the aggregation relations and the association relations. Semantic networks are widely used to represent structured knowledge. The nodes in a semantic network are concepts of a specific knowledge domain. The relations between the concepts are also formalized, typically making use of the following relation types: equivalence relations, hierarchic relations, scope notes, and associations.

In each of these application domains, object oriented databases are constructed that make use of a small number of relationship types with different semantics. This semantic difference requires these relations to be visualized in a clearly distinctive way. Further, nodes are meaningless without labeling. In product design modeling, edges also need labeling (on top of the visualisation of their relationship type).

One possible visualization solution would be to use a standard form of graph drawing, e.g. a spring-embedder model, and add labels on nodes and edges, and polish it up with color and shape to distinguish object and relation types. However, for the applications we have in mind, relation type differences are very important. Therefore, such a simple color or line-shape polishing is not sufficient. We want the user to be well aware of the semantic difference between browsing into, i.e., an object's specification list and browsing into that object's component list. Relation type differences are emphasized much better when they are reflected in the spatial layout, and not only in color and linetypes.

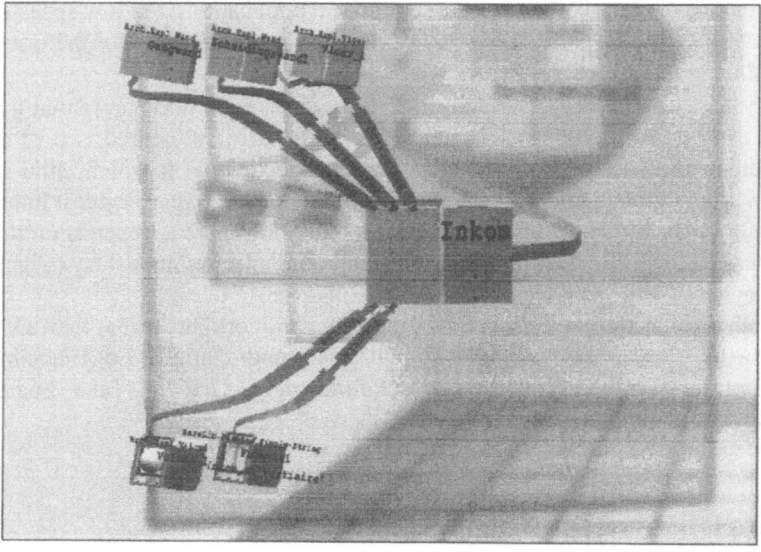

Fig. 2. The visualization of relation types in DDDiver, applied in a CAD system. The central object represents a room "Inkom". It has 2 specification-relations (directed to the bottom-left) that specify the volume and the function of the room. It also has 3 association-relations (directed to the top) with 3 other building components. Finally, it has a single aggregation-relation (horizontal link) with an object in the back (which is the building object).

In DDDiver, the relations of a node are grouped by type, and each group is given a characteristic direction. The recognizability of a relation's type is further improved by distinct arrow styles and colors. The relations themselves are straight by default, but bent into S-like shapes when the connected nodes are not in line. When bent, the relation-end-arrows preserve the characteristic direction and the rest of the relation forms a fluidly connecting spline. See figure 2.

In DDDiver, both relations (edges) and objects (nodes) can have multiple tags. The visualization of a single object with all direct relations can therefore already consume a rather big screen area. Nevertheless, we wanted to be able to visualize a maximum of context for any node in focus. We developed a network browsing technique using multiple transparent layers. Lieberman demonstrated how multiple translucent layers can be used to provide both context and focus views on huge maps [9]. We adapted Lieberman's principle towards a browsing mechanism for graphs. Unlike Lieberman's application, DDDiver visualizes the transparent layers as 3D shells in 3D workspace. Shells popup in front of a 3D cubic box that represent the database. Data objects retrieved from the database are always located on one of the shells.

Each shell can be divided into multiple fields. Each field can host a single data object and its directly related objects. (See figure 3.) When the user asks to see additional data of a related object, this object first moves to a new empty field of its own. This new field is by default located on an empty layer that is laid on top. On this new layer, the additional data objects are displayed around the central object that the user pointed out. The relations between the objects on the different layers stay intact through the spline deformation of the relation-bands, as explained above.

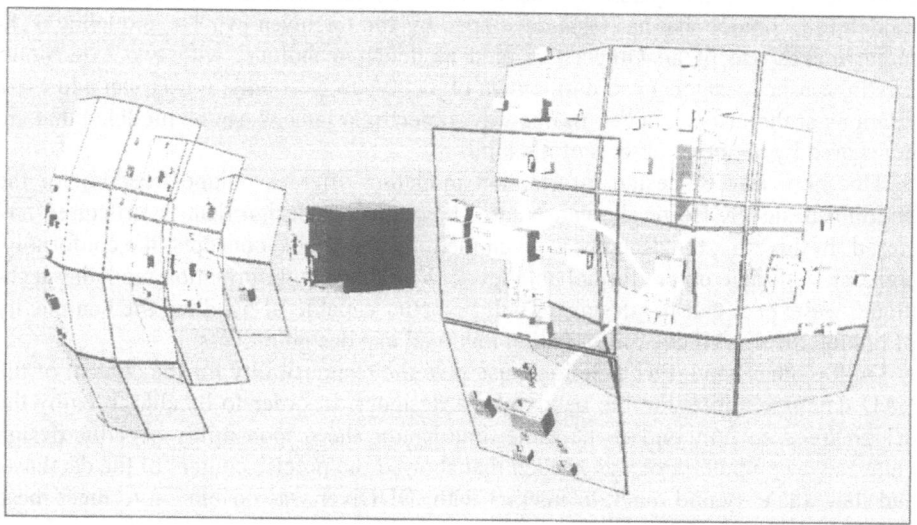

Fig. 3. DDDiver's transparant layers in front of the database box. In this example spherical layers were used with an increasing number of data fields

After multiple browsing steps, a pile of layers is formed. When a new layer is added, existing layers automatically move one step backwards. The bottom layers are automatically cleaned up and removed when the maximum number of layers is reached. We found that 5 to 6 layers is the maximum number of layers that is practically useful.

The top layer always shows the lastly retrieved data objects; layers underneath contain context data. When the user wants to return to data on a lower level, the top layers and the data on it can be removed again. The layers can also individually be moved side- and upward in a side by side position. In our approach, the order of the layers can not be changed.

DDDiver can be configured to create both planar and spherical layers (Figure 3 shows an example of spherical layers). The two layer shapes have corresponding manipulation styles: respectively sliding and rotating.

The number of field subdivisions can be chosen in function of the available screen area. The maximum number of usable fields is mainly restricted by the readability of the text labels on relations and objects. On a screen resolution of 1280x1024 pixels, we experienced that a subdivision of 3 by 2 fields is the practical maximum when the links are labeled. For unlabeled links, the numbers can be doubled. The use of super-scene anti-aliased rendering would also reduce the required screen area per field.

4 A CAD application

DDDiver is implemented as an application-independent tool-set for database visualization. However, the original specifications were formulated on the basis of the needs for a new Computer Aided Design (CAD) system for the building and construction industrie. This CAD system is characterized by an innovative design-information modeling technique that has been developed by van Leeuwen [8]. The modeling technique makes use of an Object Oriented modeling technique, with two extensions. Firstly, a user (designer) can define own object types. Secondly, a user can add extra relations at the instance level. In this way, object relationsips can be modeled that are not shared by the other objects of it's kind.

This new way of design information modeling offers a unique freedom for the building designer. He or she can control how abstract design data is structured and stored. In this way, the designer is given the power to model concepts like conformity, contrast, and scale on the formal data level. With this new information modeling technique, we expect that the designers will be better capable of handling the complexity of linking diverse kinds of information involved in a design process.

On the other hand, this technique also puts the responsibility for the content of the CAD database entirely in the hands of the designer. In order to be able to enjoy the design freedom fully and at the same time handle the responsibility over the design database, a computer tool was needed that showed the precise content of the database, and that was easy and quick to interact with. DDDiver was developed to meet these demands.

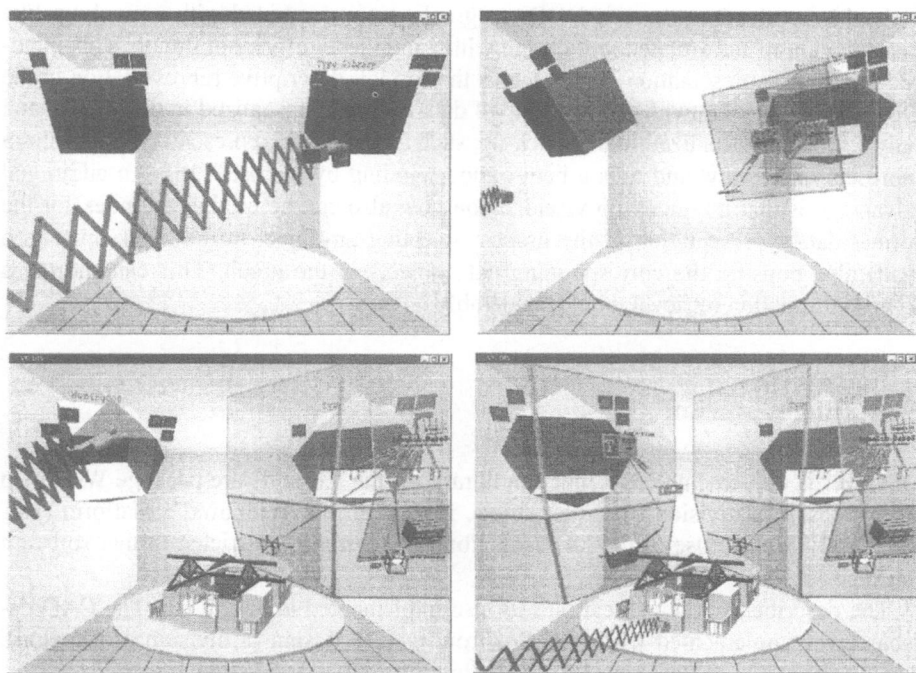

Fig. 4. In the CAD application, architects and construction engineers can directly interact with the desing database through DDDiver. Two DDDiver graphs were combined: one for the object type library (right side), a second for the object instances (left side). DDDiver's direct database visualization was further mixed with a mock-up style visualization of the design

The CAD system's database is made up of 2 parts: (1) the *type library* containing user customizable type definitions, and (2) a number of *desing models* that each describe a particular building design. A single type definition (in the type library) is small directed cyclic construct on itself, that can be composed out of 3 relation types (association, specification, component) and 9 basic node types. The design models contain the instantiations of these small constructs and form large directed, cyclic networks. These large design models can contain several thousands of nodes.

DDDiver was configured to display the type library and the instance database in 2 separate graphs that are positioned side-by-side; see figure 4, and color plate in the appendix. When looking for existing information, the browsing process works independent in both graphs. However, to instantiate objects, type nodes can be dragged from the type-graph to the instance-graph. The instantiating process is visualized as a duplication process after which the type node automatically returns to the type-graph, and the instance remains in the instance graph. To create a new user type, the inverse action can be performed: a set of instance nodes that are thought of as a prototype for the new type definition are dragged from the instance graph to the type graph. These and other user actions benefit from the built in 3D animation effects as these more clearly visualize the effects of a user's action.

In this application, the two DDDiver graphs were combined with a mock-up like visualization of the architectural design with a high level of verisimilitude. This pictorial mock-up representation complements the formal, descriptive representation in the DDDiver graphs. Notwithstanding that all data is already visualized in the graphs, and only a subset is visualized in the mock-up, such a mixed data representation provides a more complete view and thus a better understanding by the user [10]. An additional advantage is that the mock-up visualization now also can be used as an index for the formal data representation in the graphs: selecting an object in the mock-up (e.g. a wall) also pops-up the corresponding data objects in the graph. This can shorten a user's information retrieval time considerably.

5 Implementation

DDDiver is implemented as a function library for the VR software package World Up from Sense8 (a division of Unigraphics). It runs on the Windows™ platform (NT, 2000 and 98). The use of the DDDiver library is currently restricted to that software package.

The described CAD application was also implemented in World Up. The CAD database was implemented with PSE Pro from Object Design (a division of eXcelon). The database is accessed from World Up through Active-X functionality.

DDDiver requires a PC with 3D accelerated graphics board with OGL support. It runs smoothly on a recent PC with a Geforce™ graphics board.

6 Conclusions

We presented DDDiver, a tool for the visualization of relational databases that are used for product data modeling, design information databases, and semantic networks. Unlike other graph visualization tools, DDDiver explicitly supports multiple relation types and object types, as well as labeling for both relations and objects.

An interactive data exploration mechanism was developed which makes use of multiple transparent layers to maintain data context. The whole is worked out as an integrated 3D interactive visualization.

Unlike other graph drawing approaches, the visualization in DDDiver not based on a computationally insensitive lay-out algorithm. As a result, the system's performance is only a linear function of the size of the displayed database part; it is independent of the total database size.

Within layers fields, one step object-to-object relationships are visualized purely on the basis of the underlying database structure, notably the relation type and relation index. As a result, the graphical layout is predictable and thus easy to interpret by the user. This predictability is partially lost after subsequent navigation steps, because the data-objects are then spread out over multiple layers. It can be argued that as a result, the user will not be able to build a mental map of the overall database structure. In the

VR-DIS application, we showed how a second data representation can be integrated to overcome this problem.

Acknowledgement

This project has been conducted in the context of the VR-DIS research of the Eindhoven University of Technology. More details can be found at the VR-DIS web site [11].

References

1. Card, S.K., Mackinlay, J.D., Scheiderman, B. (eds.): Readings in Information Visualization, Morgan Kaufmann Publishers (1999)
2. Munzner, T., "Drawing Large Graphs with H3Viewer and Site Manager", In: Proceedings of the 6th International Symposium on Graph Drawing GD'98, Montréal, August 1998, Whitesides S. (Ed.), p. 384-393, Springer-Verlag (1998)
3. Wills, G.J.: "Niche Works – Interactive Visualization of Very Large Graphs", In: Proceedings of the Symposium on Graph Drawing GD'97, Springer-Verlag (1997)
4. Herman I., Melancon G., de Ruiter M., Delest, M, "Latour – A Tree Visualization System", In: Proceedings of the 7th International Symposium on Graph Drawing GD'99, September 1999, Kratochvil, J. (ed.), p. 392-399, Springer, Berlin (1999)
5. Graph drawing, GD : yearly international symposium 1992 -1999; proceedings as of 1995 by Springer-Verlag, Berlin
6. Eades, P., F. Cohen and M.L. Huang (1997) Online Animated Graph Drawing for Web Navigation. G. Goos, J. Hartmanis and J. van Leeuwen (eds.) 1997. Graph drawing, Proceedings of the 5th International Symposium on Graph Drawing, GD'97, held in Rome, Italy, Sept. 18-20, 1997. Springer-Verlag, Berlin, pp. 330-335.
7. Graham, J.W. (1997) NicheWorks – Interactive Visualization of Very Large Graphs, G. Goos, J. Hartmanis and J. van Leeuwen (eds.) 1997. Graph drawing, Proceedings of the 5th International Symposium, GD'97, held in Rome, Italy, Sept. 18-20, 1997. Springer-Verlag, Berlin, pp. 403-414.
8. Leeuwen, J. van, Modelling architectural Design Information by Features, Eindhoven University of Technology, Eindhoven (1999)
9. Lieberman H. (1994) Powers of Ten Thousand: Navigating in Large Information Spaces. Proceedings of the ACM Symposium on User Interface Software and Technology, 1994. p.15-16
10. M.K.D. Coomans and H.H. Achten (1998) Mixed Task Domain Representation in VR-DIS. Proceedings of the 3rd Asia-Pacific Conference on Human Computer Interaction, APCHI'98, Shonan village Center, Japan, July 15-17, 1998.
11. B. de Vries, "VR-DIS research program", on the web site of the Design Systems group, http://www.ds.arch/ , 1997

Editors' Note: see Appendix, p. 360 for colored figure of this paper

Visualization of Thermal Flows in an Automotive Cabin with Volume Rendering Method

Kenji Ono[1], Hideki Matsumoto[2], and Ryutaro Himeno[3]

[1] Vehicle Research Laboratory, Nissan Research Center,
1, Natsushima-cho, Yokosuka-shi, 237-8523, Japan
kj-ono@mail.nissan.co.jp
[2] NEC Informatec Systems,
3-2-1, Sakato, Takatsu-ku, Kawasaki-shi, 213-0012, Japan
matumoto@ssd.nis.nec.co.jp
[3] The Institute of Physical and Chemical Research,
2-1, Hirosawa, Wako-shi, 351-0198, Japan
himeno@postman.riken.go.jp

Abstract. A predictive system of thermal flow with quick turnaround time in a passenger compartment has been developed. An efficient method based on the Cartesian mesh system was used to reduce the period of analysis. The computed temperature in an automotive cabin was visualized by volume rendering techniques using an *RVSLIB* software library developed by NEC. Consecutive images of the flow were converted into MPEG1 movies, which gave us an overall understanding of the flow. The visualization results indicate that the present system is capable to sufficiently predict the thermal environment in a vehicle cabin at early stage of vehicle development.

1 Introduction

Thermal environment in a passenger compartment is an important issue in the interior design for the comfort of passengers, the reduction of stress of a driver, and the visibility. The thermal environment in a cabin is strongly influenced by sunlight, radiation of heat and ventilating flow from HVAC unit. Many researchers have studied the thermal flows in passenger compartment [1-6]. Generally, the exterior and the interior design that affect the thermal environment are determined at the early stage of vehicle design and therefore, it is important to predict the temperature distribution at this stage. Also, from the viewpoints of both the reduction of the development cost and the improvement of the efficiency, application of simulation at the early stage is of great significance. On the other hand, the vehicle shape at the early stage of vehicle development is neither accurate nor detailed. Moreover, the design may be changed every day. Thus, application of the simulation at the early stage can even dominate the direction of the vehicle development instead of trial and errors. The essential issues required for the analysis are the short turnaround time and the accuracy of the computed results. In this respect, it is more desirable of qualitative judgment in short period other than high accuracy for the prediction of the vehicle performance. For actual flow applications, there is no choice of commercial packages available because the mesh generation is very much time-consuming due to the geometric complexity. For instance, in a recently published paper that reported the thermal flow prediction with realistic configuration [4], the mesh system was generated for the whole shape, from an air duct system to a passenger compartment with four occupants, resulting in about 5 million hexahedra meshes. It was said that it had taken approximately 3 months to get the whole mesh systems completed. In order to overcome the problem of mesh generation, the authors have developed a flow solver and a mesh generation system, based on the Cartesian mesh method, which is capable to perform automatic operation [7]. This system was applied to the prediction of the thermal flow in a vehicle cabin. Concerning the visualization, there are many techniques such as streamline, contour line and so forth. In order to express the temperature distribution in a three-dimension manner, the spatial distribution is represented by combining several cross-sections, which depend on conjecture. Iso-surface technique may be used, however the information of the depth direction disappears behind the closest surface. A volume rendering method can represent the depth information as like cloud. And this method has advantage to provide intuitive image of the flow structure. Although the volume rendering is a sophisticated technique, it is hardly used for the actual design because of its complicated operation and extra computation time. Fortunately, we have a highly vectorized volume rendering library [8]. In this paper we report some preliminary test cases of the volume rendering visualization and furthermore discuss its computational performance in applying the rapid simulation to evaluation of the temperature distribution in a car cabin.

2 Voxel Modeling Method

Examining the thermal flow in a cabin need to consider the location and the direction of the ventilating opening as well as the distribution of the airflow rate for each opening. Additionally, it is important to study the temperature distribution around both driver and other occupants. The complicated shape should be treated in a realistic manner. Unfortunately, we still have no method that can generate the mesh systems for the complex shape efficiently. Generally, it needs tremendous efforts and hence is very much time-consuming. In order to largely shorten the period of analysis, a voxel modeling method is hereby employed to reduce the time in mesh generation. The voxel method approximates the shapes of objects in a step form and has only information on whether the cell is included in the objects or not [9]. The shape of the object is projected onto the Cartesian mesh through a simple judgment, i.e., whether the surface data that represent the shape cut the mesh lines or not. Therefore, mesh can be generated very quickly. Another important advantage is stability. The surface data of a vehicle are not clean. For example, some data are not interconnected, representing unclosed surfaces that should be closed. In the Cartesian mesh approach, shapes under the sub-cell are ignored. This feature acts as a filter for stability and leads to robustness, making it easy to generate mesh automatically. The flow around the approximated shape is solved instead of the original shape. On the other hands, this modeling method has disadvantage that the voxel method can never represent the accurate shapes of the objects because of the approximation by means of cubes. However, it is useful to know the outline of the flow fields in the case with little influence of the flow separation. This method enables us to investigate the flow quickly at the first stage of the vehicle design. Fig. 1 shows the geometry of a production car model used in this study. The shape in the cabin is represented by many polygons as shown in Fig 2. Projecting the polygons onto the Cartesian mesh we obtain the voxel model as illustrated in Fig. 3. This voxel model has a mesh size of 10mm and the total number of meshes is about 8.4 million cells ($171 \times 349 \times 141$). In-house software is used for the generation of the voxel model. All processes of this mesh generation procedure are automated because of their simple nature. The measured time of the conversion from 10 million polygons to 8.4 million cells is just 60 seconds on a PC with an Intel Pentium3 600MHz CPU.

3   Computational Method and Visualization Method

There are many factors that influence the thermal environment in the cabin such as sunlight, radiation and airflow from ventilating opening and so forth. In this paper, the thermal flow only from the ventilating opening is considered because it is the only controllable factor and most significant. The effect of the buoyancy caused by the temperature difference is negligible. A three-dimensional unsteady in-compressible viscous flow is assumed in this study. The governing equations in non-dimensional conservation form are the Navier-Stokes equation, the continuity equation and the energy equation like the convection-diffusion equation of the passive scalar, which is derived from the incompressible assumption. These equations are discretized by a finite volume method on a staggered mesh system. A second-order-accurate QUICK scheme is used for the convection terms and other spatial terms are discretized by central differencing manner [9]. An Euler explicit scheme is employed for the time marching method and a fractional step method [10] is used for the coupling procedure between pressure and velocity fields. The outflow boundary condition is imposed on the outer boundary of the computational domain. At wall, no-slip condition, the Neumann condition and an adiabatic condition are used for velocity, pressure and temperature, respectively. At the boundary of the ventilating opening, both velocity and temperature are given. Fortran77 and C that provide the function of the dynamic memory allocation and an efficient file I/O interface write the program. The vector ratio of the code is over 99.6% and the performances are shown as in Table 1. The measured performance is about 1.8GFLOPS on an NEC SX4 and 3.7GFLOPS on an NEC SX5. The memory size is about 666MB(single precision) for the flow and temperature calculation using the voxel model as in Fig. 3 because this program use 26 words per cell. And if we use the *RVSLIB*, additional 1,038MB will be requested in a double precision format. The flow fields were visualized using the *RVSLIB* developed by NEC [8]. The *RVSLIB* works on network environment as the server-client type application, i.e. the server runs on a high-performance computer and the clients are operated on the user PC. Functions of stream-line, pressure distribution, contour line, and volume rendering are provided by a software library on the server. For users what to do is only to call the library interface from the user program. This library visualizes the flow at the same time and generates an image as a result. This is the reason why this visualization system is suitable for the visualization with large-scale data because the data size is almost independent of the data scale. Further information about the *RVSLIB* is available in the literatures [8, 11]. Firstly, the voxel model is assembled on the local PC using the surface data based on the CAD system, and the resultant file of the voxel model is compressed by GZIP. The compressed file becomes rather small, for example, the size of the voxel model of Fig. 3 is only 217 kB. Then, the compressed files are sent to the remote server machine via the Internet. Secondly, the flow computation and the visualization are performed on the server. After the calculation, the stored time sequential images are converted into an MPEG1 movie file. Then, the movie file is transmitted to the local PC and the flow behavior is observed. For the volume rendering visualization, it is required that the set up of a frame, a color map, an intensity of the permeability of the volume and so forth should be generated in advance. Moreover, the parameter survey must be given in order to represent the characteristics of the flow fields. In this respect, the volume rendering visualization needs twice runs in the case of the first try for the target flow.

4   Results and Discussion

Before we discuss about the temperature field, the computed accuracy should be demonstrated. As the flow considered in this paper is incompressible, the passive scalar is strongly influenced by the velocity distribution. The time-averaged distribution of the flow velocity was compared in a preliminary computation as shown in Fig. 4. The flow was blowing from a duct under the front seat in this calculation. The velocity profile is measured just in front of the rear seat. It was found that the peak of the jet flow is well simulated and the overall agreement is very good. Next, we studied the changes of the flow fields and the temperature fields in the cabin according to the ventilating velocity. The interested parameter is the velocity at four ventilating openings on the instrument panel as shown in Fig. 5. The velocities at the openings were changed for the comparison of the flow field. The time dependent temperature fields were examined in this simulation. The influence of the sunlight and the radiation is not considered but only the influence of the convection from the ventilating openings. The Reynolds number of the computed flow is about 2×10^4 and the Pecret number is about 1.44×10^4. The flow fields are scaled by a reference length $L_0 = 0.07$ (m) and a reference velocity $U_0 = 5$ (m/s). Two cases in Table 2 were computed until 15 seconds in real time and the results were compared. In the case A, the flow velocities at the center ventilating openings are high while the velocities of the both sides are higher than others in the case B. The initial temperature in the cabin is 45 degrees in Celsius and the airflow tempera-ture of four ventilating openings is assumed 15 degrees. This situation is for summer cool-down case. The computed rendering images are demonstrated from Fig. 6 to 11. The color shows the temperature distribution in those images. The color for the highest temperature, i.e. 45 degrees, is transparent. If the temperature is slightly low, the color becomes red. On the contrary, the color for the lowest temperature, i.e. 15 degrees, is blue. The time evolution of the temperature distribution for case A is shown in Figs. 6 - 8 and case B is shown in Figs. 9 - 11. These snapshots show that the temperature around the left person at the rear seat was quite different from each other. From the movie shown in Fig. 11, it was observed that the flow circulates from outside to inside in the cabin. Finally, the turnaround time of the analysis period is referred. The computation time for the flow calculation is about 36 hours on the SX5 (2CPUs) including the generation of the visualized image. In the present case, the image was written to the file every 30 steps during the flow computation. The ratio of the computation time and the visualization time was about 2:1. The time to make the movie is about 30 minutes on an SGI Onyx2 reality monster 16CPUs but it may use only one CPU for its own program *dm-convert*. The file size of the movie with rather high quality is ranging from 6 to 9 MB, it depends on the rendered image. The transfer time of the movie via Internet is about 10 minutes. Consequently, the total time for this analysis is less than 40 hours because the assemble time of the voxel model is very short. This turn-around time is short enough for the educational purpose although we still have many parameters for the thermal flow computation in the passenger compartment. Further reduction of the computation time may be desired.

5 Concluding Remarks

This paper describes a method of thermal flow computation with a voxel model so as to reduce the turnaround time of the analysis. Volume rendering techniques are employed by using the *RVSLIB* visualization system. The present method was applied to compute thermal flows in a passenger compartment to investigate the thermal environment in the cabin. The results indicate that the present visualization system has capability to reasonably predict the thermal environment in vehicle cabin at early stage of vehicle development where the shortening turnaround time is greatly required. For further improvement of the present system, the following issues may need to be resolved:

- The set-up procedure is still complicated and simpler interface is expected.
- Current version of *RVSLIB* requests large memory space.
- Parallel computation of the volume rendering is attractive feature.

This system will be used in a vehicle design in the near future in our company and be deployed by the end of 2001.

References

1. Hara, J., Fujitani, K., and Kuwahara, K.: Computer Simulation of Passenger Compartment Airflow, SAE paper 881749(1988)
2. Lin, C.H., et al.: An Experimental and Computational Study of Cooling in a Simplified GM-10 Pas senger Compartment, SAE paper 910216(1991)
3. Currle, J.: Numerical Simulation of the Flow in a Passenger Compartment and Evaluation of Thermal Comfort of the Occupants, SAE paper 970529(1997)
4. Currle, J., et al.: Evaluation of the HVAC System of Passenger Cars and Prediction of the Microclimate in the Passenger Compartment by Application of Numerical Flow Analysis, SAE paper 971788(1997)
5. Currle, J and Maue, J.: Numerical Study of the Influence of Air Vent Area and Air Mass Flux of the Thermal Comfort of Car Occupants, SAE paper 2000-01-0980(2000)
6. Aronson, D., et al.: Comparison Between CFD and PIV Measurements in a Passenger Compartment, SAE paper 2000-01-0977(2000)
7. Ono, K., Fujitani, K., and Fujita, H: Applications of CFD Using Voxel Modeling to Vehicle Development, Proc. of the 3rd ASME/JSME Joint Fluids Engineering Conference, FEDSM99-7323 (1999)
8. Muramatsu, K., Matsumoto, H., Takei T. and Doi, S.: A Real-Time Visualization System for Computational Fluid Dynamics on Parallel Computers, Proc. Parallel CFD '98, Elsevier Science(1998)
9. Ono, K., Akabane, K., Shiozawa, H., Fujitani, K.: Prediction of Cooling Flow Rate through the Front Grille Using Flow Analysis with a Multi-Level Mesh System, FISITA World Automotive Congress, F2000H201 (2000)
10. Chorin, A.J.: Numerical Solution of the Navier-Stokes Equations, Math. Comput. Vol. 22 (1968) 745–762
11. Takei, T., Matsumoto, H., Muramatsu, K., Doi, S.: Parallel Vector Performance of Concur-rent Visualization System *RVSLIB* on SX-4, the 3rd pacific symposium on flow visualization and image processing, March 18-21, Maui, Hawaii (2001)

Table 1. Measured performance of the solver on high-performance servers.

	Peak Performance	Measured Performance
NEC SX4 (2CPUs)	4 GFLOPS	1.8 GFLOPS
NEC SX5 (2CPUs)	8 GFLOPS	3.7 GFLOPS

Table 2. Specified parameters of the non-dimensional velocity for each computation cases.

Opening location	Case A	Case B
Left	0.8	1.6
Center left	1.4	1.0
Center right	1.4	1.0
Right	0.8	1.6

Fig. 1. Geometry of a production car model. This surface data is based on a CAD system and consists of over 15 million polygons.

Fig. 2. The extracted surface of the passenger compartment is rendered. The total number of polygon is over 10 million.

Fig. 3. The voxel model converted from Fig. 2. The total number of voxel is about 8.4 million with 10mm in size. The upper half area is invisible for the display.

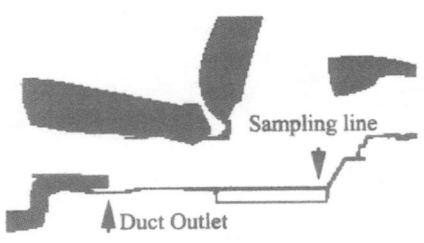

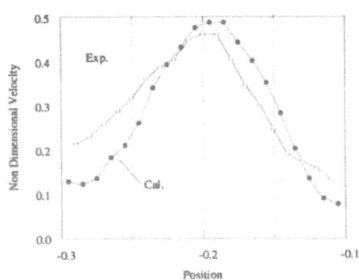

Fig. 4. The left panel shows a cross-section from the left view and measured position. The right panel shows the comparison of velocities. The velocity is normalized by the velocity at the outlet.

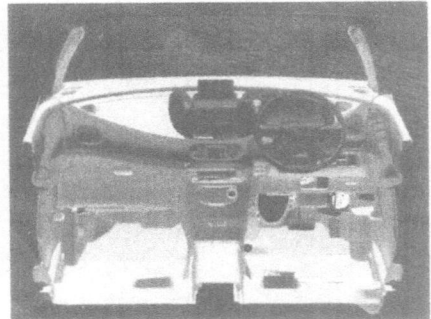

Fig. 5. The front instrument panel from the rear view. Two ventilating openings are located at center and other two openings are at both sides near the front pillar. The front seat is removed for convenience.

Editors' Note: see Appendix, p. 361f. for colored figures of this paper

Automotive Soiling Simulation Based On Massive Particle Tracing

Stefan Roettger [†], Martin Schulz [‡], Wolf Bartelheimer [*], Thomas Ertl [†]

[†] Visualization and Interactive Systems Group, IfI, University of Stuttgart
http://wwwvis.informatik.uni-stuttgart.de
[‡] Science+Computing GmbH, Tübingen
[*] BMW AG, München

Abstract. In the automotive industry Lattice-Boltzmann type flow solvers like PowerFlow from Exa Corporation are becoming increasingly important. In contrast to the traditional finite volume approach PowerFlow utilizes a hierachical cartesian grid for flow simulation. In this case study we show how to take advantage of these hierarchical grids in order to extend an existing Lattice-Boltzmann CFD environment with an automotive soiling simulation system. To achieve this, we chose to constantly generate a huge number of massive particles in user manipulable particle emitters. The process of tracing these particles step by step thus creates evolving particle streams, which can be displayed interactively by our visualization system. Each particle is created with stochastically varying diameter, specific mass and initial velocity, whereas already existing particles may decay because of aging, when leaving the simulation domain or when colliding with the vehicle's surface. On the one hand the display of these animated particles is a very natural and intuitive way to explore a CFD data set. On the other hand animated massive particles can be easily utilized for driving an automotive soiling simulation just by coloring the particles' hit points on the vehicle's surface.

1 Introduction and Motivation

In the flow visualization community many CFD simulation and visualization environments have been presented in the past [6, 5, 3, 1, 2, 10]. In order to solve the Navier-Stokes equations most of todays available simulation applications are based on the finite volume approach, which is well known and established in the automotive industry. More recent approaches such as PowerFlow from Exa Corporation [4] use a Lattice-Boltzmann simulation algorithm based on hierarchically refined cartesian grids (see Figure 1). With PowerFlow now being used as a standard flow solver in many development projects especially in automotive aerodynamics, there had been the demand for a visualization system that could take advantage of the special properties of the hierarchical grids. In cooperation with the BMW Group, Munich, such a visualization system was developed leading to a visualization application that allows interactive and intuitive exploration of a CFD data set [7]. A variety of well known flow visualization techniques like particle probes and cutting planes have been incorporated into the visualization system (see Figure 2), which can be configured also for immersive environments like the PowerWall or the CAVE.

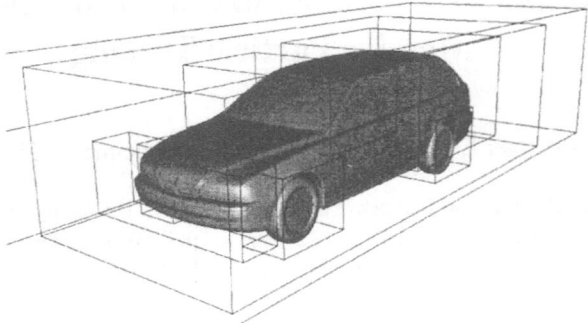

Fig. 1. The wireframes show the boundaries of a hierarchical cartesian grid, which is refined at the most interesting regions of a BMW 5 series for a Lattice-Boltzmann type CFD simulation.

Starting with this approach, the following observations lead to the development and inclusion of a soiling simulation module into the visualization system: In the traditional development process a vehicle cannot be checked for its soiling behaviour until an operable full-scale model is available. Since this is the case only at the end of the design process, our goal was to extend the existing Lattice-Boltzmann flow simulation and visualization environment in such a way that the flow engineer would be able to predict and understand the soiling behaviour at a much earlier design stage. Therefore, soiling problems could be detected before the expensive full-scale prototypes are being built. In order to achieve this we developed a particle animation system, which is able to interactively simulate the evolution of particle streams consisting of a large number of dust particles. The sustained creation and simulation of massive particles with stochastically varying diameter, initial velocity, initial position and specific mass produces a growing and evolving particle stream, which mimics the properties of dust. Each particle of the stream eventually decays either because its age is exceeding the allowed life time, or when a particle leaves the domain of the simulation, or because a collision with the vehicle's surface occured. Then the accumulation of dust on a car can be visualized simply by means of coloring each particle's hit point on the vehicle surface.

2   Physical Properties of Dust Particles

The first and most important step towards a realistic soiling simulation is to model the physical properties of dust correctly. We choose to restrict our model to dry driving conditions, because there are numerous less understood effects that influence soiling in a wet environment. Without great loss of accuracy dust particles can be idealized as spheres with a diameter D_p and a specific mass ρ_p. The drag coefficient c_{d_p}, which characterizes the forces induced on a dust particle by the fluid flow, basically depends on the particle's diameter and its velocity Δv_p relative to the flow. In general, the drag coefficient has to be measured experimentally, but for the low *Reynolds* numbers (with typical values of R_e up to 10) we en-

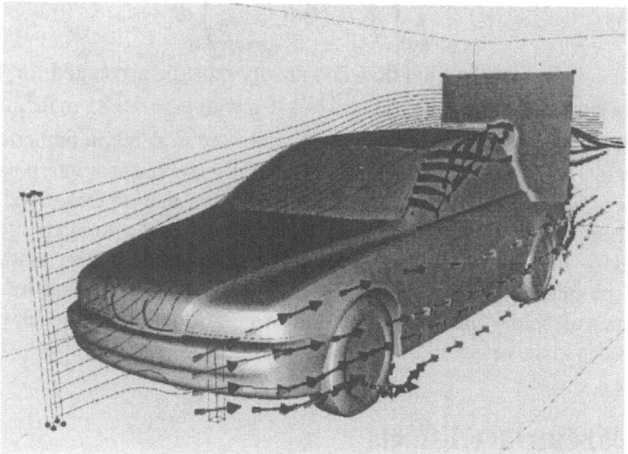

Fig. 2. An example of traditional flow visualization techniques: A cutting plane showing pressure values and several unanimated particles probes like stream lines, stream ribbons and glyphs.

counter in automotive CFD simulations the drag coefficient can be approximated with sufficient accuracy by the formula of *O'Seen*:

$$c_{d_p} = \frac{24}{R_e}\left(1 + \frac{3}{16}R_e\right), \quad R_e < 10 \quad (O'Seen)$$

Rewriting the formula of *O'Seen*, so that R_e is eliminated, we arrive at the following equation for the drag coefficient c_{d_p} in an air flow with density $\rho_{air} = 1.3\ \mathrm{kg/m^3}$ and viscosity $\nu_{air} = 10^{-5}\ \mathrm{m^2/s}$:

$$R_e = \frac{D_p \cdot \Delta v_p}{\nu_{air}} \quad \rightarrow \quad c_{d_p} = \frac{24 \cdot \nu_{air}}{D_p \cdot \Delta v_p} + \frac{9}{2}$$

Next we substitute the drag coefficient c_{d_p} in the formula, which describes the forces induced on the particles by the fluid flow and we derive the following equation for a force F_p that drives a particle of diameter D_p, flow-effective area $A_p = \frac{1}{4}\pi D_p{}^2$, mass $m_p = \frac{1}{6}\rho_p \pi D_p{}^3$ and relative velocity Δv_p:

$$F_p = \frac{1}{2}\rho_{air} A_p c_{d_p} \Delta v_p{}^2 = \frac{1}{2}\rho_{air} A_p \Delta v_p \left(\frac{24 \cdot \nu_{air}}{D_p} + \frac{9}{2}\Delta v_p\right)$$

3 Massive Particle Tracing on Cartesian Grids

Now each dust particle that was emitted with stochastically varying diameter D_p, specific mass ρ_p, initial velocity v_{p_0} and initial position x_{p_0} is treated as a point mass. According to the following second order differential equation the integration of a particle's path is accomplished by an adaptive, embedded Runge-Kutta particle tracer, which integrates the acceleration $a_p(t)$ of a dust particle to compute the particle's position after a determined period of time [9].

$$x_p(t) = \int \left(\int a_p(t)dt + v_{p_0} \right) dt + x_{p_0}$$

Both the influence of the fluid flow and gravity must be accounted for, thus the total acceleration adds up to $a_p(t) = \frac{F_p(t)}{m_p} + g$ with $g = -9.81$ m/s^2. Adaptive embedded Runge-Kutta tracers have been evaluated in depth in numerical analysis. As the domain of the integration is basically a trilinearly interpolated and hierarchically refined cartesian grid it can be shown that an embedded Runge-Kutta integrator of order 4(3) is fully sufficient with respect to integration accuracy. Taking this into account we reimplemented a particle tracer as suggested in [8]. The integration step size, however, tends to be rather small for particle diameters well below 3μm. Fortunately, the average diameter of the relevant dust particles lies in the range of 5 to several 100μm.

4 Near-Surface Effects

For soiling simulations the correct near-surface behaviour of the simulated dust particles plays an important role with respect to simulation accuracy. The adhesion probability of a particle depends mainly on the speed by which it is approaching the surface. Faster particles are more likely to hit the surface than slower ones. Since flow velocity is zero on the vehicle's surface, the adhesion of each particle is driven by its momentum and by electrostatic effects (see Figure 3). At the time of writing the latter effect is not yet completely understood and will be subject to further research activities. For now we assume that the electrostatic forces are of inverse quadratic order ($F_e = c_e \cdot \frac{1}{r^2}$ with c_e being an electrostatic field coefficient and r being the Hausdorff distance to the vehicle's surface).

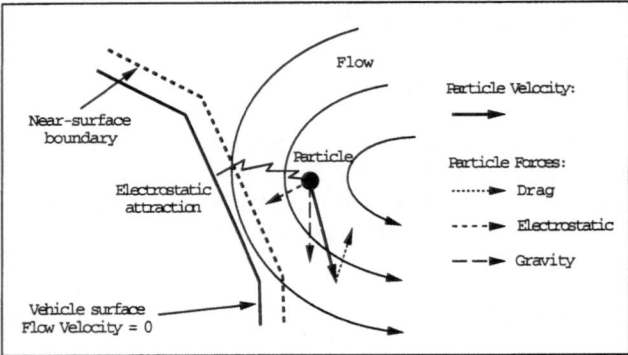

Fig. 3. A dust particle that is accelerated by its movement relative to the direction of the flow and by electrostatic and gravitational forces.

The cartesian representation of the flow field implies several restrictions. In principle, the vehicle geometry has to be stored and handled explicitly, since the simulation grid is derived by voxelization. For the same reason special care has to be taken when modeling any near-surface effect. In a correct physical setting the flow velocity, for example, is zero on the vehicle's surface as said before. This condition is not guaranteed when using a cartesian flow representation, because of the finite resolution of the grid. To compensate for that we define a so called near-surface zone. Its thickness equals the size of the smallest voxels in the grid (see Figure 4). Inside this near-surface zone the flow velocity is attenuated by the square root of the normalized Hausdorff distance to the vehicle's surface. These distances are calculated on-the-fly by utilizing an octree representation of the explicit set of surface triangles.

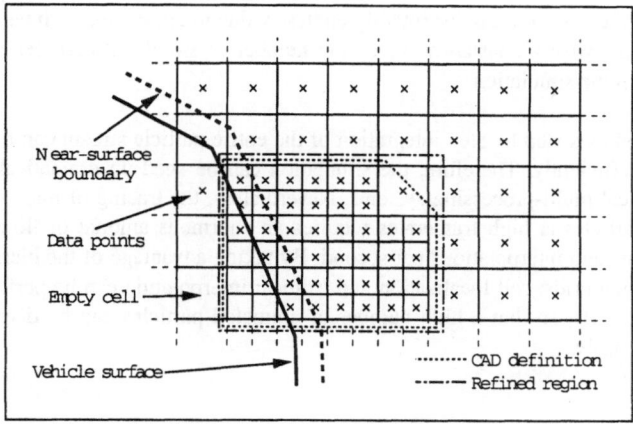

Fig. 4. Hierarchical grid representation with explicit vehicle surface and near-surface zone.

5 Massive Particle Visualization

With a particle tracer as described above a cloud of dust particles can be animated by continuously proceeding from keyframe k_n to keyframe k_{n+1}. During that task new particles are being created continuously in the extent of the particle emitters employing a stochastical process as mentioned before. Likewise, a fraction of the particles is decaying constantly because of their age, when colliding with the surface of the vehicle, or when leaving the domain of the simulation. The emerging stream of dust particles is visualized by a virtual camera with constant exposure time, such that particles with higher velocity result in longer traces on each image of the animation than slower ones (see Figure 5). This provides the viewer with an improved physical and three-dimensional insight into the flow properties, since the velocity of each particle is implicitly visible. By utilizing SGI's *Cosmo3D/OpenGL-Optimizer* the particle animation is embedded into an immersive environment, which provides the flow engineer with the ability to navigate and manipulate the particle emitters in an intuitive way. This procedure is very similar to the handling of smoke probes in a real-world wind tunnel, hence the acceptance of such a visualization technique is high.

One the one hand we mentioned that we wanted to trace the particles at real-time, but for an approaching flow with a typical velocity of $u_\infty = 20$ m/s a particle is passing the vehicle in circa 0.3 seconds on the average, which is far too fast for a human viewer to catch any details. The solution is to interactively trace the particle cloud in slow motion. On the other hand a particle is very likely to leave the simulation domain before it has reached a particle age of typically 2 to 4 seconds, which mainly depends on the mass of the particles. Under normal conditions the total number of traced particles then stabilizes on a level that is determined by the frequency of particles being created and the average time the particles spend in the domain of the simulation. In our case, however, we have found it necessary to set the maximum life time of a particle to about 5 seconds to reliably prevent the rare case of particles rotating endlessly due to simulation or integration errors. Otherwise a constantly increasing number of simulated particles would slow down the simulation.

The continuous step by step integration of the entire particle stream can be parallelized efficiently. Therefore, the simulation can be sped up dramatically on symmetrical multi-processing systems. Nevertheless, the tracing of massive animated particles at high framerates requires an enormous amount of flow field evaluations and interpolations per second. By taking advantage of the hierarchical cartesian grids, cell localization and trilinear interpolation can be performed very efficiently, so that a high number of animated particles can be displayed simultaneously.

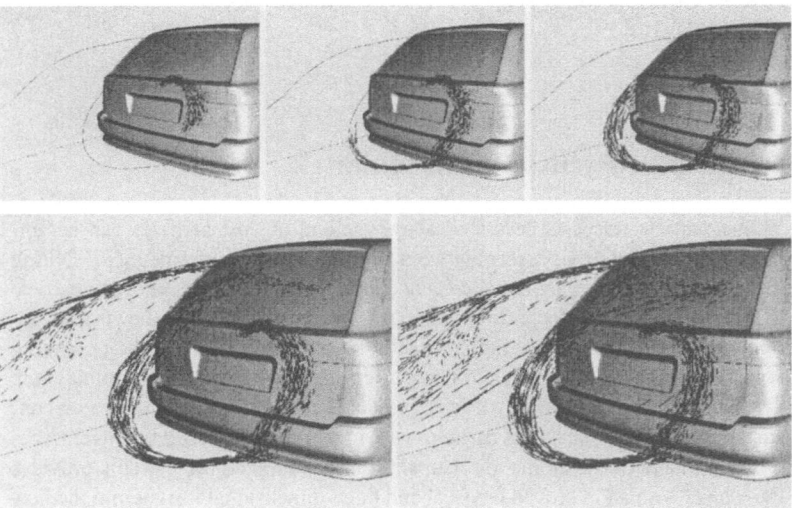

Fig. 5. Steps 1 to 5 of an interactive particle animation. Dust particles are emitted from a small user definable wireframe box behind the car. The emerging massive particle stream is first moving down and then up the back window. Eventually, it is diverging at the top of the car due to shearing forces.

Now the accumulation of dirt on a vehicle's surface can be visualized easily by coloring the hit points on the surface of the car. In our case the nearest vertices on the car surface are searched. The color of each of the vertices is determined by colorcoding the number of encountered hits. For that purpose hardware-accelerated 1D-textures are employed, which also allow smooth, Gouraud-shaded transitions between the vertices. In order to accelerate the detection of the hit points we are taking advantage of an octree representation of the explicit car surface, which reduces the total number of visited triangles from 70.000 of the original CAD model to an average of 8 to 15 triangles per octree search. By storing a single bit per voxel, which denotes the presence of geometry, we can speed up collision detection even further.

6 Results and Conclusion

A series of screenshots from a soiling simulation employing the techniques mentioned above is shown in Figure 6. The total run time was approximately 24 hours of continuous particle animation with screenshots taken after 15 minutes, 45 minutes, 2 hours, 12 hours and finally 24 hours. During the simulation run our particle animation system was simultaneously tracing an average number of about 930 dust particles on an SGI Octane MXI. With its dual 250 MHz MIPS R10K processors the refresh rate of the multi-threaded animation was 7 Hz, which corresponds to a speedup of about 85% in comparison to a single processor system. Under normal dry driving conditions the probability for a single dust particle in the air to hit the car is fairly low. When using time-averaged flow fields, which smooth out the transient flow components, the probability by which a particle hits the surface is reduced even further. Therefore the main problem of a soiling simulation as discussed above is the huge number of emitted particles that must be traced in order to achieve a sufficiently smooth dust distribution on the vehicle's surface. To improve this we are planning to use time-dependent flow fields in the future. Besides this, a variety of less understood physical effects and environmental influences makes it very difficult to compare the result of a soiling simulation to the real-world evaluations conducted by BMW. Nevertheless, we have been able to verify that our physical model of the dust behaviour is sufficiently exact. In order to improve our knowledge about the involved electrostatic effects near the vehicle's surface further research activities are needed. In the mean time our efforts will serve as a basis for further convergence towards the complex real-world soiling situation and as an accompanying visualization tool for the flow engineers.

7 Acknowledgments

This work has been financed by FORTWIHR, the Bavarian Consortium for High Performance Scientific Computing.

Fig. 6. Steps 1 to 5 of a soiling simulation running for approximately 24 hours on a dual processor SGI Octane MXI with 2x250MHz MIPS R10K. The colorcoding shows the hit rate of particles colliding with the car body. Red hot spots indicate areas with a high degree of soiling.

References

1. S. Bryson and C. Levit. The Virtual Windtunnel. In *IEEE Computer Graphics and Applications, 12(4):25-34*, 1992.
2. Steve Bryson. Approaches to the Successful Design and Implementation of VR Applications. In *Proc. SIGGRAPH'94*, 1994.
3. Steve Bryson and Steven Feiner. Virtual Environments in Scientific Visualization. In *Virtual Reality for Visualization, Course Notes of Tutorial 5 at Visualization 95*, 1995.
4. Exa Corporation. *http://www.exa.com*.
5. D. A. Lane. Scientific Visualization of Large-Scale Unsteady Fluid Flows. In G. Nielson, H. Hagen, and H. Mueller, editors, *Scientific Visualization*, pages 125–145. IEEE Computer Society, 1997.
6. F. Post and T. van Walsum. Fluid Flow Visualization. In H. Hagen, H. Mueller, and G. Nielson, editors, *Focus on Scientific Visualization*, pages 1–40. Springer Berlin, 1997.
7. M. Schulz, F. Reck, W. Bartelheimer, and Th. Ertl. Interactive Visualization of Fluid Dynamics Simulations in Locally Refined Cartesian Grids. In *Proc. Visualization '99*. IEEE, 1999.
8. C. Teitzel, R. Grosso, and T. Ertl. Efficient and Reliable Integration Methods for Particle Tracing in Unsteady Flows on Discrete Meshes. In W. Lefer and M. Grave, editors, *Visualization in Scientific Computing '97*, pages 31–41, Wien, 1997. Springer.
9. S.K. Ueng, C. Sikorski, and K.L. Ma. Efficient Construction of Streamlines, Streamribbons and Streamtubes on Unstructured Grids. *IEEE Transactions on Visualization and Graphics*, 2(2):100–109, June 1996.
10. S. P. Uselton. exVis: Developing A Wind Tunnel Data Visualization Tool. In R. Yagel and H. Hagen, editors, *Proc. Visualization '97*. IEEE, 1997.

Editors' Note: see Appendix, p. 363 for colored figures of this paper

Comparative Visualization of Instabilities in Crash-Worthiness Simulations

Ove Sommer and Thomas Ertl

University of Stuttgart, IfI, Visualization and Interactive Systems Group
{sommer, ertl}@informatik.uni-stuttgart.de
http://wwwvis.informatik.uni-stuttgart.de

Abstract. Since crash-worthiness simulations get more and more important as part of the car development process in order to reduce the cost of development, enhance the product quality, and minimize the time-to-market, the reliability of the simulation results plays a decisive role concerning their significance. Recently the simulation departments of several automotive companies started investigating the quantity and reason for deviations during a number of simulation runs on the same input model.

In this case study we discuss different measurements for instability and present a texture-based visualization method which allows the engineers to efficiently explore the simulation results by interactively hiding finite element structures with nearly constant crash performance. Furthermore, we describe those parts of our prototype which use a CORBA layer for providing the same view on a set of simulation results and allowing the visual comparison by using the marker functionality.

1 Introduction

In recent years simulation has become more and more important for the development of new cars. It supports the testing with hardware-prototypes by delivering simulation results which are close to test results. This makes the reduction of hardware-prototype tests possible and therefore allows the development at a lower price. Furthermore, the shorter cycle of simulation allows the evaluation of more iterations of variants, and thus better or safer car body parts which improves the product quality.

In the field of crash simulation the continuously increasing CPU power of high-end simulation servers and the parallelization of the simulation software leads to

- **models of finer mesh resolutions**. Finer models map the crash-worthiness of a car body more exactly.
- **extensive tracking of more model parameters**. The chance to correlate the temporal behavior of different simulation variables by using state-of-the-art visualization techniques allows the engineers to come to a deeper understanding.
- **more simulations runs**. The more iterations can be computed the more improvements can be done to the structure of car body parts.

Recently the stability of the simulation process is investigated in order to ensure the reliability of simulation results. For this purpose one and the same model will be simulated several times and the results are compared to each other.

In this case study we describe the statistical methods that are used to compare the simulation output and to evaluate the achievable stability. We will discuss two categories of comparison functions. We present a visualization method which allows the engineers to detect regions of instability even in complex models. And finally, the direct visual comparison of multiple simulation runs using marker functionality together with a CORBA connection layer will be presented.

2 Stability calculation

Today a finite element mesh of a whole car body model consists of about 500.000 elements and nodes. For crash-worthiness simulations the first 120 milliseconds of an impact are computed and the coordinates are stored in 60 time steps (*states*) together with tracked variables like velocities, forces or strains, which takes about 50 hours on 6 CPUs of a modern simulation server. 2000 simulation iterations are calculated before the next state is appended to the result file. During simulation several kinds of ramifications will cause differing results. They originate as well from the limited precision of the numerical process as from the structure of the finite element mesh. For example, if one shell element A is pressed against another element B which has a normal that lies in the plane of A then the simulator has to determine the direction in which A will slide on B. Those ramifications are called the *instability* of a simulation process.

In order to evaluate the effects of the replacement of any car body part by a variant it is absolutely necessary to be able to reduce the impact of instabilities caused by the design of the meshed car body model. At the time the engineers try to spot those regions which are responsible for unstable crash dynamics. Multiple runs for the same model with the same boundary conditions are computed. The simulation results are compared against each other by using appropriate evaluation functions.

The set of evaluation functions can be split in different classes: if they use the output data directly or proceed on a previously computed measure, if they represent a local or a global criterion, and if they use one- or multi-dimensional data. In the following three examples of geometric comparison functions are illustrated.

2.1 Global measurement functions

The function $V(p, t, r) = p(t, r) - p(0, r)$ measures the displacement of the mesh node p in the t'th state from its original position in the initial state for one simulation run r. This is done per component or Euclidean. The length of the displacement vector is compared over all simulation runs R.

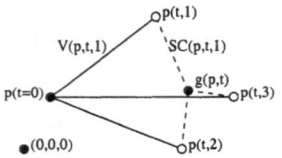

Fig. 1. The outline shows the position of node p in the initial state ($p(t = 0)$) and after t time steps of three different simulation runs. The solid lines mark the displacement function $V(p, t, r)$, the dashed lines the scatter function $SC(p, t, r)$.

The scatter function $SC(p, t, r)$ expresses the distance of node p to the centroid $g_c(p, t)$ which is calculated as $g_c(p, t) = \frac{1}{R} \sum_{r=1}^{R} p(t, r)$, where R is the number of runs. Here, the projection to one main axis could also be investigated to focus on differences in the specified dimension.

The drawback of both, the displacement and the scatter function is, that they are global measures. For example, during a front-crash simulation an instability at some finite elements of the engine mount will influence the values of a wide area of adjacent car body parts and will even force deviations in the rear part of the car. The largest differences between simulation runs occur in regions of intense deformation as Fig. 2 shows. The deviation of corresponding mesh coordinates becomes less for larger distances from the center of most buckling.

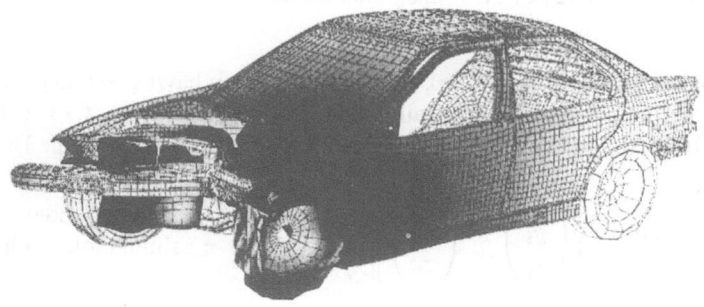

Fig. 2. The grey-scale (see Appendix for color version) maps the length of difference vectors of corresponding nodes. After 80 milliseconds the largest deviation can be found in the left front side (dark regions) and smaller values in the rear. However, using a global measurement it is hard to spot regions where different crash behavior originates.

In order to spot the origin of instability the engineers need another criterion because these global measurement functions detect large regions without determining if the deformation of a set of finite elements is the reason or the result of instability. A more adequate measurement is provided by a local criterion.

2.2 Local measurement function

At the Institute for Algorithms and Scientific Computing (SCAI) of the German National Research Center for Information Technology (GMD) a local deformation criterion [2] has been developed within the Autobench project, a research project driven by some of the leading automotive industry companies and financed by the German Bundesministerium für Bildung und Forschung [1]. This criterion considers the displacement of a node at time step t with regard to its neighborhood nodes in comparison to the initial state, and thus the deformation of its adjacent finite elements (Fig. 3).

322

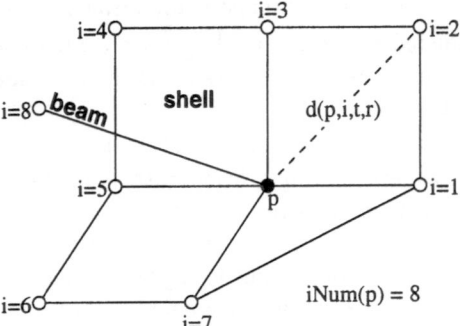

Fig. 3. $d(p, i, t, r)$ is computed as the Euclidean distance (see Eq. (1)) of node p to its *neighbor node* i in state t of the r'th simulation run.

The mesh deformation $\text{DNM}(p, t, r)$ around node p having $\text{iNum}(p)$ neighbors is calculated as the averaged sum of all distance differences $d(p, i, t, r)$ to their initial distances $d(p, i, 0, r)$ which are the same for each simulation run r (Fig. 3):

$$d(p, i, t, r) = \left\| \begin{pmatrix} x_i \\ y_i \\ z_i \end{pmatrix} - \begin{pmatrix} x_p \\ y_p \\ z_p \end{pmatrix} \right\|_{t,r} \quad , \qquad \begin{array}{l} t : \text{time step index} \\ r : \text{simulation run index} \end{array} \qquad (1)$$

$$\text{DNM}(p, t, r) = \frac{1}{N} \sum_{i=1}^{N} |d(p, i, t, r) - d(p, i, 0, r)| \quad , \qquad N := \text{iNum}(p) \quad (2)$$

The deformation $\text{DNM}(p, t, r)$ is calculated for each node p in each time step t and over all simulation runs r. This scalar quantity is only influenced by the local neighborhood which contrasts to the global measurement functions. Now, the expected value of the deformation

$$\text{DNMAV}(p, t) = \frac{1}{R} \sum_{r=1}^{R} \text{DNM}(p, t, r) \quad , \qquad R : \# \text{ simulation runs} \quad (3)$$

over all R simulation runs is determined and its standard deviation is evaluated as a measurement for the local instability of the simulation around node p.

2.3 Efficient calculation

Since large data sets consisting of about half a million finite elements and nodes have to be handled, it is impossible to store all the data in main memory. For an efficient calculation of $\text{DNMAV}(p, t)$ we generate a table which represents the neighborhood of each node. This table holds entries of index pairs. One points to the neighbor node and the other one to the corresponding field in the node distance array. The table structure makes sure that the distance of each node pair is computed just once and it provides quick access to the pre-calculated node distances $d(p, i, t, r)$ of the current and the initial time step.

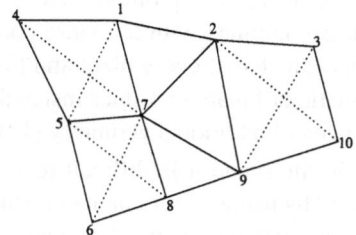

Node	Node/Index pair list						
1	(2,0)	(4,1)	(5,2)	(7,3)			
2	(1,0)	(3,4)	(7,5)	(9,6)	(10,7)		
3	(2,4)	(9,8)	(10,9)				
4	(1,1)	(5,10)	(7,11)				
5	(1,2)	(4,10)	(6,12)	(7,13)	(8,14)		
6	(5,12)	(7,15)	(8,16)				
7	(1,3)	(2,5)	(4,11)	(5,13)	(6,15)	**(8,17)**	**(9,18)**
8	(5,14)	(6,16)	(7,17)	**(9,19)**			
9	(2,6)	(3,8)	(7,18)	(8,19)	**(10,20)**		
10	(2,7)	(3,9)	(9,20)				

Fig. 4. This example outlines the structure of the node/index pair table which contains a list (row) for each node. The list stores pairs of an adjacent node and an index to an array where the corresponding node distance is stored.

First of all the node/index pair table (Fig. 4) is initialized evaluating the mesh connectivity which is constant over all states. Then the node distances for the initial state are computed and stored in an array which is used for all simulation runs. Starting with the first simulation run the node distances of the current state are calculated and the difference to the corresponding pre-calculated distance of the initial state is stored in a second array. Each time the table is traversed from bottom to top and the lists in the rows are traversed from tail to head as long as the node index of the entry is larger than the node index of the current row (bold entries in Fig. 4). Now, $DNM(p, t, r)$ is the sum of each referenced value in row p divided by the number of entries in the row. A third array holds the accumulated sum of $DNM(p, t, r)$ in order to get the expected value $DNMAV(p, t)$ at the end of all runs.

After all states of one simulation run have been processed the $DNM(p, t, r)$ is temporarily written to disk. After we have generated this file for each simulation run and divided the values in the third array by the number of simulation runs, the values are read back in and the standard deviation of the local deformation can be computed and stored to disk as a measure for instability. Later on the instability can be mapped onto the geometry of one simulation run using the technique described in the next section.

Furthermore, the span between the minimum and maximum deformation is of interest. Hence, for each state t and each node p the extreme values of $DNM(p, t, r)$ are stored together with the index r so that the most different simulation runs can be determined later on.

3 Visualization using index texture maps

The advantage of mapping scalar data as colors directly onto geometry is that the data is visualized where it appears and thus the causal relationship between geometry and mapped data is more comprehensible. In the field of CAE flat shading can be used for element-based data visualization. As the data is node-based in the majority of cases Gouraud shading will not lead to meaningful images because the colors are assigned to vertices and interpolated in RGB color space inside the polygon during rasterization. Instead the visualization could be enhanced by adding geometry and assigning appropriate colors to the subdividing vertices, but that will increase the load of the graphics pipeline.

The best way to visualize node-based data is the utilization of a one-dimensional texture which is defined as color band. Each vertex is combined with a texture coordinate. During rasterization the texture coordinate is evaluated at every pixel and then the color is looked up in the texture. Hence, high deviation of mapped values inside the same polygon will result in color-bands without the need of additional geometry [12].

In complex models with many occluding parts in the scene it is difficult to spot regions with critical values. This problem can be solved by using a four channel texture map. The additional alpha channel provides the opportunity to restrict the data mapping or the geometry rendering depending on the texture environment setting in the context of OpenGL [11]. If GL_DECAL is used, the resulting color is composed as $C_{out} = (1 - \alpha_{tex})C_{frag} + \alpha_{tex}C_{tex}$ while the transparency is not modified by the texture ($\alpha_{out} = \alpha_{frag}$). Provided we set the α component of each texel either to 0.0 or 1.0, the data visualization is only visible for those values, where the corresponding texel has an α component of 1.0. Otherwise the geometry is rendered in the original color.

If we switch the texture environment to GL_REPLACE and enable the alpha test the geometry rendering is controlled by the mapped values. In [10] boolean textures were already used to clip geometry during the rasterization stage. We use this clipping functionality of the texture subsystem in correlation with the values simulated at the geometry. While the texture defines the outgoing color, the relation of the texel's α component to the alpha test reference value decides, if the fragment is rendered or not. Thus, this technique can be used to restrict the geometry rendering to interesting data value ranges as already associated with, for example, the visualization of potential flanges [4]. The alpha test has to be enabled to avoid z-buffer pollution; otherwise the invisible geometry could hide other geometry which lies behind the transparent parts and therefore will fail the z-buffer test.

For visual data exploration and analysis the interactive modification of the mapping has turned out as very useful. It allows the engineers to interactively restrict the color mapping or the geometry rendering to the regions of interesting values. In order to provide high interactivity the texture map does not contain RGBα quadruples but indices. These indices are used to reference the color and transparency of the texel in a hardware-supported texture color lookup table. The contents of this table represents a transfer function which can be modified in a color editor dialog.

For the investigation of instability this technique allows an interactive search for regions where different crash behavior originates. First the standard deviation of the element deformation as described in section 2.3 is loaded from disk and mapped to the indices of the texture color lookup table. By switching to the GL_REPLACE/*alpha test* mode and adjusting the alpha transfer function the engineer can hide all geometry that behaves constant or shows only a small standard deviation across all simulation runs. Then the user can zoom into a remaining area, lock the camera to the geometry, and analyze the reason for the instable performance in several simulation runs activating the time animation. A semi-transparent rendering instead of hiding that geometry with low deviation may help to orientate oneself in a complex model (Fig. 5).

The described visualization methods have been integrated into *crash Viewer*, a prototype for pre- and post-processing functionality [8,9] in the area of crash-worthiness simulations using the PAM-CRASH code [5]. The application has been developed in co-

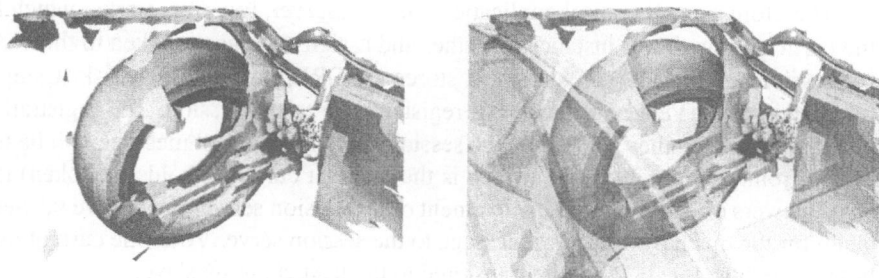

Fig. 5. To detect regions of primary instability it turned out to be very useful if the transfer function of the alpha channel is set lower than the alpha test reference value for values of small deviation. For a better orientation the user can interactively fade in the neighborhood as shown in the right image. (see also color plate in Appendix)

operation with the BMW Group and is in productive use. It uses OpenGL Optimizer [7], a tool set for large model visualization which is based on Cosmo3D [6], a scene graph layer on top of OpenGL.

4 Comparing geometry using synchronized viewers

If the most differing simulation runs have been determined the engineer could get an impression of the real deformation deviation only if it is possible to visually compare both finite element meshes in detail. A CORBA connection layer which has originally been implemented to support collaboration of two or more distant engineers evaluating simulation results [3] can be used for this task to synchronize multiple viewers on the same display.

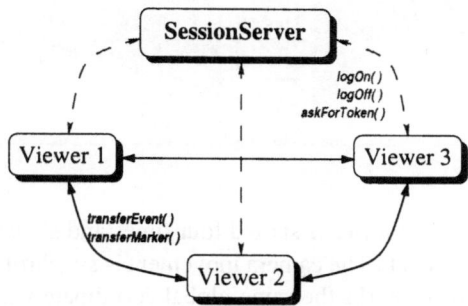

Fig. 6. The session server acts as a controler of a multi-viewer session and controls the master token. The viewer which currently holds the token sends event messages directly to other participating viewers.

Therefore a small control application (*SessionServer*, Fig. 6) is started which links the participating viewer instances together and assigns the master token to them. After the session server has been started, it stores a CORBA reference to disk. Using this reference a *crashViewer* instance can register itself to the session. The registration is propagated to the other viewers by the session server. Any event message will be transmitted from the master viewer (which is the one that currently holds the token) to the slave viewers directly without involvement of the session server. Each slave viewer can claim for the token by sending a message to the session server. After the current master has released the token, it will be transfered to the next claiming slave.

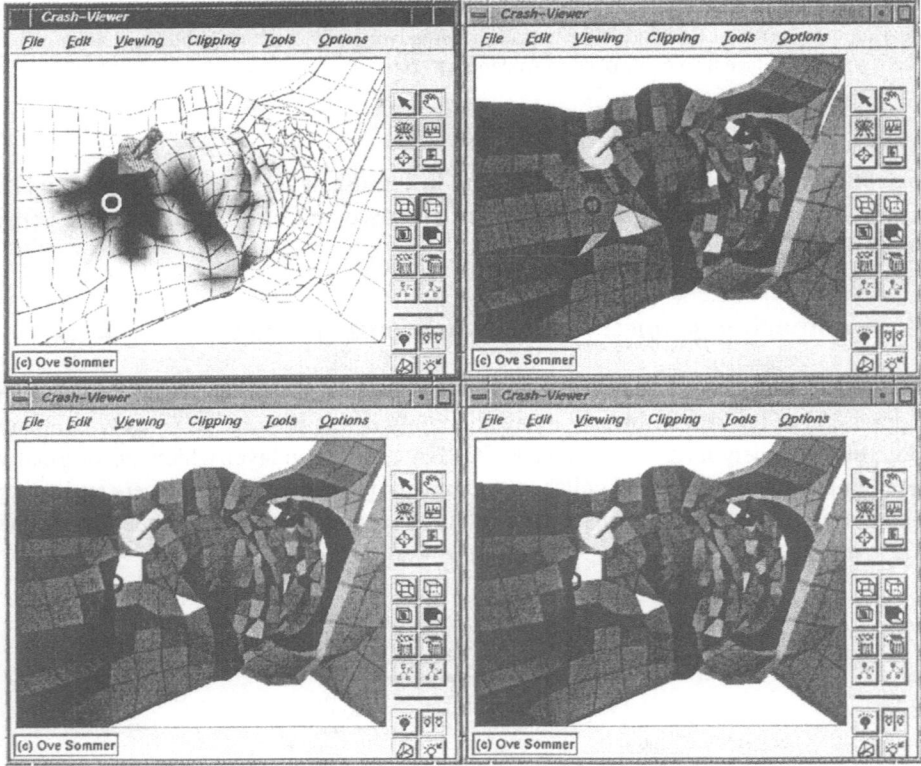

Fig. 7. Our prototype *crashViewer* is started four times and shows different simulation results of the same input deck. The camera movement is synchronized using a CORBA connection. The 3D arrow marks the same global coordinate while the circle tags the same finite element mesh node. The mapped standard deviation in the upper left window points out the different deformation behavior.

The camera position is sent by the master as a transformation matrix. Furthermore, markers can be inserted into the scene to define reference points. Fig. 7 shows four different simulation results. The upper left viewer visualizes the standard deviation of all results. The 3D arrow marks the same global coordinate in each viewer and points out the geometric difference between the simulation runs. The upper right window shows a completely different deformation around the circled node which marks the same mesh node in each window.

Of course this functionality can also be used to compare the crash performance of variants when the reasons for instability have been removed. This would enhance the car development process significantly because the differences between constructive variants and their effects to the whole model regarding crash dynamics would directly be visible.

5 Results

The calculation of a measure for the instability of crash-worthiness simulations is construed to be time and memory efficient. The test data set that can be shown here contains about 60.000 shell elements and nearly 55.000 nodes. Each of the 15 result files store 81 simulated time steps of the same source model. On a SGI Octane with one R12k/300MHz CPU the standard deviation of the local deformation as described in section 2.2 is computed for all result files in 4.5 minutes. The process needs about 45MB main memory. The memory consumption depends on the number of nodes and the number of states but it is independant of the number of simulation runs. Simulation results with 500.000 nodes over 60 time steps should require less than 500MB. For larger models or result files with more time steps it is possible to make the memory consumption also independant of the number of states which would lower the performance.

With the described methods integrated our prototype *crashViewer* allows for the first time the comparative visualization of instability in crash-worthiness simulations. The interactive modification of transfer functions used by the index texture map provides value-based geometry clipping. The engineer is visually guided to regions in the finite element model where different crash behavior originates. The detailed investigation of such areas is supported by several functions like the camera locking mechanism. (See Appendix for additional color plate.)

6 Conclusion

We introduced a method to determine and visualize the instability across multiple crash-worthiness simulations of the same source model. The integration of the presented techniques into our prototype *crashViewer*, allows the engineers of the crash simulation department to explore the origins of instability. Finally, the use of multiple synchronized viewers displaying different simulation results makes a direct comparison possible. Only the combination of advanced rendering techniques and exploiting graphics hardware allows an innovative visualization application which is in productive use at BMW.

7 Acknowledgements

We thank the Institute for Algorithms and Scientific Computing (SCAI) of the German National Research Center for Information Technology and the crash department of the BMW Group for providing the simulation results. This work was partially funded by the Bundesministerium für Bildung und Forschung in the context of the Autobench project.

References

1. Autobench – An Integrated Construction Environment for Virtual Prototypes in Automotive Industry. http://www.autobench.de, 1998–2001.
2. Jürgen Bendisch and Hartmut von Trotha. Stabilitätsuntersuchungen mit Mitteln der Statistik. Internal report of GMD, Autobench project, April 2000.
3. Klaus Engel, Ove Sommer, and Thomas Ertl. A Framework for Interactive Hardware Accelerated Remote 3D-Visualization. In *Proc. of EG/IEEE TCVG Symposium on Visualization VisSym 2000*, pages 167–177,291. Springer Wien/New York, May 2000.
4. Norbert Frisch, Dirc Rose, Ove Sommer, and Thomas Ertl. Pre-processing of Car Geometry Data for Crash Simulation and Visualization. In Vaclav Skala, editor, *WSCG 2001 - The Ninth International Conference in Central Europe on Computer Graphics and Visualization*, pages 25–32, February 2001.
5. E. Haug, A. Dagba, J. Clinckemaillie, F. Aberlenc, A. Pickett, R. Hoffman, and D. Ulrich. Industrial Crash Simulations using the PAM-CRASH code. In *Supercomputing in Engineering Structures*, pages 171–196. Computational Mechanics Publications, 1989.
6. Silicon Graphics Inc. Cosmo3D^{TM} Programmer's Guide. Silicon Graphics Inc., IRIS Insight Library, 1998. http://techpubs.sgi.com/.
7. Silicon Graphics Inc. OpenGL Optimizer^{TM} Programmer's Guide: An Open API for Large-Model Visualization. Silicon Graphics Inc., IRIS Insight Library, 1998. http://techpubs.sgi.com/.
8. Sven Kuschfeldt, Thomas Ertl, and Michael Holzner. Efficient Visualization of Physical and Structural Properties in Crash-Worthiness Simulations. In Yagel and Hagen, editors, *Proc. IEEE Visualization '97*, pages 487–490,583. IEEE Computer Society Press, October 1997. ISBN 1-58113-011-2.
9. Sven Kuschfeldt, Ove Sommer, and Thomas Ertl. Efficient Visualization of Crash-Worthiness Simulations. *IEEE Computer Graphics and Applications*, 18(4):60–65, July/August 1998.
10. William E. Lorensen. Geometric Clipping Using Boolean Textures. In *Proceedings Visualization '93*, pages 268–274. IEEE, 1993.
11. Ove Sommer and Thomas Ertl. Geometry and Rendering Optimizations for the Interactive Visualization of Crash-Worthiness Simultations. In *Proceedings of IT&T/SPIE Electronic Imaging, Visual Data Exploration and Analysis VII*, volume 3960, pages 124–134, January 2000.
12. Michael Teschner and Christian Henn. Texture Mapping in Technical, Scientific, and Engineering Visualization. http://www.sgi.com/chembio/resources/texture/index.html, 1995. Technical Report, Silicon Graphics Inc.

Editors' Note: see Appendix, p. 364 for colored figures of this paper

Authors Index

Color Plates

Color Plates

Harding et al. (pp. 3–14)

Fig. 8. Structural model created with GDIS

Fig. 5. GDIS uses stereographics, haptic force feed-back and sonification

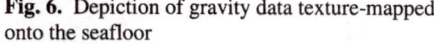

Fig. 6. Depiction of gravity data texture-mapped onto the seafloor

Fig. 7. Depiction of the seafloor data with curvature coloration

334

Jiang et al. (pp. 15–24)

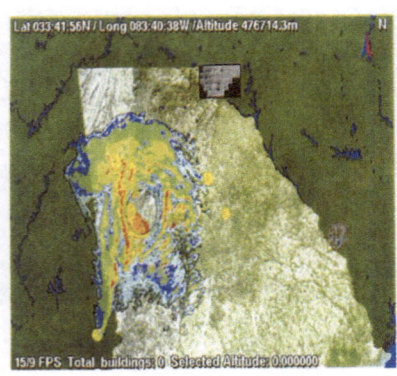

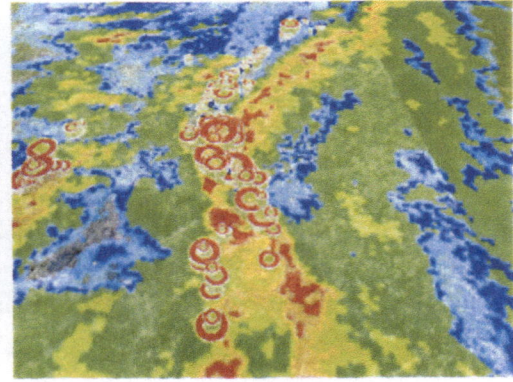

Color Plate 1. Overview of combined raw and analyzed data for a given time step

Color Plate 2. Close-up of combined data at same timestep. Reddish icons are analyzed mesocyclones

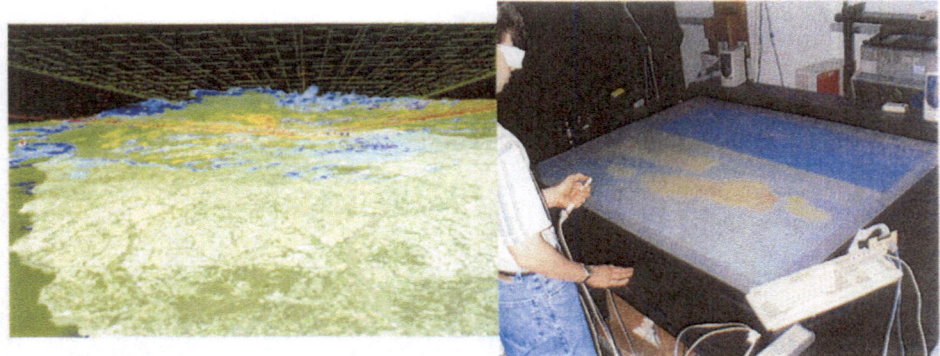

Color Plate 3. View of raw 3D data in fly mode

Color Plate 4. Weather visualization using the virtual workbench interface

Color Plate 5. Overhead view of emergency response area and simulated toxic gas plume

Color Plate 6. Fly mode view of simulated toxic gas plume. Positions of first responders are indicated by the red icons

Weber et al. (pp. 25–34)

(i) Isosurface extracted using two out of seven levels of the AMR hierarchy. Generating the stitch cells required approximately 55ms, generating the isosurface approximately 250ms

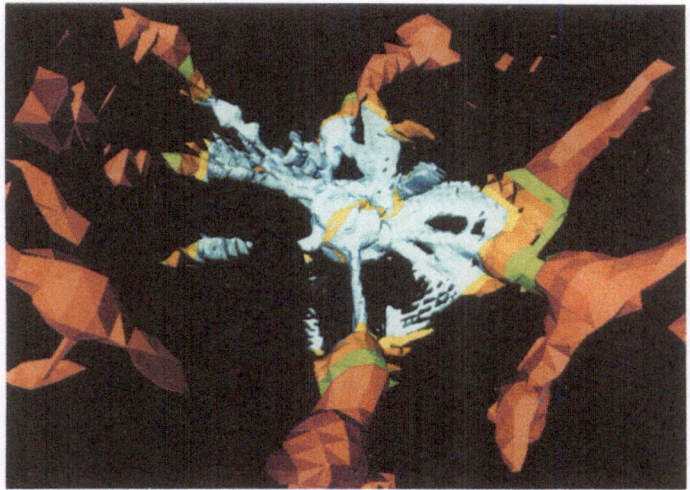

(ii) Isosurface extracted using three out of seven levels of the AMR hierarchy. Generating the stitch cells required approximately 340ms, generating the isosurface approximately 600ms

Fig.7. Isosurface extracted from AMR hierarchy (data set courtesy of Greg Bryan, Massachusetts Institute of Technology, Theoretical Cosmology Group, Cambridge, Massachusetts)

Gerstner (pp. 35–44)

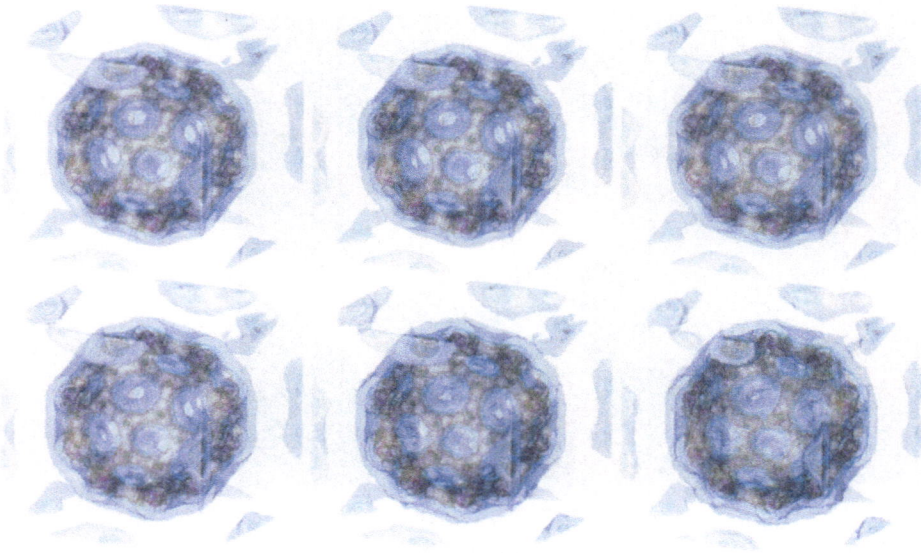

Fig. 5. Multiple transparent isosurfaces of the buckyball data set for varying error thresholds

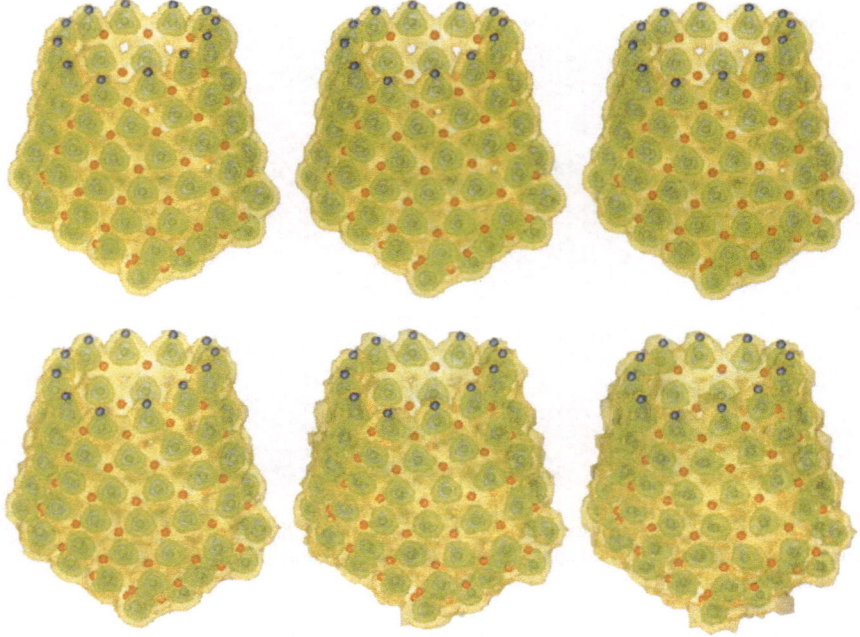

Fig. 6. Multiple transparent isosurfaces of a boron-nitrite nanotube cap with corresponding atomic positions for varying error thresholds

Bertram and Hagen (pp. 55–63)

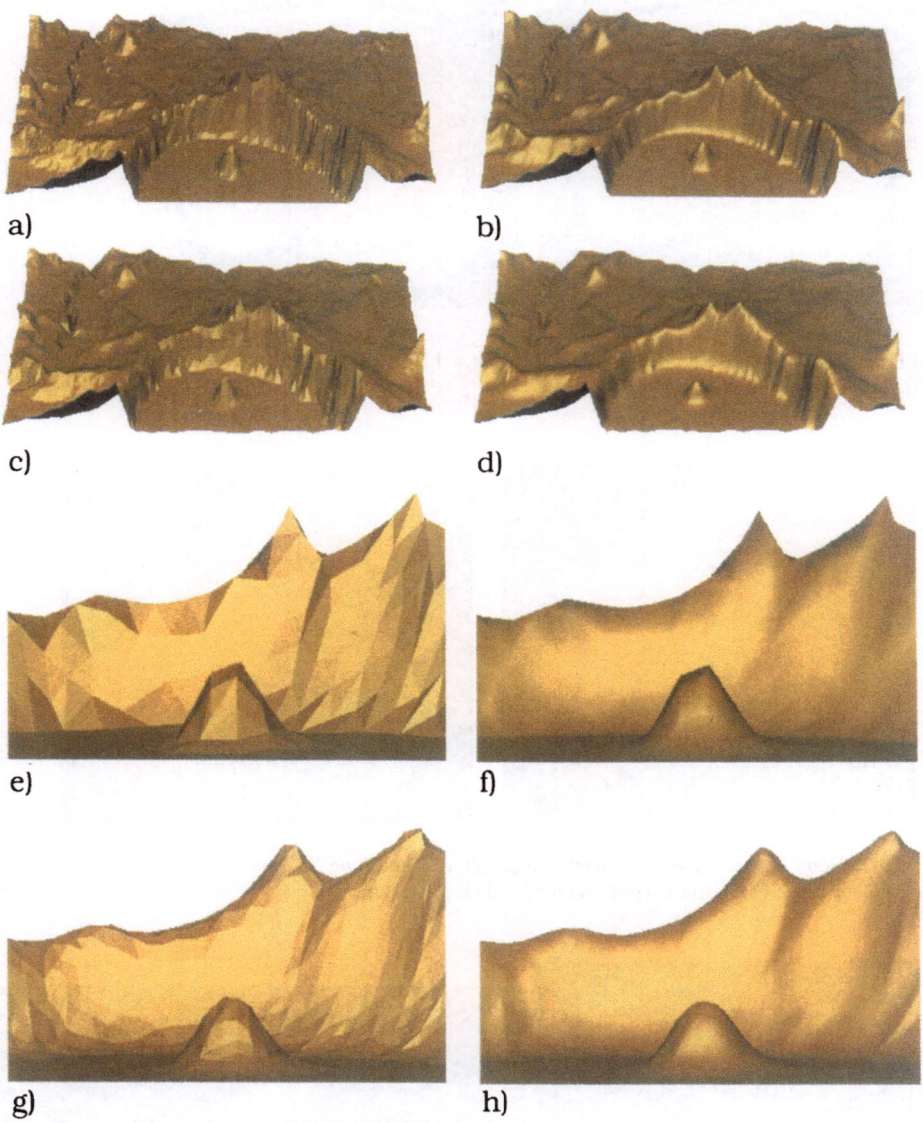

Fig. 7. Modified Loop subdivision for "Crater-Lake" terrain model. **a** Original model (19380 triangles); **b** third subdivision (1240320 triangles); **c** simplified model (5000 triangles); **d** third subdivision (320000 triangles); **e** local view of **c**; **f** Gouraud shaded; **g** first subdivision; **h** fourth subdivision

Westermann (pp. 65–74)

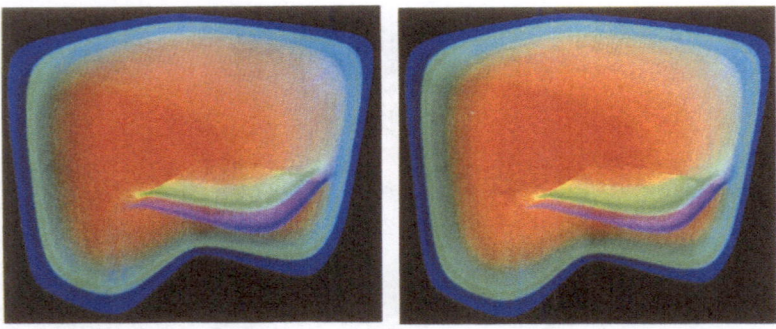

Fig. 6. The heat-sink data set was resampled on a 128^3 (left) and on a 256^3 (right) Cartesian grid and rendered via 3D textures

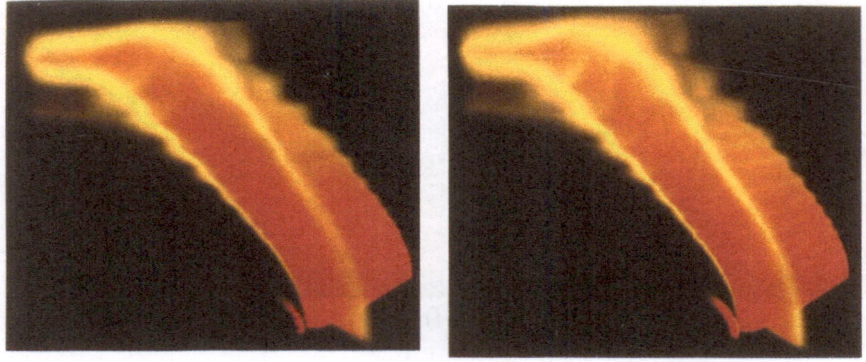

Fig. 7. Both images show the bluntfin data set resampled on a 256×128×64 (left) and a 512×256×128 (right) Cartesian grid and rendered via 3D textures

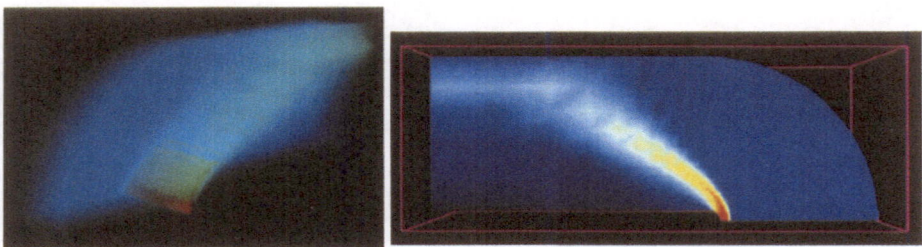

Fig. 8. On the left the bluntfin data set was resampled on a 512×256×128 grid. On the right our method was employed to resample the data on an arbitrary slice. The resampling time was 0.18 seconds.

Zaugg and Egbert (pp. 85–93)

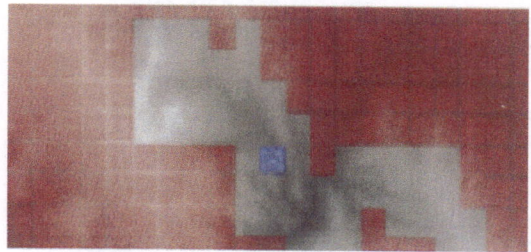

Fig. 5. Visualization of visibility. The viewpoint voxel is colored blue, and occluded tiles are shown in red

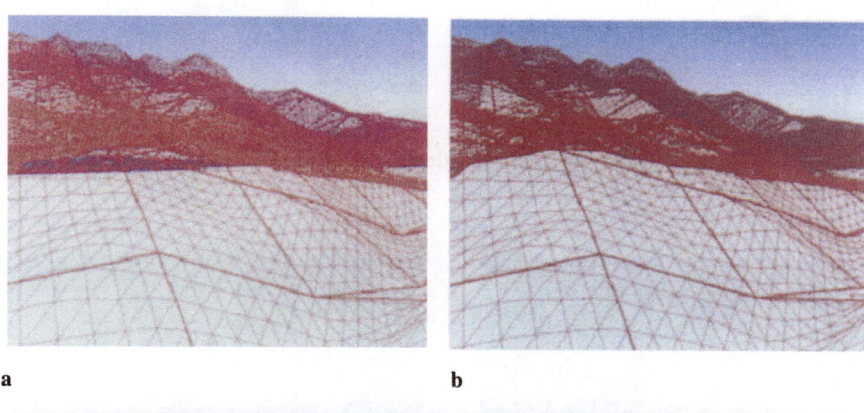

a   b

c

Fig. 6. A frame rendered from the viewpoint shown in blue in Fig. 6. **a** Wireframe with culling disabled, **b** wireframe with culling enabled, **c** textured view

Van Gelder (pp. 95–105)

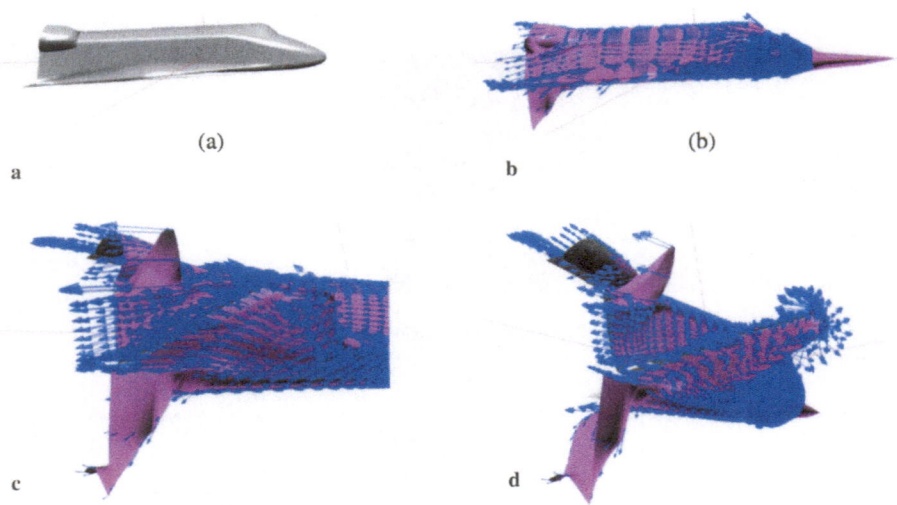

(a)

(b)

a

b

c

d

Fig. 3. Stream surface for simulation of the space shuttle launch vehicle, zone 3 of a nine-zone grid; see text for discussion

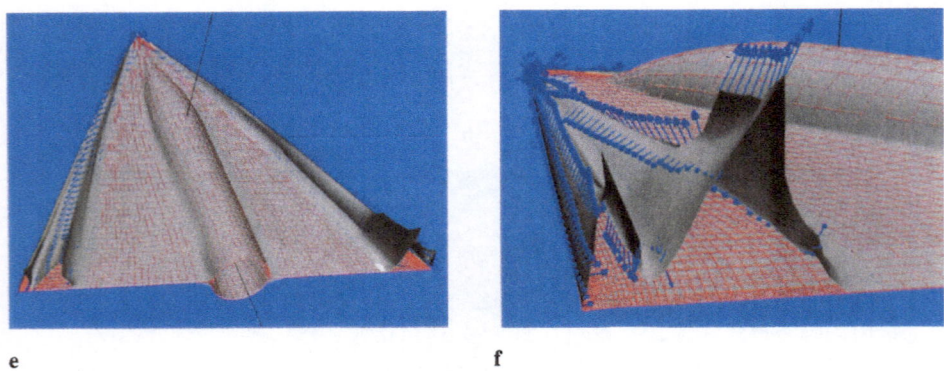

e

f

Fig. 4. Stream surface (smooth-shaded) in steady-state delta wing simulation, showing lines of separation and attachment near the outher edges. **e** Overview. **f** Closeup

Tricoche et al. (pp. 107–116)

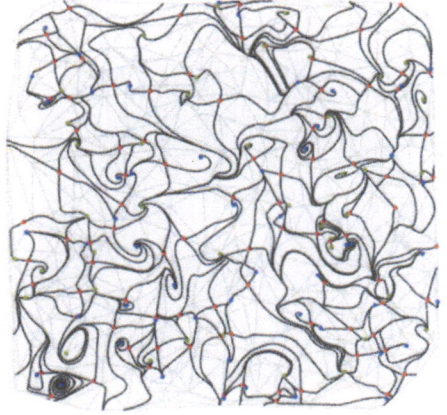 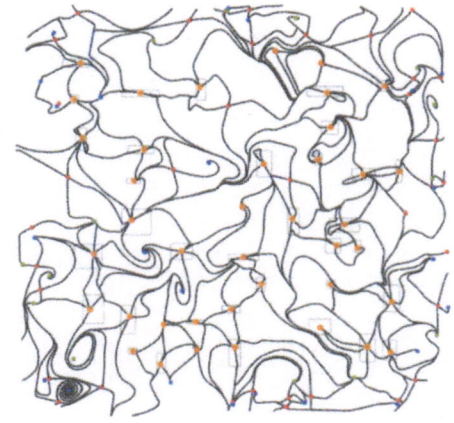

a Original topology   b Simplified topology (0.5)

Color Plate 1. First example: Vector case

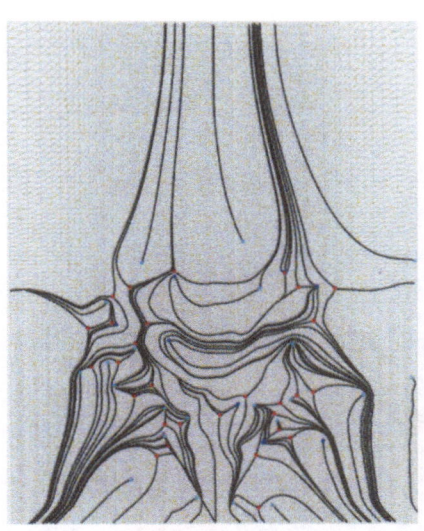 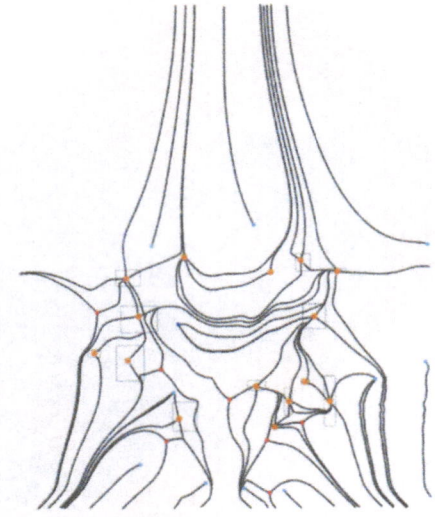

a Original topology   b Simplified topology (0.5)

Color Plate 2. Second example: Tensor case

Tricoche et al. (pp. 117–126)

Color Plate 1. Singular paths and bifurcations

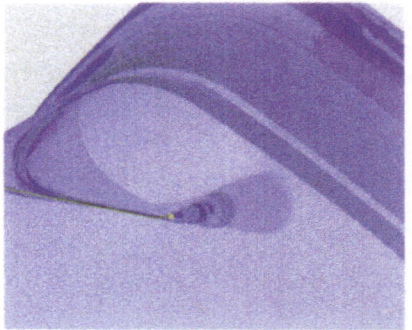

Color Plate 2. Hopf bifurcation (surfaces)

Color Plate 3. Overview of the topology evolution

Bartroli et al. (pp. 127–136)

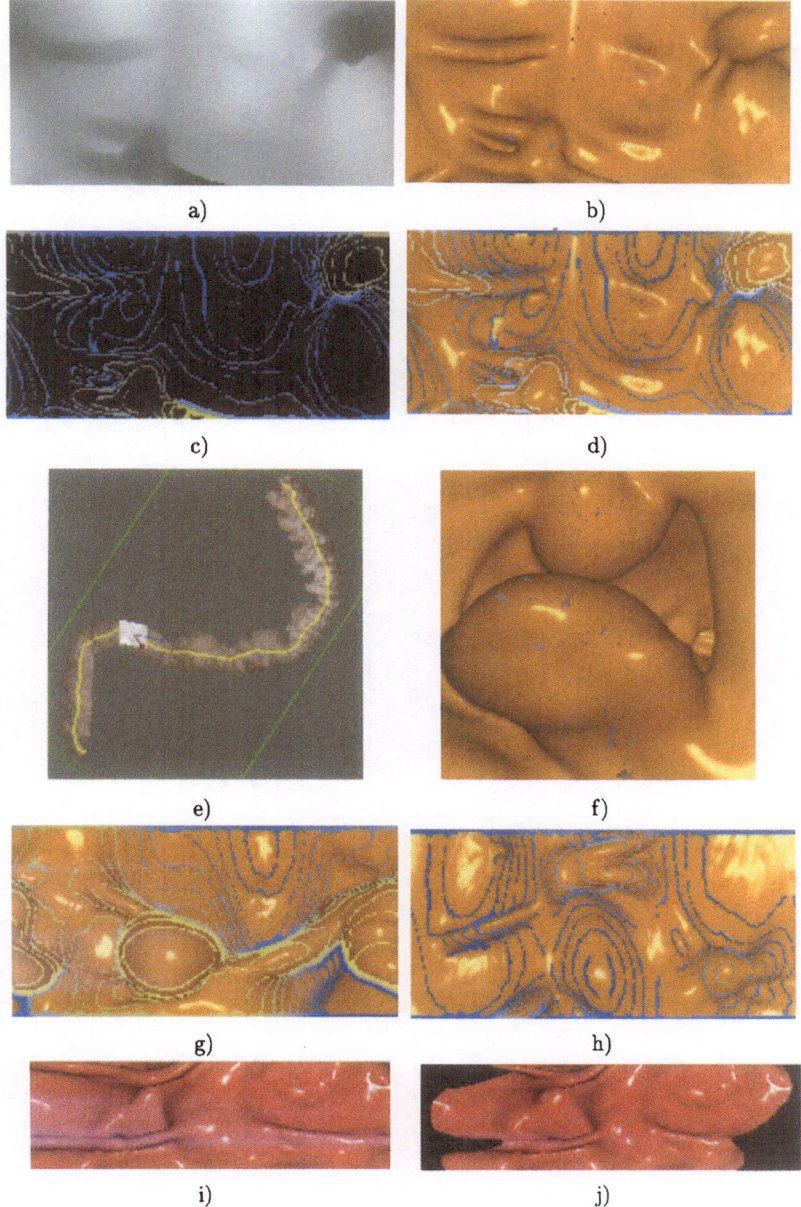

Fig. 6. Cadaveric colon CT data set 381×120×632. For the same camera position and using constant angle sampling: **a** depth image, **b** shaded image, **c** level lines with hue shift color coded, **d** combination of the level lines and shaded image. **e** Outside view and camera position for **g**. **f** Endoscopy view moving the camera in **e** a bit backwards. **g** Constant angle sampling showing 2 polyps with level lines enhancement. **h** Constant angle sampling from another camera position showing 2 polyps. Colon CT dataset 256×256×311: **i** Constant angle sampling. **j** The same camera position as **i** but with perimeter sampling

Serlie et al. (pp. 137–146)

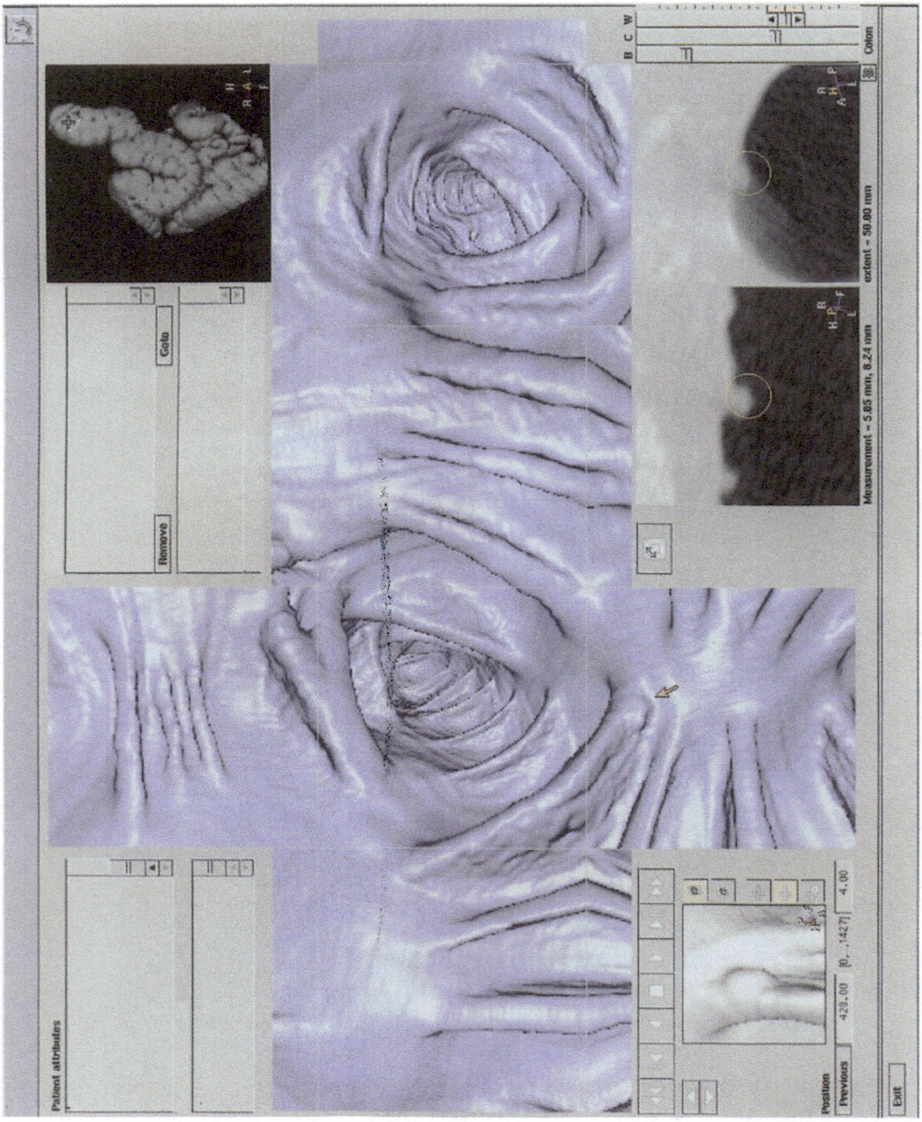

Demelio et al. (pp. 147–156)

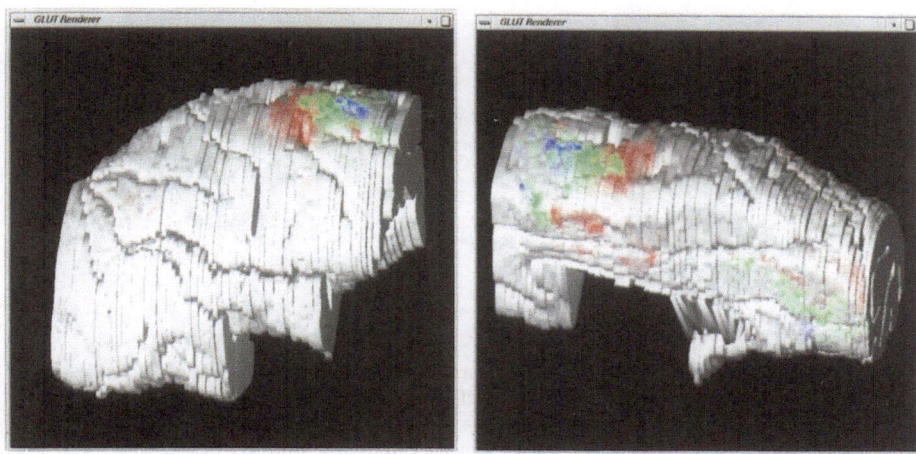

Fig. 6. Lateral (left) and mesial (right) view of the right hemisphere of the brain of a macaque monkey. All the major cortical sulci are clearly distinguishable and the same is true for finer macroscopical details as for example small cortical dimples. Neurons labeled with different tracers are shown with dots of different colors and the color intensity is attenuated depending of the distance of the neurons from the surface in the viewing direction

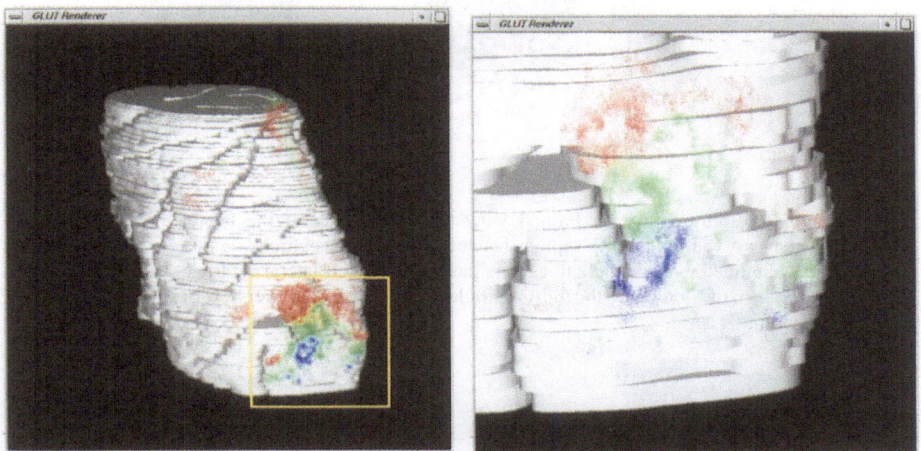

Fig. 7. Dorsal view of the right hemisphere of the brain of a macaque monkey. Three dense aggregates of labeling surround the injections of three neural tracers. The sites of the injections are clearly identified in the zoomed image as the regions empty of labeling within each of the three clouds of labeling

Bartz et al. (pp. 157–164)

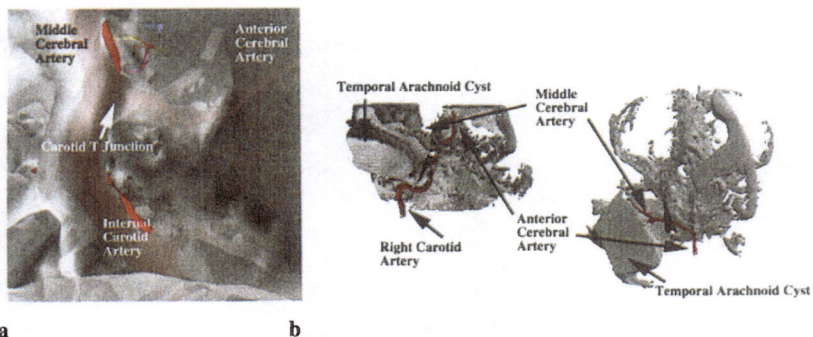

Fig. 4. Temporal Arachnoid Cyst dataset: **a** View on to carotid T junction, where the internal carotid artery branches into the lateral middle cerebral artery and the frontal anterior cerebral artery. The red vascular geometry – extracted from MRI Angiography – penetrates through the visible vascular geometry extracted from MRI TSE. **b** Frontal and top overview of temporal arachnoid cyst

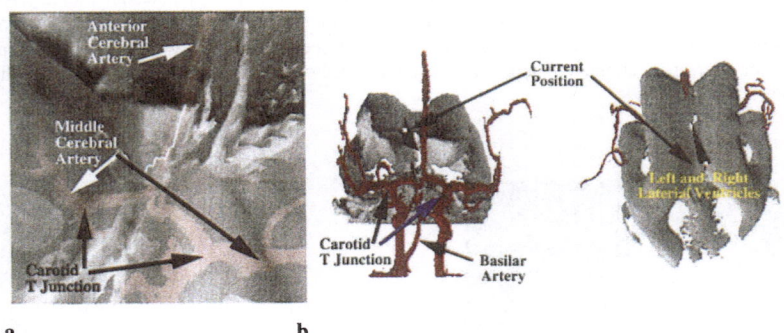

Fig. 5. Ventriculostomy dataset: **a** Frontal view from the center of the lateral ventricles (first two ventricles); the septum between the lateral ventricles is dissolved. The white line marks the default camera path from the left lateral ventricle through the foramen of Monro into the third ventricle. **b** Frontal and top overview of ventricular system

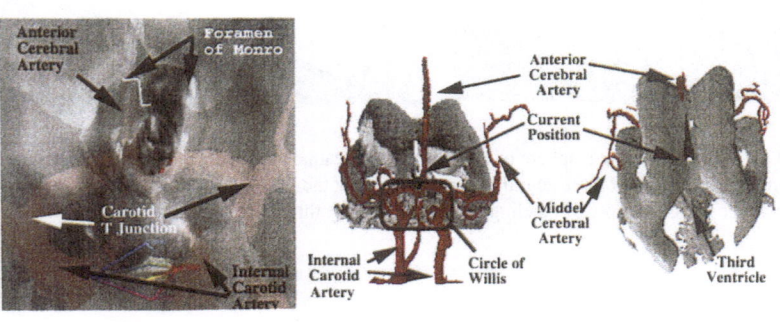

Fig. 6. Ventriculostomy dataset: **a** Frontal view from the cerebral aqueduct entrance in the third ventricle. The floor of the third ventricle – the potential location for a new CSF drain – is bounded by the arterial *circle of Willis*, a potential cause for mass bleeding. **b** Frontal and top overview of ventricular system

Telea and van Wijk (pp. 165–174)

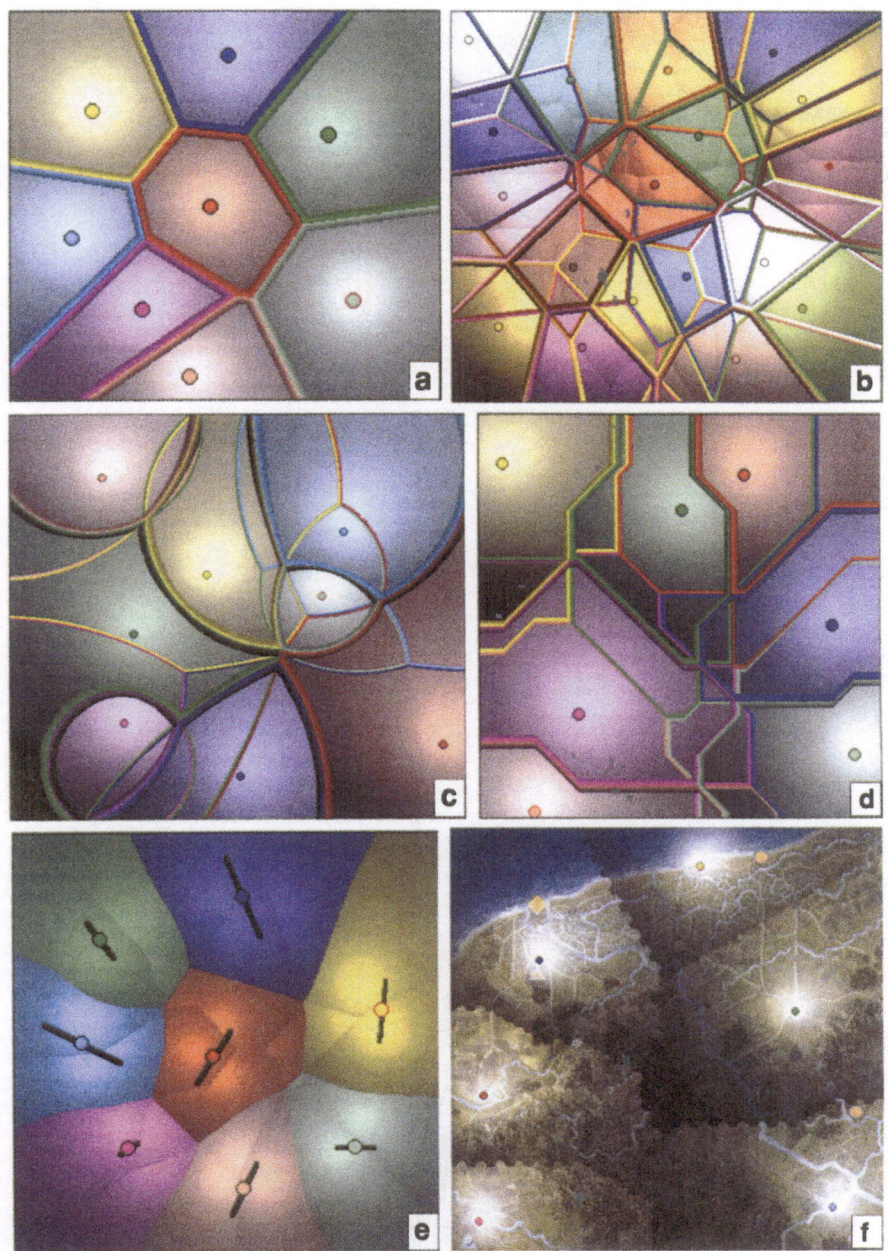

Fig. 8. Voronoi diagram renderings for Euclidian distance, order-1 and order-3 (**a, b**), multiplicative-weighted distance, order-2 (**c**), and Manhttan (**d**) distance order-2, line sites order-2 (**e**), and hexagonal grid-based discrete distance (**f**)

Diehl et al. (pp. 175–184)

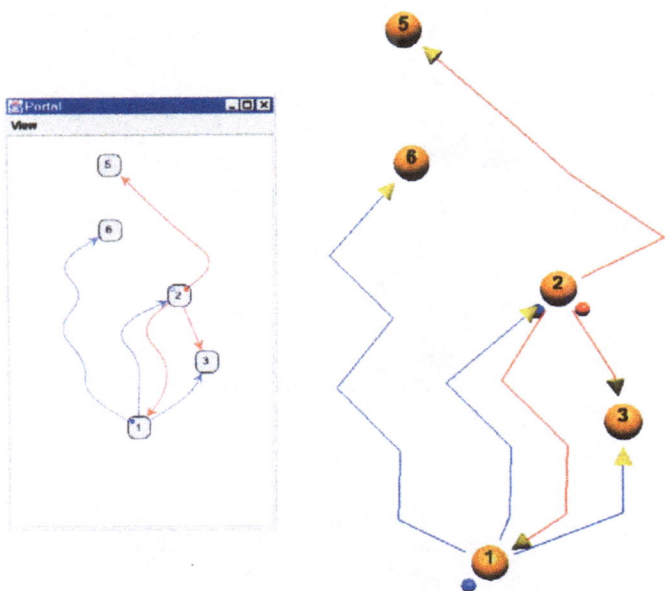

Fig. 2. A graph in 2D and 3D view

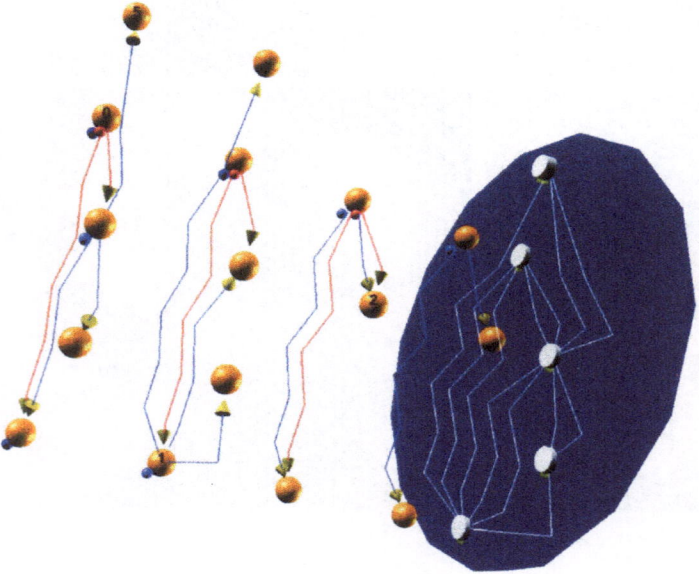

Fig. 3. A history view of navigation by two users

Hao et al. (pp. 185–192)

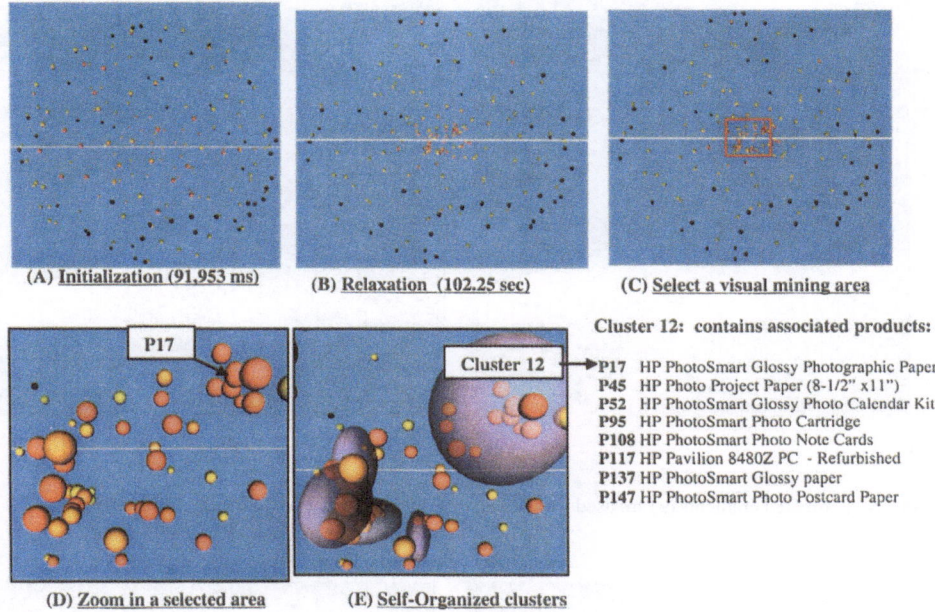

(A) <u>Initialization (91,953 ms)</u>   (B) <u>Relaxation (102.25 sec)</u>   (C) <u>Select a visual mining area</u>

Cluster 12: contains associated products:

P17 HP PhotoSmart Glossy Photographic Paper
P45 HP Photo Project Paper (8-1/2" x11")
P52 HP PhotoSmart Glossy Photo Calendar Kit
P95 HP PhotoSmart Photo Cartridge
P108 HP PhotoSmart Photo Note Cards
P117 HP Pavilion 8480Z PC - Refurbished
P137 HP PhotoSmart Glossy paper
P147 HP PhotoSmart Photo Postcard Paper

(D) <u>Zoom in a selected area</u>   (E) <u>Self-Organized clusters</u>

Fig. 4. An example of market basket analysis (Hewlett Packard E-Store)

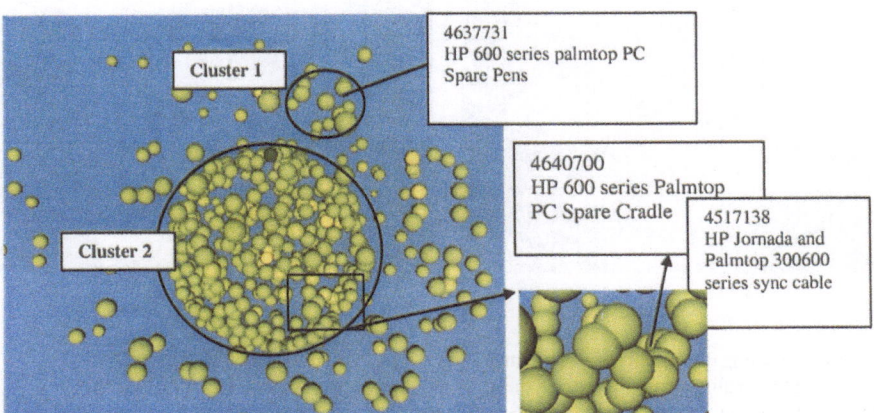

4637731
HP 600 series palmtop PC
Spare Pens

4640700
HP 600 series Palmtop
PC Spare Cradle

4517138
HP Jornada and
Palmtop 300600
series sync cable

Fig. 5. An example of customer profiling (purchasing similar products) Grouping 478,000 customers into 12 clusters from 171,000 real-time transactions

Mroz and Hauser (pp. 193–202)

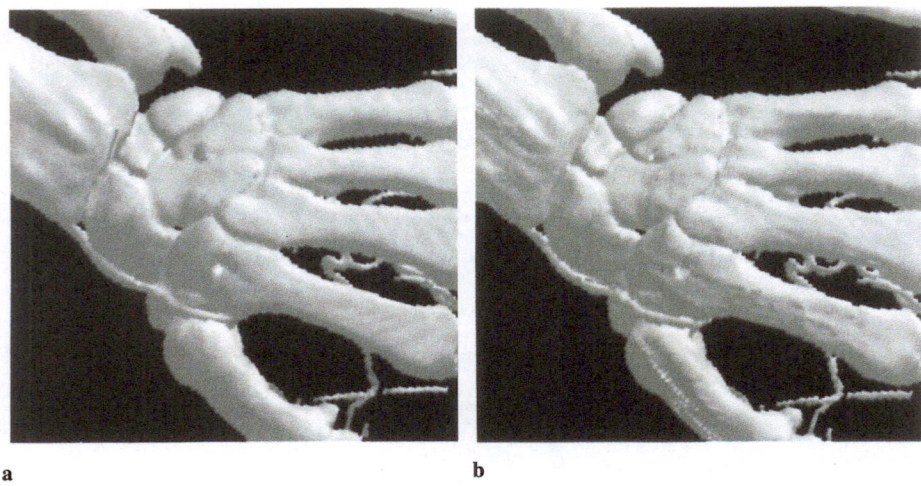

a   b

Fig. 3. Estimated gradients (**a**) are used for shading until the original gradient data has arrived (**b**)

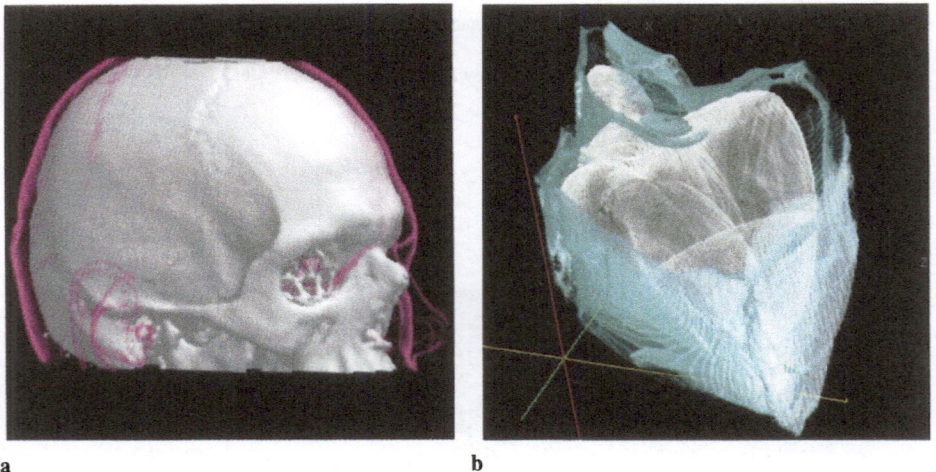

a   b

Fig. 4. a By adjusting the visualization mappings at the client, the skin surface has been rendered using a non-photorealistic technique over the conventionally shaded skull. **b** A data channel containing distance information has been used to modulate opacity of the basin surface to emphasize areas of almost-contact between the surface and the attractor contained within

Hladùvka et al. (pp. 203–211)

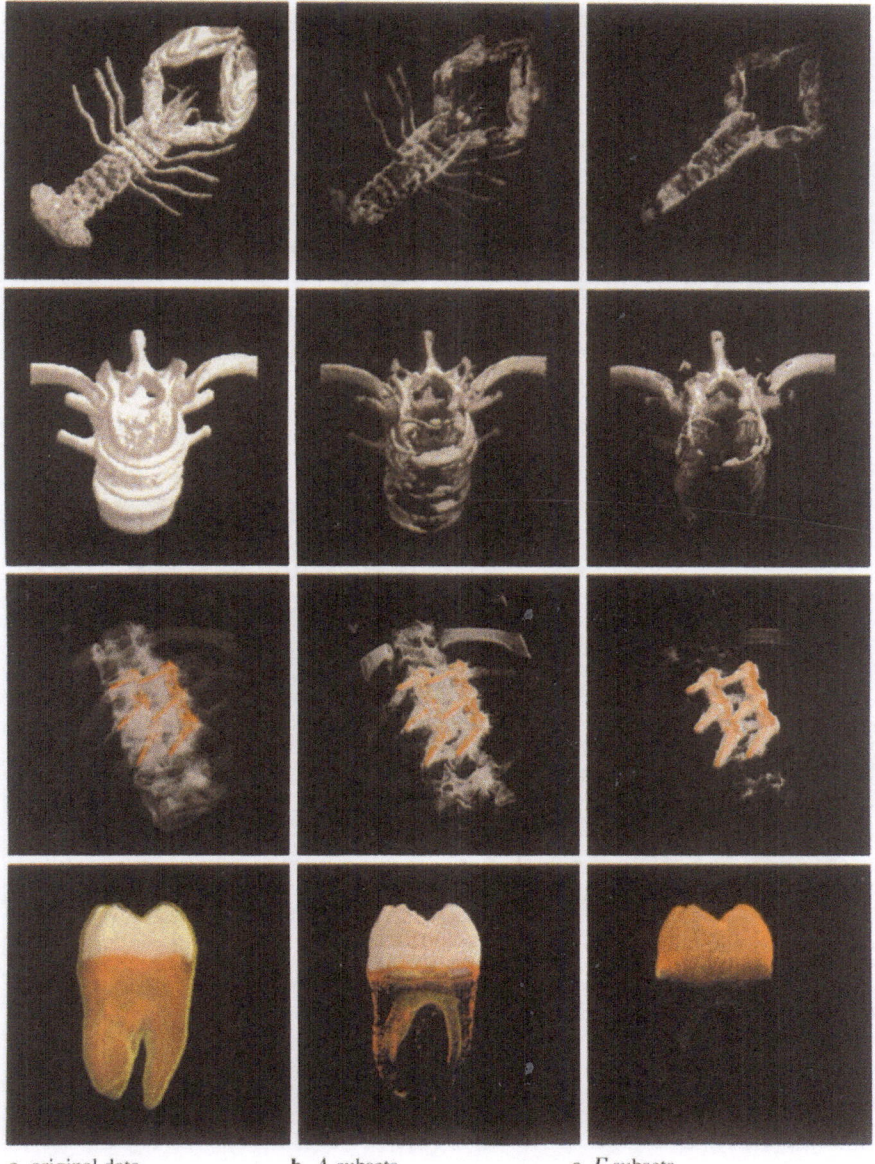

a original data   b Λ-subsets   c Γ-subsets

Fig. 2. Direct volume rendering of Lobster, Vertebra1, Vertebra2, and Tooth data sets (**a**) and their representations due to salience provided by our method (**b**) and by detection due to gradient magnitude (**c**). The Lobster subsets consist of 2.01% voxels of the original data set, Vertebra1 and Verteba2 of 4.03% voxels, and the Tooth subsets of 0.67% voxels

Laney et al. (pp. 213–222)

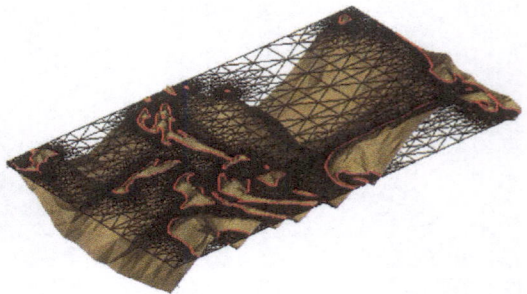

Fig. 4. A slice of the entropy field showing the isocontour used in the color figures

Fig. 5. An adaptive distance transform with $\lambda = 1$ and $\varepsilon_{min} = 0.5$. The adaptive distance triangulation is shown

Fig. 6. An adaptive distance transform with $\lambda = 1$ and $\varepsilon_{min} = 0.5$

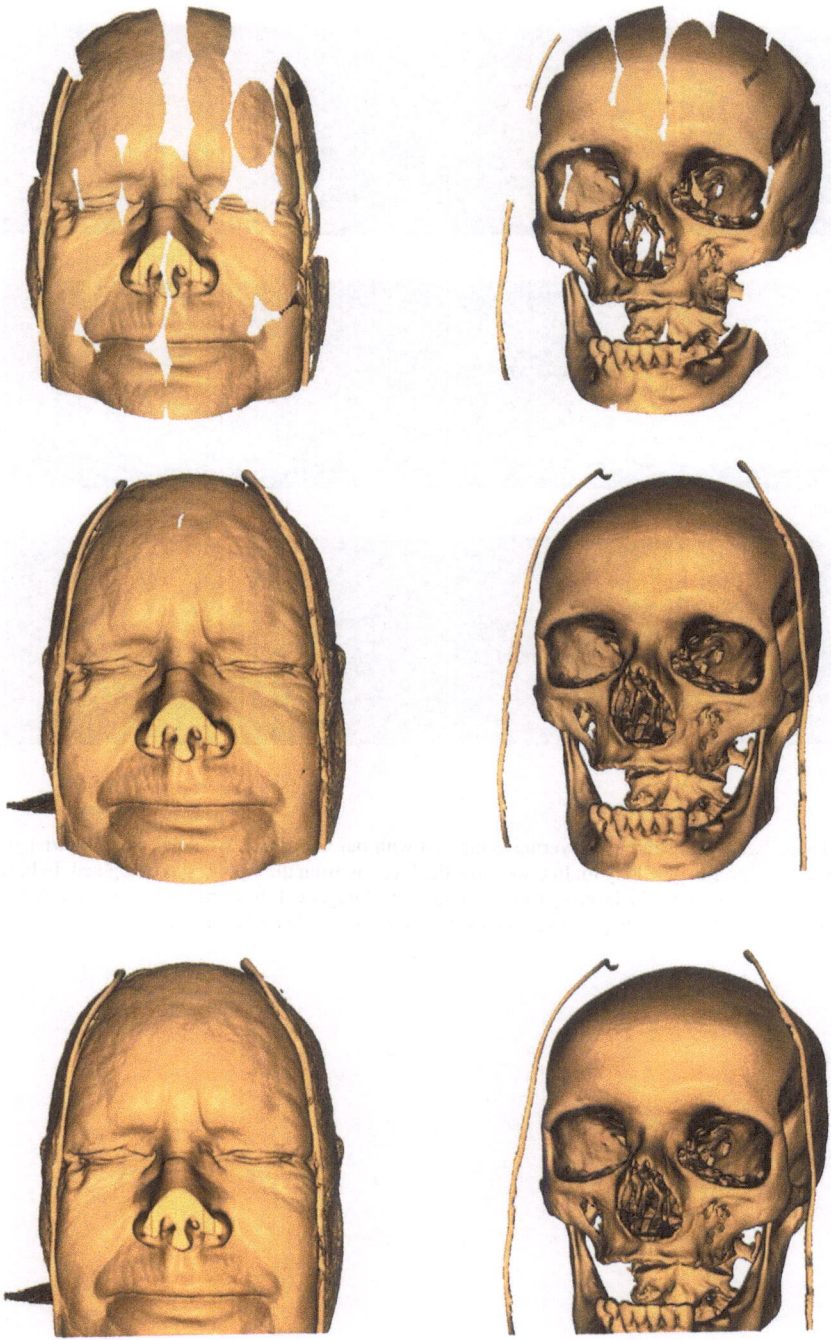

Color Plate 1. The left column shows three progressively refined images of the skin surface generated at points A, B and C in Graph 1. The right column shows three images of the bone surface generated at points A, B and C in Graph 2. In the middle row, after less than 1% of the rays have been cast, the resulting images are almost indistinguishable from the final images in the bottom row

Krishnan et al. (pp. 233–242)

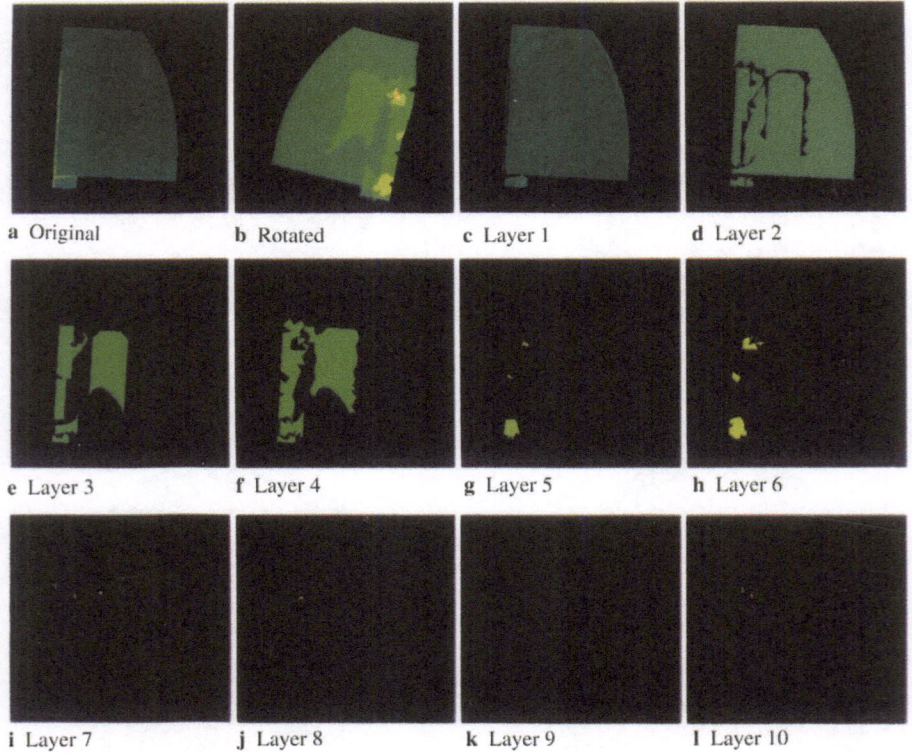

a Original   b Rotated   c Layer 1   d Layer 2

e Layer 3   f Layer 4   g Layer 5   h Layer 6

i Layer 7   j Layer 8   k Layer 9   l Layer 10

Fig. 3. These figures illustrate the layering computed with our algorithm. We color code the triangles according to the layer they belong to. In **a** we show the layering from the view it was computed. In **b**, we rotated the object as to show the layering from the other side. Images **c–l** show the ten layers computed for this particular view. Note how the 2D footprint of the layers get smaller and smaller

Djurcilov et al. (pp. 243–252)

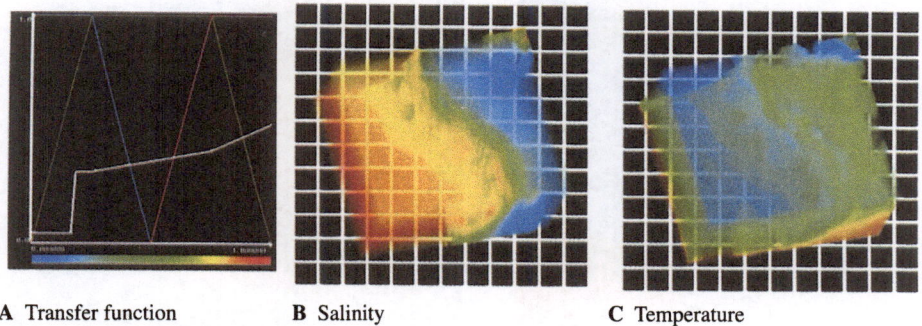

A Transfer function   **B** Salinity   **C** Temperature

Fig. 10. High contrast transfer function

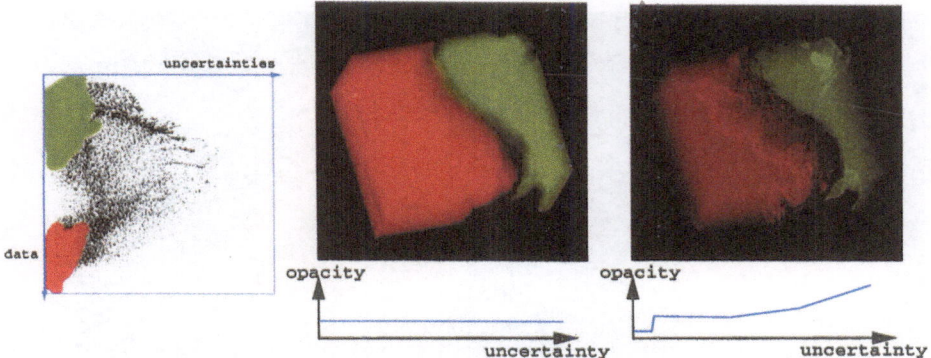

Fig. 11. The scatter plot in Fig. 5 is used as a 2D transfer function. Good (low uncertainty) data with low values are mapped to green, while good data with high values are mapped to red. Rest are mapped to gray

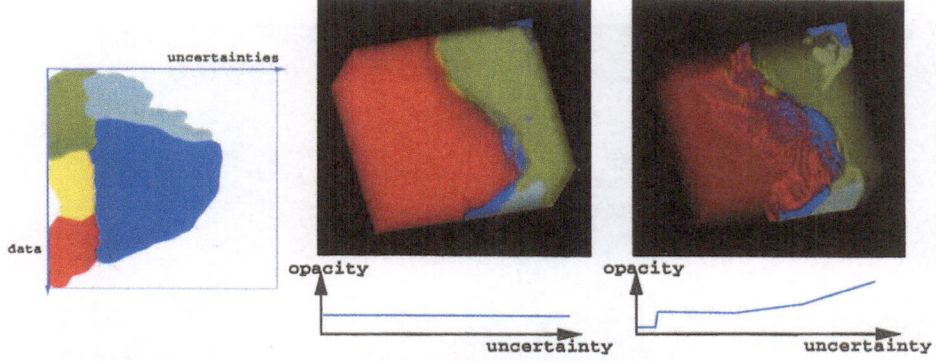

Fig. 12. The 2D transfer function identifies 5 regions instead of just 2

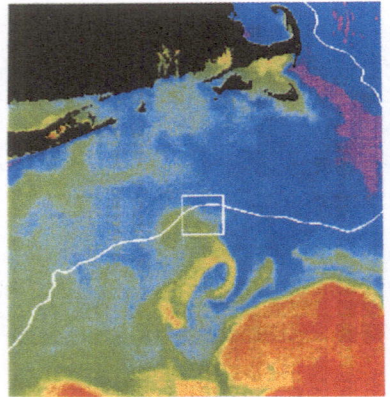

Fig. 13. Surface temperature

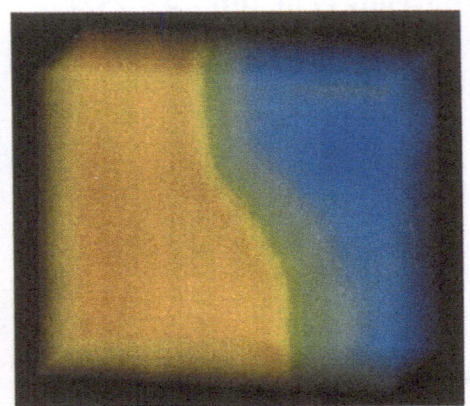

Fig. 14. Rendering of mean salinity

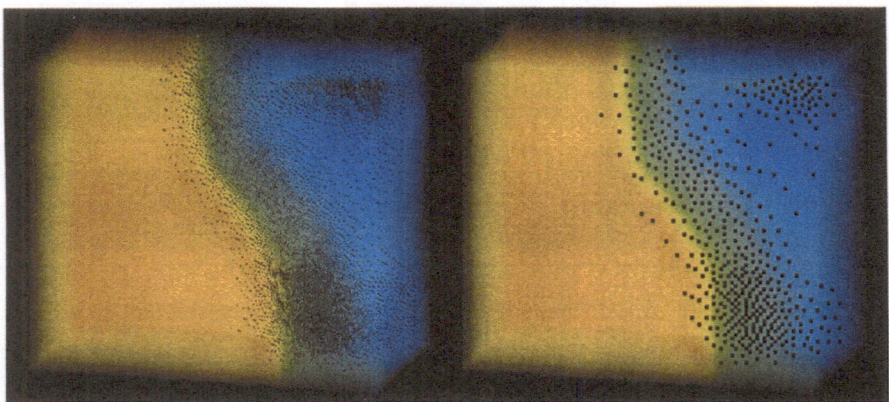

Fig. 15. Rendering with speckles   **Fig. 16.** Larger speckles emphasize holes

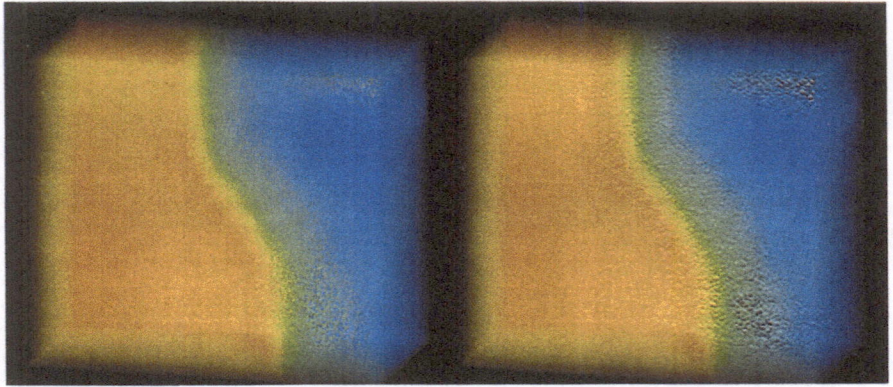

Fig. 17. Noise in high uncertainty areas   **Fig. 18.** Texture is another option

Li and Shen (pp. 253–262)

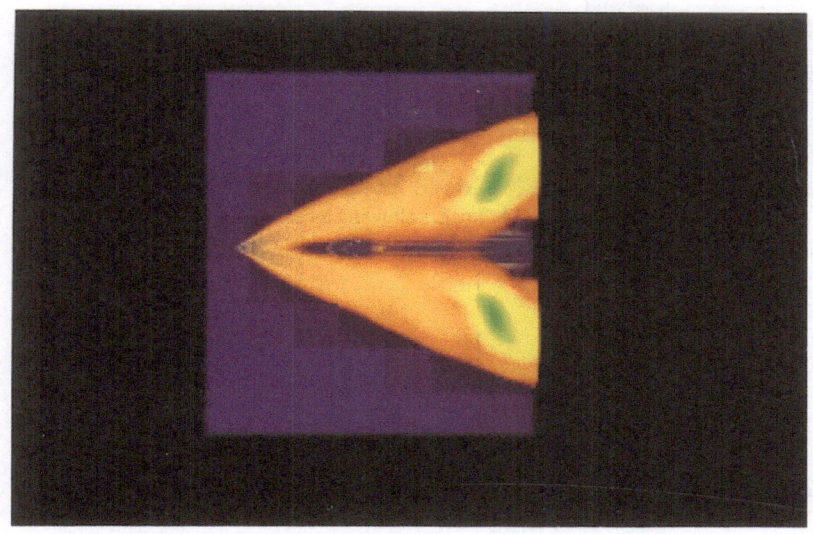

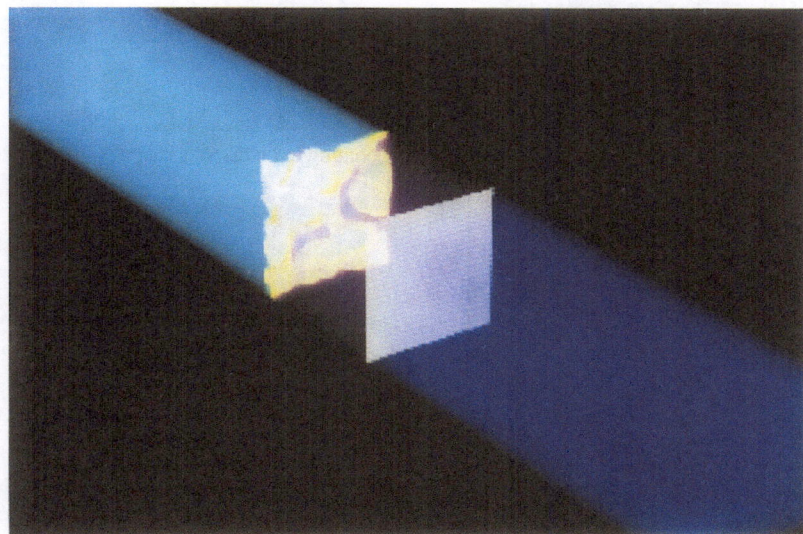

Fig. 8. Images for the delta wing and shockwave data sets

358

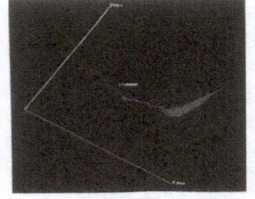

Fig. 10. A Bar Chart in 3D, showing price, time and volume.

Fig. 11. Images showing a user rotating the 3D graph model on the 'Workbench' at GMD.

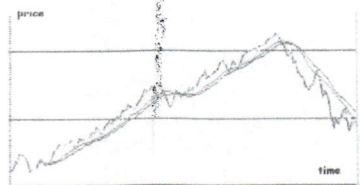

Fig. 12. Displaying multiple moving averages in 2D.

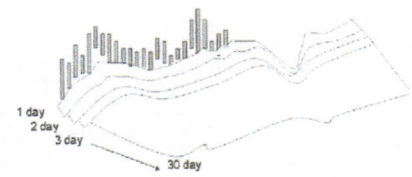

Fig. 13. Constructing a surface of multiple moving averages.

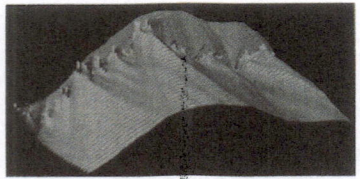

Fig. 14. The moving average surface.

Fig. 15. Zooming in on the Workbench.

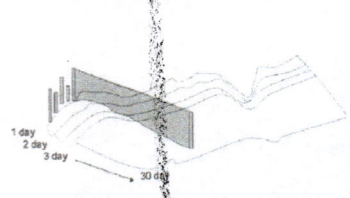

Fig. 16. Extending the price bars to intersect with the moving average plane.

Fig. 17. The model showing the moving average surface and intersecting price bars.

Fig. 18. Interacting with the moving average surface and extended price bars using the workbench & pen stylus.

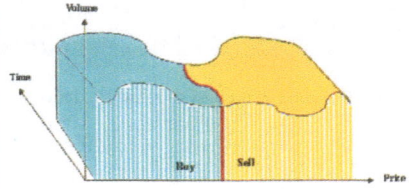

Fig. 19. The Depth of Market landscape, which consists of a series of surface strips representing a volume and price histogram at different time steps.

Atkison et al. (pp. 283–290)

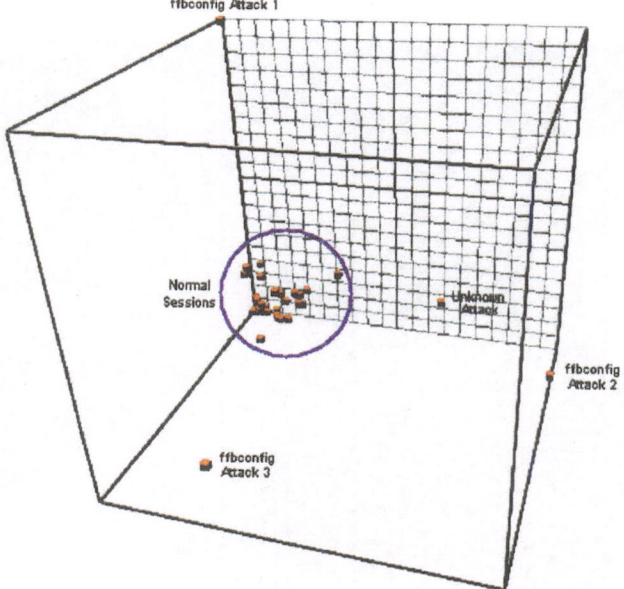

Fig. 1. Discovery of the same type of attack as the query shown

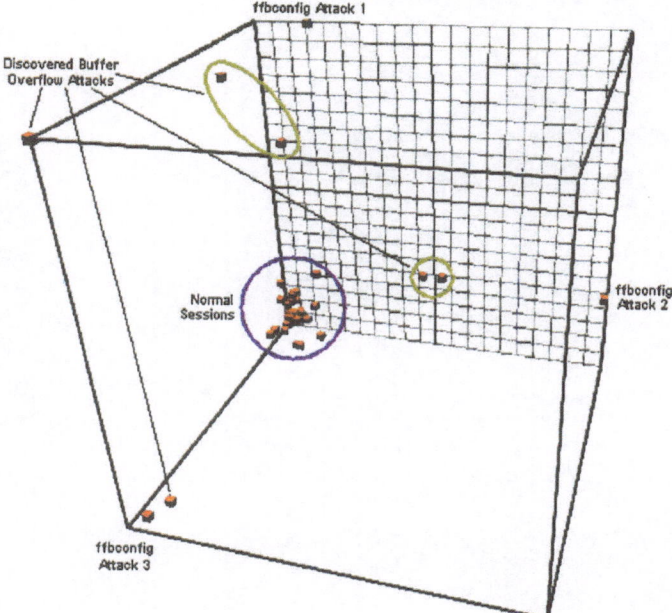

Fig. 2. Discovery of the attack of the same family as the query shown

Coomans and Timmermans (pp. 291–299)

Ono et al. (pp. 301–308)

Fig. 6. The temperature distribution of case A after 5 seconds past from the initial state

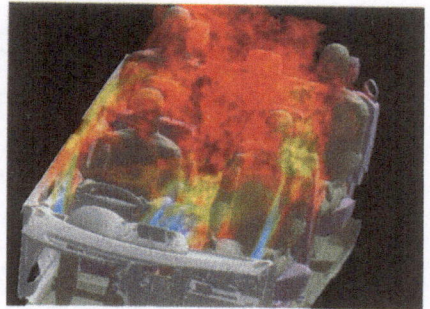

Fig. 7. The temperature distribution of case A after 10 seconds past from the initial state

Fig. 8. The temperature distribution of case A after 15 seconds past from the initial state

362

Fig. 9. The temperature distribution of case B after 5 seconds past from the initial state

Fig. 10. The temperature distribution of case B after 10 seconds past from the initial state

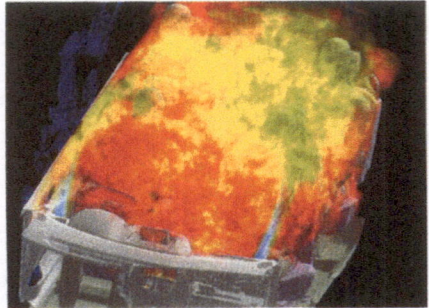

Fig. 11. The temperature distribution of case B after 15 seconds past from the initial state. Click figure, then you will find the thermal flow behavior in the passenger compartment

Roettger et al. (pp. 309–317)

Steps 1 to 3 of an interactive particle animation. Dust particles are emitted from a small user definable wireframe box behind the car. The emerging massive particle stream is first moving down and then up the back window. Eventually, it is diverging at the top of the car due to shearing forces

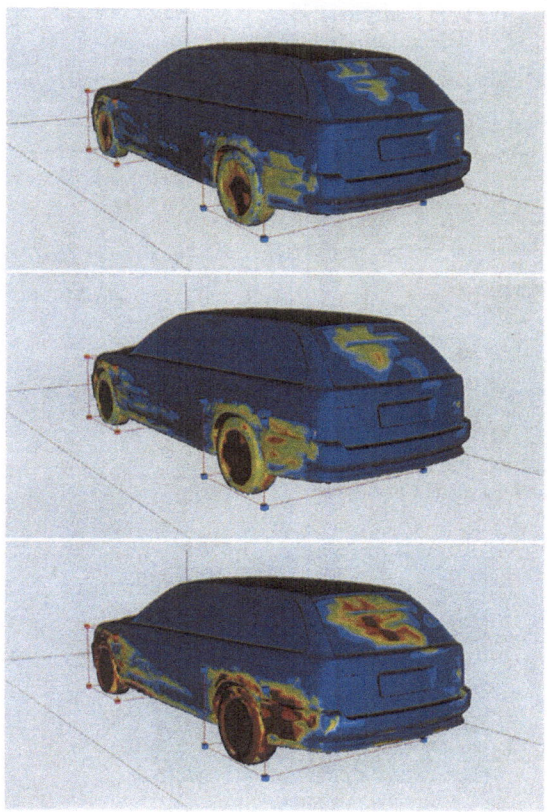

Steps 1 to 3 of a soiling simulation running for approximately 24 hours on a dual processor SGI Octane MXI with 2×250MHz MIPS R 10K. The colorcoding shows the hit rate particles colliding with the car body. Red hot spots indicate areas with a high degree of soiling

Sommer and Ertl (pp. 319–328)

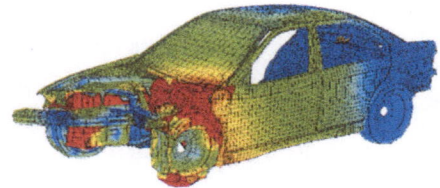

Fig. 1. The color maps the length of difference vectors of corresponding nodes

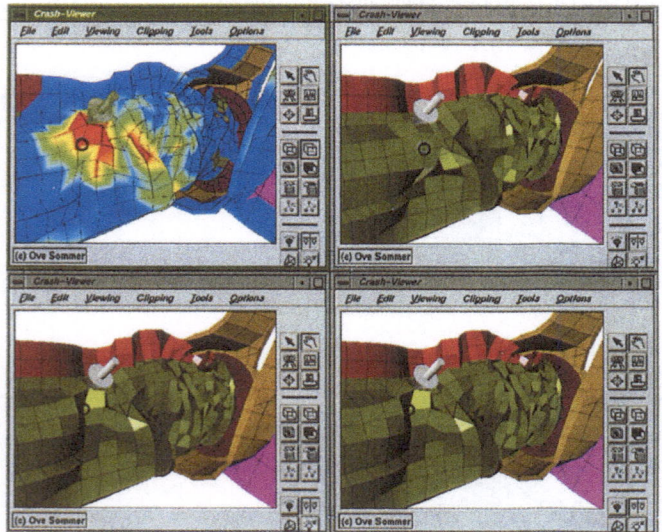

Fig. 2. Comparative visualization of instabilities using *Crash-Viewer*

Fig. 3. For a better orientation the neighborhood of primary instability regions can be faded in interactively by modifying the α transfer function

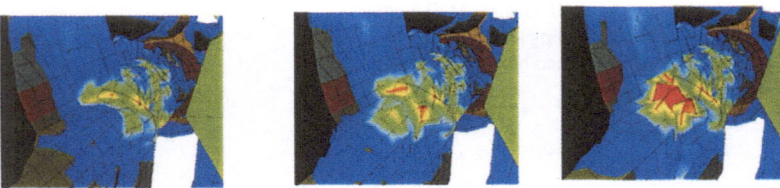

Fig. 4. Development of an instability over three time steps

SpringerEurographics

Bernd Fröhlich,
Joachim Deisinger,
Hans-Jörg Bullinger (eds.)

Immersive Projection Technology and Virtual Environments 2001

Proceedings of the Eurographics Workshop,
Stuttgart, Germany, May 16-18, 2001

2001. XI, 284 pages. 150 partly coloured figures.
Softcover DM 121,–, öS 847,–
(recommended retail price)
ISBN 3-211-83671-3
Eurographics

17 papers report on the latest scientific advances in the fields of immersive projection technology and virtual environments. The main topics included here are human computer interaction (user interfaces, interaction techniques), software developments (virtual environment applications, rendering techniques), and input/output devices.

Please visit our new website: **www.springer.at**

 **Springer**WienNewYork

A-1201 Wien, Sachsenplatz 4–6, P.O. Box 89, Fax +43.1.330 24 26, e-mail: books@springer.at, Internet: **www.springer.at**
D-69126 Heidelberg, Haberstraße 7, Fax +49.6221.345-229, e-mail: orders@springer.de
USA, Secaucus, NJ 07096-2485, P.O. Box 2485, Fax +1.201.348-4505, e-mail: orders@springer-ny.com
Eastern Book Service, Japan, Tokyo 113, 3–13, Hongo 3-chome, Bunkyo-ku, Fax +81.3.38 18 08 64, e-mail: orders@svt-ebs.co.jp

SpringerEurographics

Nadia Magnenat-Thalmann,
Daniel Thalmann,
Bruno Arnaldi (eds.)

Computer Animation and Simulation 2000

Proceedings of the Eurographics
Workshop in Interlaken, Switzerland,
August 21–22, 2000

2000. X, 211 pages.
110 partly coloured figures.
Softcover DM 89,–, öS 625,–
ISBN 3-211-83549-0

This book presents state-of-the-art
methods in computer animation and
simulation. This collection of papers
covers current research in human animation, physically-based modeling, motion
control, animation systems, and other
key aspects.

View table of contents at:
www.springer.at

Bernard Péroche,
Holly Rushmeier (eds.)

Rendering Techniques 2000

Proceedings of the Eurographics
Workshop in Brno, Czech Republic,
June 26–28, 2000

2000. XIII, 422 pages.
250 partly coloured figures.
Softcover DM 128,–, öS 896,–
ISBN 3-211-83535-0

This book presents state-of-the-art
methods in computer graphics rendering.
The papers included in this volume were
selected after careful review by an international committee of experts. Included
are various topics related to the generation of synthetic images: visibility, global
illumination, perception, hardware assisted rendering, reflectance models, real-
time rendering, antialiasing, compression and image-based rendering.

All prices are recommended retail prices.

 SpringerWienNewYork

A-1201 Wien, Sachsenplatz 4–6, P.O. Box 89, Fax +43.1.330 24 26, e-mail: books@springer.at, Internet: **www.springer.at**
D-69126 Heidelberg, Haberstraße 7, Fax +49.6221.345-229, e-mail: orders@springer.de
USA, Secaucus, NJ 07096-2485, P.O. Box 2485, Fax +1.201.348-4505, e-mail: orders@springer-ny.com
Eastern Book Service, Japan, Tokyo 113, 3–13, Hongo 3-chome, Bunkyo-ku, Fax +81.3.38 18 08 64, e-mail: orders@svt-ebs.co.jp

SpringerEurographics

J. D. Mulder,
R. van Liere (eds.)

Virtual Environments 2000

Proceedings of the Eurographics
Workshop in Amsterdam,
The Netherlands, June 1–2, 2000

2000. X, 217 pages. 95 partly coloured figures.
Softcover DM 98,–, öS 686,–. ISBN 3-211-83516-4

D. J. Duke, A. Puerta (eds.)

Design, Specification and Verification of Interactive Systems '99

Proceedings of the Eurographics Workshop in Braga, Portugal, June 2–4, 1999

1999. IX, 280 pages. 89 figures.
Softcover DM 118,–, öS 826,–. ISBN 3-211-83405-2

W. C. de Leeuw,
R. van Liere (eds.)

Data Visualization 2000

Proceedings of the Joint EURO-
GRAPHICS and IEEE TCVG Symposium
on Visualization in Amsterdam,
The Netherlands, May 29–31, 2000

2000. XI, 296 pages. 166 partly coloured figures.
Softcover DM 118,–, öS 826,–. ISBN 3-211-83515-6

N. Magnenat-Thalmann,
D. Thalmann (eds.)

Computer Animation and Simulation '99

Proceedings of the Eurographics
Workshop in Milano, Italy,
September 7–8, 1999

1999. X, 230 pages. 148 partly coloured figures.
Softcover DM 89,–, öS 625,–. ISBN 3-211-83392-7

N. Correia, T. Chambel,
G. Davenport (eds.)

Multimedia '99

Proceedings of the Eurographics
Workshop in Milano, Italy,
September 7–8, 1999

2000. IX, 222 pages. 85 figures.
Softcover DM 85,–, öS 595,–. ISBN 3-211-83437-0

M. Gervautz, A. Hildebrand,
D. Schmalstieg (eds.)

Virtual Environments '99

Proceedings of the Eurographics
Workshop in Vienna, Austria,
May 31–June 1, 1999

1999. X, 191 pages. 78 figures.
Softcover DM 85,–, öS 595,–. ISBN 3-211-83347-1

All prices are recommended retail prices

 SpringerWienNewYork

A-1201 Wien, Sachsenplatz 4–6, P.O. Box 89, Fax +43.1.330 24 26, e-mail: books@springer.at, Internet: **www.springer.at**
D-69126 Heidelberg, Haberstraße 7, Fax +49.6221.345-229, e-mail: orders@springer.de
USA, Secaucus, NJ 07096-2485, P.O. Box 2485, Fax +1.201.348-4505, e-mail: orders@springer-ny.com
Eastern Book Service, Japan, Tokyo 113, 3–13, Hongo 3-chome, Bunkyo-ku, Fax +81.3.38 18 08 64, e-mail: orders@svt-ebs.co.jp

SpringerComputerScience

Pauline J. Sheldon,
Karl W. Wöber,
Daniel R. Fesenmaier (eds.)

Information and Communication Technologies in Tourism 2001

Proceedings of the International Conference in Montreal, Canada, 2001

2001. XII, 386 pages. 83 figures.
Softcover DM 138,–, öS 966,–
(recommended retail price)
ISBN 3-211-83649-7

This collection comprises the papers presented at the International Conference on Information and Communication Technologies in Tourism held in Montreal, Canada. The role played by information and communication technologies in the management and marketing of destinations, the convergence of communication systems and the integration of heterogeneous information systems, the designs of intelligent recommendation systems, the use of destination web sites by consumers, and the implementation of information and communication technologies in small and medium-sized tourism enterprises are addressed.

 **Springer**WienNewYork

A-1201 Wien, Sachsenplatz 4–6, P.O. Box 89, Fax +43.1.330 24 26, e-mail: books@springer.at, Internet: **www.springer.at**
D-69126 Heidelberg, Haberstraße 7, Fax +49.6221.345-229, e-mail: orders@springer.de
USA, Secaucus, NJ 07096-2485, P.O. Box 2485, Fax +1.201.348-4505, e-mail: orders@springer-ny.com
Eastern Book Service, Japan, Tokyo 113, 3–13, Hongo 3-chome, Bunkyo-ku, Fax +81.3.38 18 08 64, e-mail: orders@svt-ebs.co.jp